Student Solutions Manual
for Yandl's

Finite Mathematics

Mary B. Ehlers
Seattle University

Brooks/Cole Publishing Company
Pacific Grove, California

Brooks/Cole Publishing Company
A Division of Wadsworth, Inc.

Printed in the United States of America

10   9   8   7   6   5   4   3   2   1

ISBN 0-534-14192-7

Sponsoring Editor: *Faith B. Stoddard*
Editorial Assistant: *Nancy Champlin*
Production Coordinator: *Dorothy Bell*
Cover Design: *Katherine Minerva*
Cover Photo: *Lee Hocker*
Art Coordinator: *Lisa Torri*
Printing and Binding: *Malloy Lithographing, Inc.*

PREFACE

This manual contains the detailed solutions to the more than **950** odd-numbered problems in the text **Finite Mathematics** by Andre L. Yandl.

In order to develop the ability to solve problems, the student should first attempt the problem on his/her own before comparing the solution with the solution given in this manual. The solution process is more important than the answer, and understanding is more important than memorization. Many problems can be solved by more than one method and sometimes the answer may be expressed in different forms. Thus some solutions may differ from the one in the manual and still be correct.

I wish to thank the members of the team that helped in the preparation and production for this project. Mike Allen, Bob Heighton, and Jim Riester assisted with the typing of the various versions and revisions of the manuscript. Professor Yandl checked many of the solutions for correctness and prepared the complete solutions for many of the problems in Chapter 7. At Brooks/Cole, Dorothy Bell served as Production Coordinator and Lisa Torri as Art Coordinator, and Faith Stoddard, Assistant Mathematics Editor, coordinated the entire project and provided editorial supervision and encouragement. I especially thank Frank Arena, who also assisted with the typing of the manuscript, assembled it for production, and remained with the project from beginning to end.

The manual was prepared using EXP: The Scientific Word Processor for use on MS-DOS computers. EXP is available from Brooks/Cole Publishing Company, Pacific Grove, California 93950.

Mary B. Ehlers
Seattle University
Seattle, Washington 98122

CONTENTS

GETTING STARTED

1. $x^7 - y^7 = (x - y)(x^6 + x^5 y + x^4 y^2 + x^3 y^3 + x^2 y^4 + x y^5 + y^6)$

3. $8x^3 - 125 y^3 = (2x)^3 - (5y^3) = (2x - 5y)(4x^2 + 10xy + 25y^2)$

5. $32 a^5 - 243 c^5 = (2a)^5 - (3c)^5 = (2a - 3c)[(2a)^4 + (2a)^3 (3c) + (2a)^2 (3c)^2$
      $+ (2a)(3c)^3 + (3c)^4] = (2a - 3c)(16 a^4 + 24 a^3 c + 36 a^2 c^2 + 54 a c^3 + 81 c^4)$

7. $x^3 + y^3 = \left(x^3 - (-y)^3\right) = (x + y)(x^2 - xy + y^2)$

9. $\dfrac{x^3}{8} + 27 y^3 = \left(\dfrac{x}{2}\right)^3 - \left(-3y\right)^3 = \left(\dfrac{x}{2} + 3y\right)\left(\dfrac{x^2}{4} - \dfrac{3}{2}xy + 9y^2\right)$

11.

$$
\begin{array}{ccccccccccccc}
 & & & & & & 1 & & & & & & \\
 & & & & & 1 & & 1 & & & & & \\
 & & & & 1 & & 2 & & 1 & & & & \\
 & & & 1 & & 3 & & 3 & & 1 & & & \\
 & & 1 & & 4 & & 6 & & 4 & & 1 & & \\
 & 1 & & 5 & & 10 & & 10 & & 5 & & 1 & \\
1 & & 6 & & 15 & & 20 & & 15 & & 6 & & 1
\end{array}
$$

13. $(x + y)^6 = x^6 + 6x^5 y + 15 x^4 y^2 + 20 x^3 y^3 + 15 x^2 y^4 + 6 x y^5 + y^6$

15. $(x + y)^7 = x^7 + 7 x^6 y + 21 x^5 y^2 + 35 x^4 y^3 + 35 x^3 y^4 + 21 x^2 y^5 + 7 x y^6 + y^7$

17. $(2x + 3y)^3 = (2x)^3 + 3(2x)^2 (3y) + 3(2x)(3y)^2 + (3y)^3$
      $= 8x^3 + 36 x^2 y + 54 x y^2 + 27 y^3$

19. $(3a - 2b)^4 = (3a)^4 - 4(3a)^3 (2b) + 6(3a)^2 (2b)^2 - 4(3a)(2b)^3 + (2b)^4$
      $= 81 a^4 - 216 a^3 b + 216 a^2 b^2 - 96 a b^3 + 16 b^4$

21.
$$\left(2x+\frac{y}{2}\right)^8 = (2x)^8 + 8(2x)^7\left(\frac{y}{2}\right) + 28(2x)^6\left(\frac{y}{2}\right)^2 + 56(2x)^5\left(\frac{y}{2}\right)^3 + 70(2x)^4\left(\frac{y}{2}\right)^4$$
$$+ 56(2x)^3\left(\frac{y}{2}\right)^5 + 28(2x)^2\left(\frac{y}{2}\right)^6 + 8(2x)\left(\frac{y}{2}\right)^7 + \left(\frac{y}{2}\right)^8 = 256x^8 + 512x^7y + 448x^6y^2$$
$$+ 224x^5y^3 + 70x^4y^4 + 14x^3y^5 + \frac{7}{4}x^2y^6 + \frac{xy^7}{8} + \frac{y^8}{256}$$

23. (a) $n = 8 + 4 = 12$

(b) $\frac{495}{5}(8)a^7b^5 = 792\,a^7b^5$

(c) The preceeding term is ka^9b^3 where $\frac{9k}{3+1} = 495$ and $k = 220$. The preceeding term is $220\,a^9b^3$.

25. (a) $n = 4 + 10 = 14$

(b) $\frac{1001(4)}{11}a^3b^{11} = 364\,a^3b^{11}$

(c) $\frac{1001(10)}{5}a^5b^9 = 2002\,a^5b^9$

27. (a) $n = 5 + 12 = 17$

(b) $6188\,a^4b^{13}\frac{(5)}{13} = 2380\,a^4b^{13}$

(c) $6188\left(\frac{12}{6}\right)a^6b^{11} = 12376\,a^6b^{11}$

29. (a) $n = 16 + 4 = 20$

(b) $\frac{4845(16)}{5}a^{15}b^5 = 15504\,a^{15}b^5$

(c) $\frac{4845(4)}{17}a^{17}b^3 = 1140\,a^{17}b^3$

31. (a) $n = 7 + 8 = 15$

(b) $\frac{6435(7)}{9}a^6b^9 = 5005\,a^6b^9$

(c) $\frac{6435(8)}{8}a^8b^7 = 6435\,a^8b^7$

33. $\frac{(3+101)}{2}(50) = \frac{104}{2}(50) = 52(50) = 2600$

35. $1 + 2 + 3 + \cdots + 50 = \frac{(1+50)}{2}(50) = (51)(25) = 1275$

37. (a) $1^3 + 2^3 = 1 + 8 = 9, \quad 1 + 2 = 3$

(b) $1^3 + 2^3 + 3^3 = 36, \quad 1 + 2 + 3 = 6$

(c) $1^3 + 2^3 + 3^3 + 4^3 = 100, \quad 1 + 2 + 3 + 4 = 10$

(d) $1^3 + 2^3 + 3^3 + 4^3 + 5^3 = 225, \quad 1 + 2 + 3 + 4 + 5 = 15$

(e) $1^3 + 2^3 + 3^3 + 4^3 + 5^3 + 6^3 = 441, \quad 1 + 2 + 3 + 4 + 5 + 6 = 21$

(f) $1^3 + 2^3 + 3^3 + 4^3 + 5^3 + 6^3 + 7^3 = 784, \quad 1 + 2 + 3 + 4 + 5 + 6 + 7 = 28$

Noting $3^2 = 9$, $6^2 = 36$, $10^2 = 100$, $15^2 = 225$, etc., we guess

$$1^3 + 2^3 + \cdots + n^3 = (1 + 2 + \cdots + n)^2 = \left(\frac{n(n+1)}{2}\right)^2 = \frac{n^2(n+1)^2}{4}$$

39. To find $53(47) + 36(72) = 5083$, press the following calculator keys in order: (53 × 47) + (36 × 72) = 5083

41. Press the following calculator keys in order: (72 × 23.5) + (64 × 90.2) − (13.3 × 14.7) = 7269.29

43. ((23.5 × 45.3) − (17.6 × 34.8)) ÷ 17.5 = 25.83257143

45. Press the following calculator keys in order: ((59.5 × 23.6) + (27.5 × 41.7)) ÷ (19.8 + 64.2) = 30.36845238

47. Calculator keys in order are (45 × (13.2 + 17.8)) − (16 × (14.9 + 13.1)) = 947

49. Calculator keys in order are 32 x^2 + 53 x^2 = 3833

51. By calculator, 45 x^2 − 78 x^2 = − 4059

53. By calculator, 5.2 y^x 5 − 7.4 y^x 4 = 803.3827

55. By calculator, (31.2 x^2 + 14.7 y^x 3 =) × (6.2 y^x 3) = 989052.3819

57. By calculator, (8.3 x^2 + 63.2 y^x 3 =) × (51.3 x^2 + 63.2 y^x 3 − 17.4 y^x 4) = 41260314020

59. By calculator, (1.04 y^x 40 − 1.03 y^x 38 =) × (1.02 y^x 12 − 23 y^x 3) = − 21000.93812

61. By calculator, the operations (5.1 y^x 5 + 6.2 y^x 4 =) ÷ (3.2 y^x 2 + 4.1 y^x 3 =) give 62.25143833

63. $\dfrac{32^3 + .43^4}{5.2^2 + 4.5^4} = (32^3 + .43^4)/(5.2^2 + 4.5^4)$. The operations (32 y^x 3 + .43 y^x 4 =) ÷ (5.2 y^x 2 + 4.5 y^x 4 =) give 74.96647626

65. $\dfrac{4^{12} + 1.02^{42}}{1.04^{25} - 13.2} = (4^{12} + 1.02^{42})/(1.04^{25} - 13.2)$. The operations

$(\ 4\ y^x\ 12\ +\ 1.02\ y^x\ 42\ =\ )\div(\ 1.04\ y^x\ 25\ -\ 13.2\ =\ )$ give -1592648.342

67. The operations $(\ 1.6\ y^x\ 20\ +\ 1.05\ y^x\ 30\ =\ )\div(\ 1.015\ y^x\ 20$ $+\ 1.025\ y^x\ 10\ -\ 2313.2\ =\ )$ give -5.234017633

69. (a) To find $3500(1.08)^{15}$ by calculator the operations $1.08\ y^x\ 15\ =\ \times\ 3500\ =$ give \$11,102.59

(b) To find $3500(1.04)^{30}$ by calculator the operations $1.04\ y^x\ 30\ =\ \times\ 3500\ =$ give \$11,351.89

(c) To find $3500(1.02)^{60}$ by calculator the operations $1.02\ y^x\ 60\ =\ \times\ 3500\ =$ give \$11,483.61

(d) To find $3500\left(1 + \dfrac{.08}{12}\right)^{15(12)}$ by calculator the operations $.08\ \div\ 12\ =\ +\ 1$ $=\ y^x\ 15\ =\ y^x\ 12\ =\ \times\ 3500\ =$ give \$11,574.23

71. With Bank A, $500\left(1 + \dfrac{.0695}{12}\right)^{96} = \$870.44$ (By calculator $.0695\ \div\ 12$ $=\ +\ 1\ =\ y^x\ 96\ =\ \times\ 500\ =$ give 870.44). With Bank B, $500\left(1 + \dfrac{.07}{4}\right)^{32}$ $= \$871.11$ (By calculator $.07\ \div\ 4\ =\ +\ 1\ =\ y^x\ 32\ =\ \times\ 500\ =$ give 871.11) Choose Bank B to obtain \$871.11

73. $\dfrac{350\left[(1 + .045)^{40} - 1\right]}{.045} = \$37460.61$ will be the balance.

(By calculator, $1\ +\ .045\ =\ y^x\ 40\ =\ -\ 1\ =\ \times\ 350\ =\ \div\ .045\ =$ give 37460.61)

75. $77,313.47 = \dfrac{R\left[\left(1 + \dfrac{.09}{12}\right)^{144} - 1\right]}{.09/12} \Rightarrow 77,313.47\left(\dfrac{.09}{12}\right)\cdot\dfrac{1}{\left[\left(1 + \dfrac{.09}{12}\right)^{144} - 1\right]}$

$= R = 300.00$. \$300.00 was deposited monthly. (By calculator, $.09\ \div\ 12$ $=\ +\ 1\ =\ y^x\ 144\ =\ -1\ =\ 1\ /\ x\ \times\ 77313.47\ =\ \times\ .09\ =\ \div\ 12$ $=$ give 300.00)

CHAPTER 1

MATHEMATICAL MODELS

EXERCISE SET 1.1 ELEMENTARY INTRODUCTION TO SETS

1. $\{1,2,3,4,5,6\}$

3. $\{1\}$

5. {Wisconsin, Washington, West Virginia, Wyoming}

7. $\{c,d,e\}$

9. {Johnson, Nixon, Ford, Carter, Reagan, Bush}

11. $2x - x = 10 - 3, \quad x = 7.$ The solution set is $\{7\}$.

13. $x = 2x - 6, \quad 6 = x.$ The solution set is $\{6\}$.

15. $10 - 6 = 5x - 3x, \quad 4 = 2x, \quad x = 2.$ The solution set is $\{2\}$.

17. $1 + 4 = 5x - 4x, \quad 5 = x$ but 5 is not in the replacement set. The solution set is $\emptyset$.

19. $8 = 8x, \quad x = 1.$ The solution set is $\{1\}$.

21. Lincoln, Jefferson, Truman

23. 3 and 4 are the only members of the set.

25. $5, 9, 233$

27. $1, 2, 4$

29. Harry Reasoner, Barbara Walters, Phil Donahue

31. False,   since $2 < 3$

33. True, $\pi \doteq 3.14 > 2$

35. False, m is not a vowel in the English alphabet.

37. $2^2 = 4$; The subsets are: $\emptyset, \{1,2\}, \{1\}, \{2\}$

39. $2^5 = 32$; The subsets are: $\emptyset,\{1\},\{2\},\{3\},\{4\},\{5\},\{1,2\},\{1,3\},\{1,4\},\{1,5\},\{2,3\},$
$\{2,4\},\{2,5\},\{3,4\},\{3,5\},\{4,5\},\{3,4,5\},\{2,4,5\},\{2,3,5\},\{2,3,4\},\{1,4,5\},\{1,3,5\},$
$\{1,3,4\},\{1,2,5\},\{1,2,4\},\{1,2,3\},\{1,2,3,4\},\{1,2,3,5\},\{1,2,4,5\},\{1,3,4,5\},$
$\{2,3,4,5\},\{1,2,3,4,5\}$

41. $2^2 = 4$, $4 - 1 = 3$. There are 3 proper subsets.

43. $2^4 = 16$, $16 - 1 = 15$. There are 15 proper subsets.

45. $2^n - 1$

47. (a) 1, (b) 6, (c) 15, (d) 20, (e) 15, (f) 6, (g) 1

$$
\begin{array}{ccccccccccccc}
 & & & & & & 1 & & & & & & \\
 & & & & & 1 & & 1 & & & & & \\
 & & & & 1 & & 2 & & 1 & & & & \\
 & & & 1 & & 3 & & 3 & & 1 & & & \\
 & & 1 & & 4 & & 6 & & 4 & & 1 & & \\
 & 1 & & 5 & & 10 & & 10 & & 5 & & 1 & \\
\hline
1 & & 6 & & 15 & & 20 & & 15 & & 6 & & 1 \\
\end{array}
$$

49. $\{0,7,8,9\}$

51. $\{0,1,2,5,6,7,9\}$

53. $A \bigcup B = \{1,2,3,4,5,6\}$
$A \bigcap B = \{2,5\}$

55. $B \bigcup C = \{2,3,4,5,8\}$
$B \bigcap C = \emptyset$

57. $(A \bigcap B)' = \{2,5\}' = \{0,1,3,4,6,7,8,9\}$
$A' \bigcup B' = \{0,7,8,9\} \bigcup \{0,1,3,4,6,7,8,9\} = \{0,1,3,4,6,7,8,9\}$

59. $(A \bigcap C)' = \{3,4\}' = \{0,1,2,5,6,7,8,9\}$
$A' \bigcup C' = \{0,7,8,9\} \bigcup \{0,1,2,5,6,7,9\} = \{0,1,2,5,6,7,8,9\}$

EXERCISE SET 1.2 VENN DIAGRAMS

1.

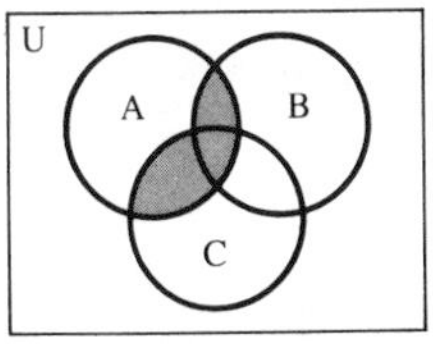

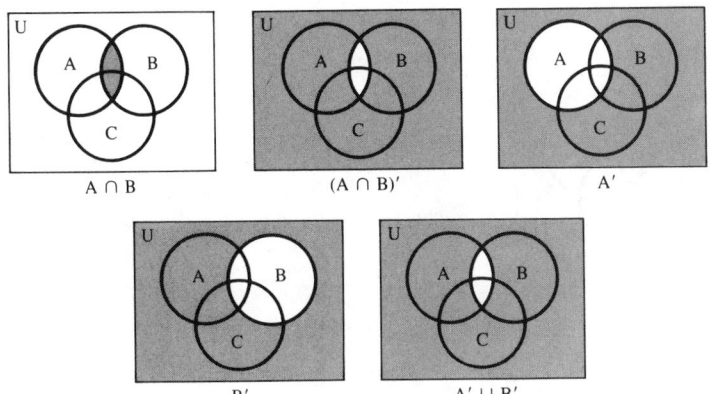

$A \bigcap (B \bigcup C)$

3.

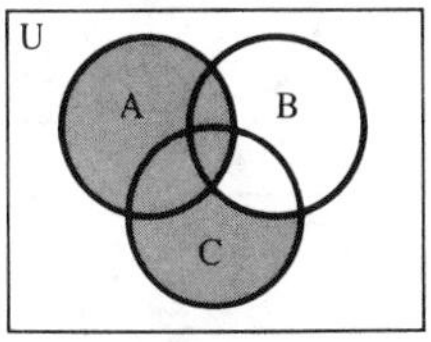

$A \bigcup (B' \bigcap C)$

Shade $B' \bigcap C$ and then shade
the region not already shaded
which lies within A.

5.

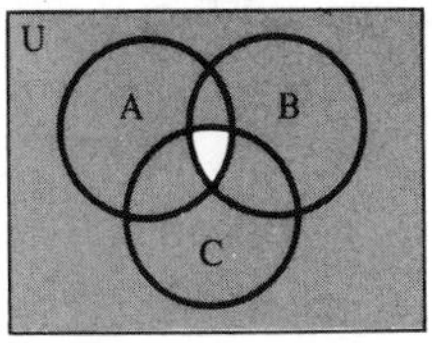

$(A \bigcap B)' \bigcup C'$

Shade the region outside $A \bigcap B$,
and then also shade all points
outside of C.

7.

$A \cap B$

$(A \cap B)'$

A'

B'

$A' \cup B'$

The second and fifth Venn diagrams are the same.

9.

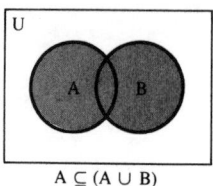

A ⊆ (A ∪ B)

A is within $A \bigcup B$

11.

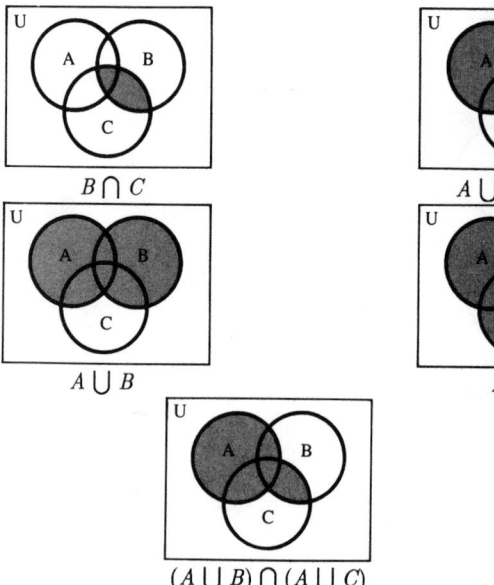

$B \bigcap C$

$A \bigcup (B \bigcap C)$

$A \bigcup B$

$A \bigcup C$

$(A \bigcup B) \bigcap (A \bigcup C)$

The second and fifth Venn diagrams are the same.

13.

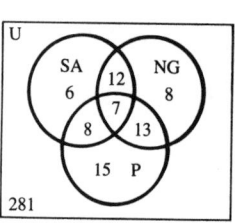

Let SA, NG and P denote the sets of students who read Scientific American, National Geographic, and Playboy respectively.

(a) 6   (b) 8   (c) $350 - 69 = 281$

15.

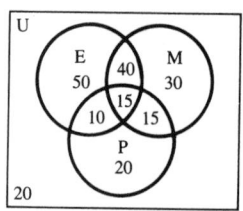

Let E, M and P denote the sets of students who took English, Mathematics and Philosophy respectively.

(a)   15
(b) $40 + 15 = 55$
(c) $50 + 40 + 10 + 15 = 115$
(d) $40 + 30 + 15 + 15 = 100$
(e) $10 + 15 + 15 + 20 = 60$

17.

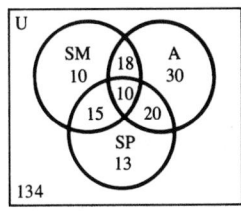

Let SM, A and SP denote the sets of students who smoked, drank alcohol, and played sports respectively.

(a)   13
(b)   20
(c) $250 - 116 = 134$

19.

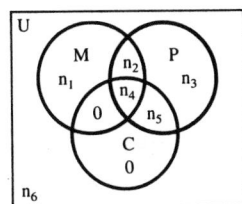 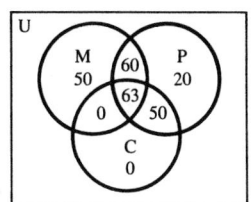

Let M be the set of male customers, P be the set of customers who made a purchase, and C be the set of customers who use a credit card. Note: One can't use a credit card unless they make a purchase.

$$n_1 + n_2 + n_3 + n_4 + n_5 + n_6 = 230$$

$$n_1 + n_2 + n_4 = 173, \quad n_2 + n_3 + n_4 + n_5 = 193$$

$$n_4 + n_5 = 113$$

$$n_2 + n_4 = 123$$

$$n_4 = 63$$

(a)   50
(b)   60
(c)   50
(d)   20

Although $50 + 60 + 63 + 50 + 20 = 243$, only 230 customers were surveyed.

EXERCISE SET 1.3 EQUATIONS IN ONE VARIABLE

1.   All real numbers except $-2, 3$ and -5.

3.   All real numbers except $-1, 1$ and -6.

5.   All real numbers greater than or equal to 4.

7. $5 - 3y = 2y + 20$
$-15 = 5y, \quad y = -3.$ The solution set is $\{-3\}$.

9. $2x - 3 - 5x = 4x - 17$
$-3x - 3 = 4x - 17$
$14 = 7x, \quad x = 2.$ The solution set is $\{2\}$.

11. $3y - 2 + 4y = 2y + 6 - 18$
$7y - 2 = 2y - 12, \quad 5y = -10, \quad y = -2.$ The solution set is $\{-2\}$.

13. $30\left(\dfrac{x+3}{5} - \dfrac{x-8}{3}\right) = \left(\dfrac{x+4}{2}\right) \cdot 30$
$6x + 18 - 10x + 80 = 15x + 60, \quad -4x + 98 = 15x + 60,$
$38 = 19x, \quad x = 2.$ The solution set is $\{2\}$.

15. $x^2 - 6x + 5 = 0$
$(x - 5)(x - 1) = 0, \quad x = 5 \text{ or } x = 1.$ The solution set is $\{1, 5\}$.

17. $z^2 + 7z - 18 = 0$
$(z + 9)(z - 2) = 0, \quad z = -9 \text{ or } z = 2.$ The solution set is $\{2, -9\}$.

19. $y^2 + 10y - 75 = 0$
$(y + 15)(y - 5) = 0, \quad y = -15 \text{ or } y = 5.$ The solution set is $\{5, -15\}$.

21. $x^2 + x - 2 = x^2 - 9 + 13$
$(x - 2) = 4, \quad x = 6.$ The solution set is $\{6\}$.

23. $2z^2 + 3z - z^2 + 1 = 11$
$z^2 + 3z - 10 = 0, \quad (z + 5)(z - 2) = 0$
$z = -5 \text{ or } z = 2.$ The solution set is $\{-5, 2\}$.

25. $5x^2 - 7x - 196 = 0$
$(5x + 28)(x - 7) = 0, \quad x = \dfrac{-28}{5} \text{ or } x = 7.$ The solution set is $\{\dfrac{-28}{5}, 7\}$.

27. $3z^2 + 11z - 60 = 0$
$(3z + 20)(z - 3) = 0, \quad z = \dfrac{-20}{3} \text{ or } z = 3.$ The solution set is $\{\dfrac{-20}{3}, 3\}$.

29. $6(x + 3) + 5(x + 1) = 3(x + 1)(x + 3), \quad 6x + 18 + 5x + 5 = 3x^2 + 12x + 9,$
$11x + 23 = 3x^2 + 12x + 9, \quad 0 = 3x^2 + x - 14, \quad 0 = (3x + 7)(x - 2),$
$x = \dfrac{-7}{3} \text{ or } x = 2$ both of which are in the the replacement set.

The solution set is $\{2, \dfrac{-7}{3}\}$.

31. $\sqrt{x+1} = x - 5$.   Squaring both sides of the equation,
$x + 1 = x^2 - 10x + 25$, $0 = x^2 - 11x + 24$, $0 = (x-8)(x-3)$,
$x = 8$ or 3 but only 8 satisfies the original equation. The solution set is $\{8\}$.

33. Squaring both sides of the equation we obtain $x + 6 + 6\sqrt{x+6} + 9 = 27 + x$,
$6\sqrt{x+6} = 12$, $\sqrt{x+6} = 2$, $x + 6 = 4$, $x = -2$ which satisfies the
original equation.   The solution set is $\{-2\}$.

35. $5 - x = \sqrt{5x - 1}$.   Squaring both sides of the equation,
$25 - 10x + x^2 = 5x - 1$, $x^2 - 15x + 26 = 0$, $(x-2)(x-13) = 0$,
$x = 2$ or $x = 13$ but only 2 satisfies the original equation. The solution set is $\{2\}$.

37. Let x be the amount invested at 12%. Then $40{,}000 - x$ is the amount
invested at 8%. $40000(.09) = x(.12) + .08(40000 - x)$,
$400(9) = .12x + 400(8) - .08x$, $3600 - 3200 = .04x$,
$\dfrac{400}{.04} = \dfrac{40000}{4} = \$10{,}000$. Invest $\$10{,}000$ at 12% and $\$30{,}000$ at 8%.

39. $1600 - \frac{1}{4}p^2 = \frac{37}{11}p + \frac{1}{2}p^2$, $0 = \frac{3}{4}p^2 + \frac{37}{11}p - 1600$,

$$p = \dfrac{-\dfrac{37}{11} \pm \sqrt{\left(\dfrac{37}{11}\right)^2 + 3(1600)}}{\dfrac{3}{2}} = \$44 \;(p \text{ must be positive so the other}$$

solution to the equation is not considered).

41.
number of trees	yield per tree	
30	475	
$30 + x$	$475 - 7x$	for $x \le 20$. x is the number of additional trees planted.

$(30 + x)(475 - 7x) = 16318$, $14250 + 265x - 7x^2 = 16318$,

$0 = 7x^2 - 265x + 2068$, $x = \dfrac{265 \pm \sqrt{265^2 - 4(7)(2068)}}{14} = \dfrac{265 \pm 111}{14}$

$x = 26.86$ or 11. Only 11 is a feasible solution and 41 trees should
be planted per acre.

43. If $\$p$ is the price per hour and x is the number of hours per month,
$p = 120 - 1.5x$ where $p \le 75$.
profit = income $-$ expense, $1264 = px - 100 - 5x$,
$1264 = (120 - 1.5x)x - 100 - 5x$, $1264 = 115x - 1.5x^2 - 100$,
$1.5x^2 - 115x + 1364 = 0$, $15x^2 - 1150x + 13640 = 0$,

$3x^2 - 230x + 2728 = 0$, $x = \dfrac{230 \pm \sqrt{230^2 - 4(3)(2728)}}{6}$.

$x = 62$ or $x = \frac{44}{3}$.   If $x = 62$, $p = \$27$. If $x = \frac{44}{3}$, $p = \$98$ which is not a feasible

solution. Therefore the price per hour was $\$27$ and 62 hours were taught per month.

45. Let x the number of pounds of the first kind. Then $120 - x$ is the
number of pounds of the second kind.
Value of mixture $= x(5.10) + (120 - x)4.50 = 4.90(120)$,

$5.10x + 540 - 4.5x = 588$, $\quad 6x = 48$, $\quad x = \dfrac{480}{6} = 80$. Therefore, use 80 pounds

of the first kind and 40 pounds of the second.

47. Let x be the number of quarts withdrawn. Then $30 - x$ quarts remain leaving
$.1(30 - x) = 3 - .1x$ quarts of antifreeze. Since x quarts of antifreeze are added, the
new amount of antifreeze is $x + 3 - .1x$.
We need $.2(30) = 6$ quarts of antifreeze. $\quad x + 3 - .1x = 6$, $\quad .9x = 3$,

$x = \dfrac{3}{.9} = \dfrac{30}{9} = \dfrac{10}{3} = 3\frac{1}{3}$qt. Drain $3\frac{1}{3}$ quarts.

49. Let x be the number of spark plugs sold. They cost $66\frac{2}{3}$¢ each.

$\text{profit} = \text{income} - \text{expense}$, $\quad \text{profit} = \dfrac{x}{2}(80) + \dfrac{x}{2}(60) - \dfrac{200}{3}x = 1000$,

$40x + 30x - \dfrac{200}{3}x = 1000$, $\quad 210x - 200x = 3000$, $\quad 10x = 3000$, $\quad x = 300$.

He sold 300 spark plugs.

51. Let x be the number of gallons of ginger ale. Then $18 - x$ is the number
of gallons of cranberry juice. The cost of the punch is $x(4) + 6(18 - x) = 84$,
$4x + 108 - 6x = 84$, $\quad 2x = 24$, $\quad x = 12$. 12 gallons of ginger ale were in the punch.
Hence, 2/3 gallon of ginger ale was in one gallon of punch.

53. Let x be the length as shown

$5x^2 = 18000$
$x^2 = 3600$
$x = 60$

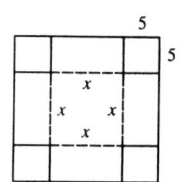

The dimensions of the sheet were 70 cm by 70 cm.

55. Let x and y be the lengths as labeled.

$x \le 35$
$xy = 500$

The cost is $(x + 2y)120 + x(40) = 9200$, $\quad 120x + 240y + 40x = 9200$,

$160x + 240\left(\dfrac{500}{x}\right) = 9200$, $\quad 160x^2 + 120{,}000 = 9200x$, $\quad 160x^2 - 9200x + 120{,}000 = 0$,

$2x^2 - 115x + 1500 = 0$, $\quad x = \dfrac{115 \pm \sqrt{115^2 - 8(1500)}}{4}$. $\quad x = 37.5$ or 20.

Since $x \le 35$, the dimensions are 20 ft (along existing wall) by 25 ft.

EXERCISE SET 1.4 INEQUALITIES

1. $2x + 13 < 5x + 28$, $\quad 13 - 28 < 5x - 2x$, $\quad -15 < 3x$, $\quad -5 < x$

The solution set is $(-5, +\infty)$.

3. $6x + 5 \leq 2x + 33$, $6x - 2x \leq 33 - 5$, $4x \leq 28$, $x \leq 7$

The solution set is $(-\infty, 7]$.

5. $\frac{x}{2} + \frac{1}{3} < \frac{3x}{5} - \frac{2}{7}$, $\frac{1}{3} + \frac{2}{7} < \frac{3x}{5} - \frac{x}{2}$, $\frac{7+6}{21} < \frac{6x - 5x}{10}$, $\frac{13}{21} < \frac{x}{10}$, $\frac{130}{21} < x$

The solution set is $\left(\frac{130}{21}, +\infty\right)$.

7. $3x + 1 < 6x + 4 < 4x + 10$, $3x + 1 < 6x + 4$ and $6x + 4 < 4x + 10$

$-3 < 3x$ and $2x < 6$, $-1 < x$ and $x < 3$, $-1 < x < 3$

The solution set is $(-1, 3)$.

9. $15 + 4x < 8x + 3 < 6x + 5$, $12 < 4x$ and $2x < 2$, $3 < x$ and $x < 2$

which is impossible. No solutions

11. $x^2 + 8x - 20 \leq 0$,

$(x + 10)(x - 2) \leq 0$,

$-10 \leq x \leq 2$

x		-10			2	
$x + 10$	$-$	0	$+$	$+$	$+$	$+$
$x - 2$	$-$	$-$	$-$	$-$	0	$+$
$(x + 10)(x - 2)$	$-$	0	$-$	$-$	0	$+$

The solution set is $[-10, 2]$.

13. $x^2 + 3x - 10 \leq 0$,

$(x - 5)(x + 2) \leq 0$,

$-2 \leq x \leq 5$

x		-2			5	
$x - 5$	$-$	$-$	$-$	0	$+$	
$x + 2$	$-$	0	$+$	$+$	$+$	
$(x - 5)(x + 2)$	$+$	0	$-$	0	$+$	

The solution set is $[-2, 5]$.

15. $8x^3 < 27$, $x^3 < \frac{27}{8} = \left(\frac{3}{2}\right)^3$, $x < \frac{3}{2}$

The solution set is $(-\infty, 1.5)$.

17. $16x^4 - 81 \leq 0$, $16x^4 \leq 81$, $x^4 \leq \frac{81}{16} = \left(\frac{3}{2}\right)^4$, $\frac{-3}{2} \leq x \leq \frac{3}{2}$

The solution set is $[-1.5, 1.5]$.

19.

$$3x^2 + 24x - 9 > 0, \quad x^2 + 8x - 3 > 0, \quad x^2 + 8x + 16 > 3 + 16,$$

$(x+4)^2 > 19, \quad x+4 > \sqrt{19} \text{ or } x+4 < -\sqrt{19}, \quad x > \sqrt{19} - 4 \text{ or } x < -\sqrt{19} - 4$

The solution set is $(-\infty, -\sqrt{19} - 4) \cup (\sqrt{19} - 4, +\infty)$.

21.

$$\frac{x^2 + 3x - 4}{x+5} = \frac{(x+4)(x-1)}{x+5} > 0$$

x		-5		-4		1	
$x+4$	$-$	$-$	$-$	0	$+$	$+$	$+$
$x-1$	$-$	$-$	$-$	$-$	$-$	0	$+$
$x+5$	$-$	0	$+$	$+$	$+$	$+$	$+$
$\dfrac{(x+4)(x-1)}{x+5}$	$-$	U	$+$	0	$-$	0	$+$

$-5 < x < -4$ or $x > 1$. The solution set is $(-5, -4) \cup (1, +\infty)$.

23.

$$\frac{(x-3)(x-2)}{5-x} > 0$$

x		2		3		5	
$x-3$	$-$	$-$	$-$	0	$+$	$+$	$+$
$x-2$	$-$	0	$+$	$+$	$+$	$+$	$+$
$5-x$	$+$	$+$	$+$	$+$	$+$	0	$-$
$\dfrac{(x-3)(x-2)}{5-x}$	$+$	0	$-$	0	$+$	U	$-$

The solution set is $(-\infty, 2) \cup (3, 5)$.

25.

$$\frac{x^2 - 16}{x^3 - 8} < 0$$

x		-4		2		4	
$x^2 - 16$	$+$	0	$-$	$-$	$-$	0	$+$
$x^3 - 8$	$-$	$-$	$-$	0	$+$	$+$	$+$
$\dfrac{x^2 - 16}{x^3 - 8}$	$-$	0	$+$	U	$-$	0	$+$

The solution set is $(-4, 2) \cup (4, +\infty)$

27.

$$\frac{x^2 + 3x + 8}{x + 5} < 0$$

$x^2 + 3x + 8 = 0$

if $x = \dfrac{-3 \pm \sqrt{9 - 32}}{2}$

x		-5	
$x^2 + 3x + 8$	$+$	$+$	$+$
$x + 5$	$-$	0	$+$
$\dfrac{x^3 + 3x + 8}{x + 5}$	$-$	U	$+$

$x^2 + 3x + 8$ has no real solutions. The solution set is $(-\infty, -5)$
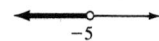
-5

29.

$$\frac{2x^2 + 4x - 4}{x + 1} < 0 \qquad \frac{x^2 + 2x - 2}{x + 1} < 0 \qquad x^2 + 2x - 2 = 0$$

if $x^2 + 2x + 1 = 3 \qquad (x + 1)^2 = 3 \qquad x + 1 = \sqrt{3} \qquad x = \sqrt{3} - 1$

or $x + 1 = -\sqrt{3} \qquad x = -\sqrt{3} - 1$

x		$-\sqrt{3} - 1$		-1		$\sqrt{3} - 1$	
$x - (\sqrt{3} - 1)$	$-$	$-$	$-$	$-$	$-$	0	$+$
$x - (-\sqrt{3} - 1)$	$-$	0	$+$	$+$	$+$	$+$	$+$
$x + 1$	$-$	$-$	$-$	0	$+$	$+$	$+$
$\dfrac{2x^2 + 4x - 4}{x + 1}$	$-$	0	$+$	U	$-$	0	$+$

Solution set is $(-\infty, -\sqrt{3} - 1) \cup (-1, \sqrt{3} - 1)$.

$-\sqrt{3} - 1 \quad -1 \quad \sqrt{3} - 1$

31.

$$\frac{x^2 + 3x - 6}{x + 3} > 0$$

$x^2 + 3x - 6 = 0$ if $x = \dfrac{-3 \pm \sqrt{9 + 24}}{2} = \dfrac{-3 \pm \sqrt{33}}{2}$

$$\frac{\left(x - \left(\frac{-3 + \sqrt{33}}{2}\right)\right)\left(x - \left(\frac{-3 - \sqrt{33}}{2}\right)\right)}{x + 3} > 0$$

x		$\dfrac{-3 - \sqrt{33}}{2}$		-3		$\dfrac{-3 + \sqrt{33}}{2}$	
$x - \left(\dfrac{-3 + \sqrt{33}}{2}\right)$	$-$	$-$	$-$	$-$	$-$	0	$+$
$x - \left(\dfrac{-3 - \sqrt{33}}{2}\right)$	$-$	0	$+$	$+$	$+$	$+$	$+$
$x + 3$	$-$	$-$	$-$	0	$+$	$+$	$+$
$\dfrac{x^2 + 3x - 6}{x + 3}$	$-$	0	$+$	U	$-$	0	$+$

The solution set is $\left(\dfrac{-3 - \sqrt{33}}{2}, -3\right) \cup \left(\dfrac{-3 + \sqrt{33}}{2}, +\infty\right)$

-3

$\dfrac{-3 - \sqrt{33}}{2} \qquad \dfrac{-3 + \sqrt{33}}{2}$

33. $\dfrac{3x^2-6x+2}{x^2-5} \le 0$

$3x^2-6x+2=0$ if $x = \dfrac{6\pm\sqrt{36-24}}{6} = \dfrac{6\pm\sqrt{12}}{6} = \dfrac{3\pm\sqrt{3}}{3}$

$x^2-5=0$ if $x = \pm\sqrt{5}$

x		$-\sqrt{5}$		$\dfrac{3-\sqrt{5}}{3}$		$\dfrac{3+\sqrt{5}}{3}$		$\sqrt{5}$	
$3x^2-6x+2$	$+$	$+$	$+$	0	$-$	0	$+$	$+$	$+$
x^2-5	$+$	0	$-$	$-$	$-$	$-$	$-$	0	$+$
$\dfrac{3x^2-6x+2}{x^2-5}$	$+$	U	$-$	0	$+$	0	$-$	U	$+$

The solution set is $\left(-\sqrt{5}, \dfrac{3-\sqrt{3}}{3}\right] \cup \left[\dfrac{3+\sqrt{3}}{3}, \sqrt{5}\right)$

$$\begin{array}{ccccc} \circ\!\!\!-\!\!\!-\!\!\!-\!\!\!\bullet & & \bullet\!\!\!-\!\!\!-\!\!\!\circ \\ -\sqrt{5} & \dfrac{3-\sqrt{3}}{3}\ \dfrac{3+\sqrt{3}}{3}\ \sqrt{5} \end{array}$$

35. $\dfrac{x^2+5x+2}{x+5} - (2x-3) > 0$

$\dfrac{x^2+5x+2-(2x-3)(x+5)}{x+5} = \dfrac{x^2+5x+2-2x^2-7x+15}{x+5} > 0$

$\dfrac{-x^2-2x+17}{x+5} > 0, \quad \dfrac{x^2+2x-17}{x+5} < 0.$

$x^2+2x-17=0$ if $x = -1 \pm 3\sqrt{2}$

x		$-1-3\sqrt{2}$		-5		$-1+3\sqrt{2}$	
$x+5$	$-$	$-$	$-$	0	$+$	$+$	$+$
$x^2+2x-17$	$+$	0	$-$	$-$	$-$	0	$+$
$\dfrac{x^2+2x-17}{x+5}$	$-$	0	$+$	U	$-$	0	$+$

The solution set is $(-\infty, -1-3\sqrt{2}) \cup (-5, -1+3\sqrt{2})$

$$\longleftarrow\!\!\!\!\circ\!\!-\!\!\!-\!\!\circ\!\!\!\qquad\!\!\circ\!\!-\!\!\!\longrightarrow$$
$$-1-3\sqrt{2}\ \ -5 \qquad -1+3\sqrt{2}$$

37. $\dfrac{3x+2}{x-3} - \dfrac{6x+4}{x-2} = \dfrac{(3x+2)(x-2)-(x-3)(6x+4)}{(x-3)(x-2)} > 0$

$\dfrac{3x^2-10x-8}{(x-3)(x-2)} = \dfrac{(3x+2)(x-4)}{(x-3)(x-2)} < 0$

x		$-\frac{2}{3}$		2		3		4	
$3x+2$	$-$	0	$+$	$+$	$+$	$+$	$+$	$+$	$+$
$x-4$	$-$	$-$	$-$	$-$	$-$	$-$	$-$	0	$+$
$x-2$	$-$	$-$	$-$	0	$+$	$+$	$+$	$+$	$+$
$x-3$	$-$	$-$	$-$	$-$	$-$	0	$+$	$+$	$+$
$\dfrac{(3x+2)(x-4)}{(x-3)(x-2)}$	$+$	0	$-$	U	$+$	U	$-$	0	$+$

The solution set is $(-2/3,2) \cup (3,4)$

39. $-x^2 + 3x - 9 = 0$ if $x = \dfrac{-3 \pm \sqrt{9-36}}{-2}$. There are no real solutions. Thus,

$-x^2 + 3x - 9$ is always positive or always negative and if $x = 0$, $-x^2 + 3x - 9$ is negative.

x		-2		-1	
$-x^2 + 3x - 9$	$-$	$-$	$-$	$-$	$-$
$x+2$	$-$	0	$+$	$+$	$+$
$x+1$	$-$	$-$	$-$	0	$+$
$\dfrac{-x^2 + 3x - 9}{(x+2)(x+1)}$	$-$	U	$+$	U	$-$

The solution set is $(-\infty,-2) \cup (-1,+\infty)$

41. Assume a, b and c are real numbers, $a < b$ and $c < 0$.
$b-a$ is positive, c is negative. Hence, $(b-a)(c)$ is negative.
$(b-a)c < 0$
$bc - ac < 0$
$bc - ac+ + ac < 0 + ac$
$bc < ac$

43. The solution shown is not correct as $x = -5$ is not a solution which as may be easily verified. In the last step, both sides of the inequality are multiplied by $\dfrac{1}{x+2}$. This step is invalid since we don't know if $\dfrac{1}{x+2}$ is positive, negative, or worse yet, undefined.

45. Let x be the hours Betty must work.

$(x + 10)6 + 7x \geq 385$

$13x \geq 385 - 60 = 325$

$x \geq \dfrac{325}{13} = 25$ hours

47. Profit $=$ revenue $-$ cost > 0

$$p(252 - 7p) - 1092 - 4q > 0$$

$$p(252 - 7p) - 4(252 - 7p) - 1092 > 0$$

$$252p - 7p^2 - 1008 + 28p > 1092$$

$$0 > 7p^2 - 280p + 2100 = 7(p^2 - 40p + 300)$$

$$0 > (p - 30)(p - 10)$$

$$\$10 < p < \$30$$

p		10		30	
$p - 10$	$-$	0	$+$	$+$	$+$
$p - 30$	$-$	$-$	$-$	0	$+$
$(p - 10)(p - 30)$	$+$	0	$+$	0	$+$

49. We want the revenue to exceed the cost.

$$p(935 - 17p) > 5100 + 15(935 - 17p)$$

$$935p - 17p^2 > 5100 + 14025 - 255p$$

$$0 > 17p^2 - 1190p + 19125$$

$$0 > 17(p^2 - 70p + 1125)$$

$$0 > (p - 45)(p - 25)$$

$$\$25 < p < \$45$$

51. Revenue$-$cost $\geq 400{,}000$

$q(1000)(30 - .25q) - (200 + 5q)(1000) \geq 400{,}000$ where q is the number of units sold in thousands.

$$q(30 - .25q) - 200 - 5q \geq 400$$

$$30q - .25q^2 - 200 - 5q \geq 400$$

$$0 \geq .25q^2 - 25q + 600$$

$$0 \geq q^2 - 100q + 2400$$

$$0 \geq (q - 40)(q - 60)$$

q		40		60	
$q - 40$	$-$	0	$+$	$+$	$+$
$q - 60$	$-$	$-$	$-$	0	$+$
$(q-40)(q-60)$	$+$	0	$-$	0	$+$

$$40 \leq q \leq 60$$

At least 40,000 units but no more than 60,000 units must be produced and sold.

53. Let x be the number of quarts to be drained.

$$.35(15) \leq \text{quarts of antifreeze} \leq .45(15)$$

$$5.25 \leq (15 - x)(.25) + x \leq 6.75$$

$$5.25 \leq \frac{15}{4} - \frac{x}{4} + x \leq 6.75$$

$$5.25 \leq \frac{15}{4} + \frac{3}{4}x \leq 6.75$$

$$21 \leq 15 + 3x \leq 27$$
$$6 \leq 3x \leq 12$$
$$2 \text{ quarts} \leq x \leq 4 \text{ quarts}$$

55. $x^2 + 2x + 6 = (x+1)^2 + 5$ and $x^2 + 8x + 20 = (x+4)^2 + 4$ are always positive. Thus, the solution is found by examining the signs of the other factors.

x			-5					-3				5	
$(x+3)^3$	$-$	$-$	$-$	$-$	$-$	$-$	0	$+$	$+$	$+$	$+$	$+$
$(x-5)^7$	$-$	$-$	$-$	$-$	$-$	$-$	$-$	$-$	$-$	$-$	0	$+$
$(x+5)^3$	$-$	$-$	0	$+$	$+$	$+$	$+$	$+$	$+$	$+$	$+$	$+$
fraction	$-$	$-$	U	$+$	$+$	$+$	0	$-$	$-$	$-$	0	$+$

The solution set is $(-\infty, -5) \cup (-3, 5)$

EXERCISE SET 1.5 FUNCTIONS AND GRAPHS

1. (a) The domain is $\{0, 1, 2, 3, 4\}$ (b) $f(1) = 2$, $f(3) = 6$, $f(4) = 8$.

3. (a) The domain is $\{0, 1, 2, 3, 4\}$ (b) $f(0) = 0$, $f(2) = 8$, $f(4) = 64$.

5. (a) The domain is S.   (b) $f(-2) = 4 + 1 = 5$, $f(2) = 4 + 1 = 5$
$f(6) = 36 + 1 = 37$.

7. $\{x \mid x \neq 4\}$

9. $\{x \mid x \neq 4 \text{ or } -4\}$

11. $x - 4 > 0 \rightarrow x > 4$.   The domain is $(4, \infty)$

13. $f(5) = \dfrac{2}{5-4} = 2, \quad f(0) = \dfrac{2}{0-4} = -.5, \quad f(-3) = \dfrac{2}{-3-4} = -\dfrac{2}{7}.$

15. $h(-2) = \dfrac{-2-2}{4-16} = \dfrac{-4}{-12} = \dfrac{1}{3}, \quad h(1) = \dfrac{1-2}{1-16} = \dfrac{-1}{-15} = \dfrac{1}{15}, \quad h(5) = \dfrac{5-2}{25-16} = \dfrac{3}{9} = \dfrac{1}{3}.$

17. $g(9) = \sqrt{9-8} = 1, \quad g(12) = \sqrt{12-8} = \sqrt{4} = 2, \quad g(5)$ is undefined.

19.

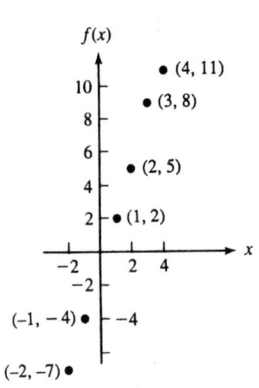

21.

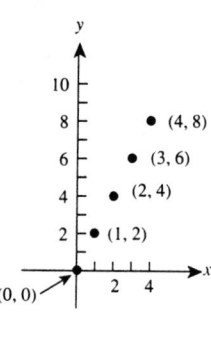

23.   The slope of the line $m = \dfrac{6-(-2)}{-3-(-5)} = \dfrac{6+2}{-3+5} = \dfrac{8}{2} = 4$

The equation of the line is $y-(-2) = 4\left(x-(-5)\right) \qquad y+2 = 4(x+5)$

$y+2 = 4x + 20 \qquad y = 4x + 18$

25.   Since the two points have the same second coordinate of 2 the line is horizontal and has the equation $y = 2$.

27.   By Exercise 23, the slope of the line is 4. The equation of the line is $y - 8 = 4(x-3) \qquad y - 8 = 4x - 12 \qquad y = 4x - 4.$

29.   By Exercise 25, the line that is perpendicular is horizontal. Hence this line is vertical and has the equation $x = 3$.

31.   The perpendicular line has equation $12x - 6y = 36$ Hence $6y = 12x - 36$
$y = 2x - 6$ and its slope is 2. Hence our line has slope $-\dfrac{1}{2}$
and since $(-7, 0)$ lies on the line its equation is

$y - 0 = -\dfrac{1}{2}\left(x - (-7)\right) \qquad 2y = -x - 7 \qquad x + 2y + 7 = 0$

33. $f(x) = x^2 + 10x - 4$ gives a parabola which is concave up. The vertex has first coordinate $x = -\frac{10}{2} = -5$. The axis of symmetry is $x = -5$
Some points on the graph are

x	-7	-6	-5	-4
$f(x)$	-25	-28	-29	-28

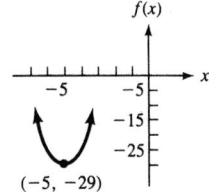

$(-5, -29)$

35. $h(x) = 3x^2 - 18x + 5$ gives a parabola which is concave up. The first coordinate of the vertex is $x = -\frac{(-18)}{6} = 3$. Some points on the graph are

x	1	2	3	4
$h(x)$	-10	-19	-22	-19

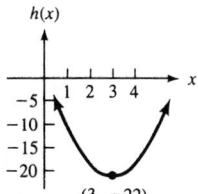

$(3, -22)$

37. $g(x) = -4x^2 + 24x - 9$. The parabola is concave down. The vertex has first coordinate $x = \frac{-24}{2(-4)} = 3$. Some points on the graph are

x	1	2	3	4
$g(x)$	11	23	27	23

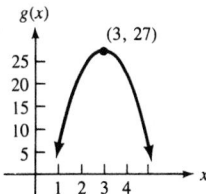

$(3, 27)$

39. (a) $g(x) = |x|$. Some points on the graph are

x	-2	-1	0	1	2
$g(x)$	2	1	0	1	2

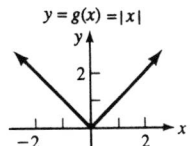

$y = g(x) = |x|$

(b) $h(x) = |x - 2|$. Some points on the graph are

x	-1	0	1	2	3	4
$h(x)$	3	2	1	0	1	2

The function shifts the graph of the absolute value function 2 units to the right.

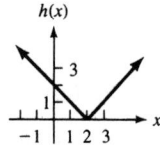

41.   Hours worked x Earnings I(x).

$\quad 0 \le x \le 40$ $14x$

$\quad 40 \le x \le 48$ $40(14) + (x - 40)21 = 21x - 280$

$\quad 168 \ge x \ge 48$ $40(14) + 8(21) + (x - 48)28$
$\qquad\qquad\qquad\qquad\qquad = 728 + 28x - 1344 = 28x - 616$

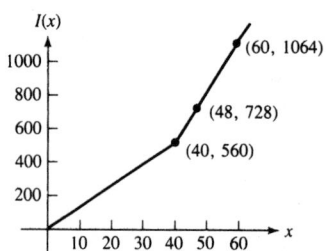

43. (a) $p = mx + b$ $(90, 45)$ and $(135, 30)$ are on the graph.

Hence $m = \dfrac{45 - 30}{90 - 135} = \dfrac{15}{-45} = -\dfrac{1}{3}$

$p - p_1 = m(x - x_1)$

$p - 30 = -\dfrac{1}{3}(x - 135)$

$p - 30 = -\dfrac{1}{3}x + 45$

$p = -\dfrac{1}{3}x + 75$

(b)   revenue = no sold · price = $x\left(-\dfrac{1}{3}x + 75 \right)$

$\qquad\qquad\qquad\qquad R = -\dfrac{1}{3}x^2 + 75x$

(c) profit = revenue − expense = $-\frac{1}{3}x^2 + 75x - (200 + 21x)$

$P(x) = -\frac{1}{3}x^2 + 54x - 200$. The graph is a parabola which is concave

down. Hence $P(x)$ has a maximum when $x = \dfrac{-54}{-\frac{2}{3}} = 27(3) = 81$ razors.

The maximum monthly profit is $-\frac{1}{3}(81)^2 + 54(81) - 200 = \$1,987$

45.

20 feet | 26.25 feet | 20 feet

26.25 feet

Let x be the length of the side costing

$13 per foot. Cost of fencing is

$$x(13) + (x + 2y)8 = 840 \qquad 21x + 16y = 840$$

$$y = \frac{840 - 21x}{16} = \frac{840}{16} - \frac{21}{16}x$$

$$\text{Area} = xy = x\left(\frac{840}{16} - \frac{21}{16}x\right) = -\frac{21}{16}x^2 + \frac{105}{2}x$$

The graph of this function is a parabola which is concave downward

and has a maximum when $x = \dfrac{-\frac{105}{2}}{-\frac{42}{16}} = \dfrac{105(16)}{42(2)} = 20$. The dimensions of the

rectangle that encloses the largest area are as shown.

47.

tuition/credit hour	credit hours taken
$160	225000
$160 + x$	$225000 - 1250x$

revenue $= (160 + x)(225000 - 1250x) = 36000000 + 25000x - 1250x^2$

The graph is a parabola which is concave downward. Hence the maximum revenue

occurs when $x = \dfrac{-25000}{-2(1250)} = \$10$. Increase tuition by $10 per credit hour.

49. (a)

Price per paddle (p)	Number sold (x)
$ 15	10
$ 5	30

$p = mx + b$ $(10, 15)$ and $(30, 5)$ are on the graph. Hence $m = \dfrac{15 - 5}{10 - 30}$

$= \dfrac{10}{-20} = -\frac{1}{2}$ $p - p_1 = -\frac{1}{2}(x - x_1)$ $p - 5 = -\frac{1}{2}(x - 30)$

$p = -\frac{1}{2}x + 15 + 5$ $p = -\frac{1}{2}x + 20$

(b) Revenue $= x\left(-\frac{1}{2}x + 20\right) = -\frac{1}{2}x^2 + 20x$

(c) Profit $=$ revenue $-$ expense $= -\frac{1}{2}x^2 + 20x - 25 - 4x = -\frac{1}{2}x^2 + 16x - 25$.

The maximum occurs when $x = \dfrac{-16}{-1} = 16$. Selling 16 paddles will

maximize the profit.

1. $5\left(8^{\frac{2}{3}}\right)\left(9^{\frac{3}{2}}\right) = 5 \cdot \left(\sqrt[3]{8}\right)^2 \cdot \left(\sqrt{9}\right)^3 = 5 \cdot 2^2 \cdot 3^3 = 5 \cdot 4 \cdot 27 = 540$

3. $7\left(\sqrt[3]{27}\right)^4 / \left(\sqrt{49}\right)^3 = 7(3^4) / 7^3 = 3^4 / 7^2 = 81 / 49$

5. $\left(\sqrt[4]{16}\right)^{-5}\left(\sqrt[3]{8}\right)^7 / \left(\sqrt[3]{27}\right)^{-5} = \dfrac{2^{-5} \cdot 2^7}{3^{-5}} = 2^2 \cdot 3^5 = 4 \cdot 243 = 972$

7. $\sqrt[5]{59049} \cdot \left(\sqrt[3]{4096}\right)^2 = 9(16)^2 = 2304$

9. $\sqrt[3]{81} \cdot \sqrt[5]{729} / \left(\sqrt[5]{2187}\right)^6 = \left(3^4\right)^{\frac{1}{3}} \cdot \left(3^6\right)^{\frac{1}{5}} / \left(3^7\right)^{\frac{6}{5}} = \dfrac{3^{\frac{4}{3}+\frac{6}{5}}}{3^{\frac{42}{5}}}$

$= 3^{\frac{4}{3}+\frac{6}{5}-\frac{42}{5}} = 3^{\frac{20+18-126}{15}} = 3^{-\frac{88}{15}}$

11. $f(x) = 3^x$. Some of the points on the graph are

x	-2	-1	0	1	2
$f(x)$	$\frac{1}{9}$	$\frac{1}{3}$	1	3	9

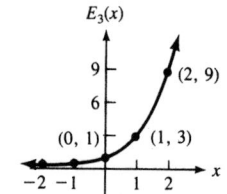

13. $f(x) = \left(\frac{1}{2}\right)^x$. Some points on the graph are

x	-3	-2	-1	0	1	2
$f(x)$	8	4	2	1	$\frac{1}{2}$	$\frac{1}{4}$

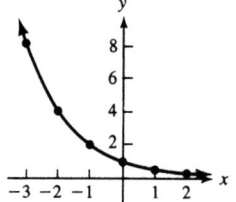

15. $f(x) = \left(\frac{1}{4}\right)^x$. Some points on the graph are

x	-2	-1	0	1	2
$f(x)$	16	4	1	$\frac{1}{4}$	$\frac{1}{16}$

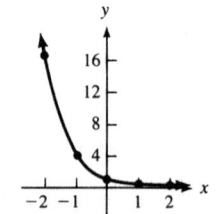

17. (a) $\left(h \circ g\right)(x) = h(2x-3) = 3^{2x-3}$.   Some points on the graph are

x	-1	0	1	2	3
$\left(h \circ g\right)(x)$	$\dfrac{1}{243}$	$\dfrac{1}{27}$	$\dfrac{1}{3}$	3	27

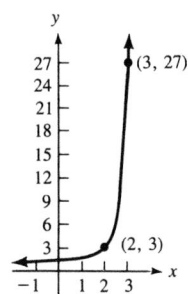

(b) $\left(g \circ h\right)(x) = g\left(3^{x}\right) = 2\left(3^{x}\right) - 3$.   Some points on the graph are

x	-1	0	1	2
$g \circ h(x)$	$-\dfrac{7}{3}$	-1	3	15

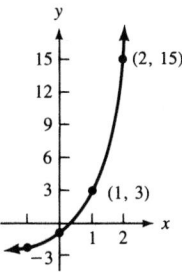

19. $A = 5000(1+.08/4)^{9 \cdot 4} = 5000(1.02)^{36} = \$10{,}199.44$. The interest is earned is $\$10{,}199.44 - \$5000 = \$5199.44$.

21. $A = 8000\left(1+\dfrac{.09}{12}\right)^{28} = \$9861.69$. Compound interest earned is $\$9861.69 - \$8000 = \$1{,}861.69$.

23.   The amounts of the deposit at the three banks after 3 months are:

Bank A: $A = 3000\left(1+\dfrac{.0795}{365}\right)^{90} = \$3060.21$

Bank B: $A = 3000\left(1+\dfrac{.08}{12}\right)^{3} = \$3060.40$

Bank C: $A = 3000\left(1+\dfrac{.081}{4}\right) = \$3060.75$

Hence choose Bank C for greatest interest.

25. Effective interest rate $= 1\left(1 + \frac{.08}{4}\right)^4 - 1 = (1.02)^4 - 1 = .0824$. Hence \$1 grows to \$1.0824 in one year and the effective interest rate is 8.24%.

27. If \$1 is invested for one year, it grows to $1\left(1 + \frac{.07}{365}\right)^{365} = \$1.0725$. Hence the effective interest rate is 7.25%.

29. $A = 4000e^{(.09)(15)} = 4000(3.8574) = \$15,429.70$

31. $P = 75000(1 + .025)^{10} = 75000(1.28008) = 96006$

33. $P = 250000(1.035)^{18} = 250000(1.8575) = 464,372$

35. Since growth is exponential and the initial population is 150,000, the population after t hours is given by $P(t) = 150,000b^t$ where b is a constant. Since the population after 23 hours is 250,000, $250,000 = 150,000b^{23}$ $\quad \frac{25}{15} = \frac{5}{3} = b^{23}, \quad b = \sqrt[23]{\frac{5}{3}}$ and $P(t) = 150,000\left(\frac{5}{3}\right)^{\frac{t}{23}}$ $\quad P(86) = 150,000\left(\frac{5}{3}\right)^{\frac{86}{23}} = 1,013,007$

37. $P(t) = P_0 b^t$. We let $P_0 = 5$ so $P(t) = 5b^t$. But $P(5730) = 2.5$ and $2.5 = 5\left(b^{5730}\right)$ Hence $(.5)^{\frac{1}{5730}} = b$ $\quad P(t) = 5(.5)^{\frac{t}{5730}}$
$P(17190) = 5(.5)^{\frac{17190}{5730}} = 5(.5)^3 = .625\,\text{gm}.$

39. $P(t) = P_0 b^t$. Let $t = 0$ at the beginning of 1975. Then $P_0 = 3$ and $P(t) = 3b^t$. We know that $P(12.3) = \frac{3}{2} = 1.5$.
Hence $1.5 = 3b^{12.3}$, $\frac{1}{2} = b^{12.3}$, $(.5)^{\frac{1}{12.3}} = b$. $P(t) = 3\left((.5)^{\frac{1}{12.3}}\right)^t$
or $P(t) = 3(.5)^{\frac{t}{12.3}}$ If observation begins in 1975 at the end of year 2000 there will be $P(t) = 3(.5)^{\frac{26}{12.3}} = .6931\,\text{gm}.$

41. (a) $W(t) = 110\left(1 - e^{-.2(3)}\right) = 49.63$ words/minute

(b) $W(t) = 110\left(1 - e^{-.2(6)}\right) = 76.87$ words/minute

(c) $W(t) = 110\left(1 - e^{-.2(12)}\right) = 100.02$ words/minute

43.   If the value is declining at the rate of 15% per year, the value

after t years is $V(t) = 30{,}000(1 - .15)^t = 30{,}000(.85)^t$

(a) $30000(1 - .15)^2 = 30000(.85)^2 = \$21{,}675$

(b) $V(4) = 30000(.85)^4 = \$15{,}660$

(c) $V(8) = 30000(.85)^8 = \$8{,}175$

45. (a) $15{,}000(1 - .18)^4 = 15{,}000(.82)^4 = \$6{,}782$

(b) $15{,}000(.82)^6 = \$4{,}560$

(c) $15{,}000(.82)^8 = \$3{,}066$

47.

m	2	5	15	50	150	500	1500
$\left(1 + \frac{1}{m}\right)^m$	2.25	2.48832	2.63288	2.69159	2.70928	2.71557	2.71738

EXERCISE SET 1.7 THE LOGARITHMIC FUNCTION

1. (a) $\log 64 = \log 2^6 = 6 \log 2 = 6(.3010) = 1.8060$

(b) $\log 36 = \log 2^2 \cdot 3^2 = \log 2^2 + \log 3^2 = 2 \log 2 + 2 \log 3$

$= 2(.3010) + 2(.4771) = 1.5562$

(c) $\log 54 = \log 27(2) = \log 3^3 + \log 2 = 3 \log 3 + \log 2 = 3(.4771) + .3010 = 1.7323$

3. (a) $\log 4^{.2} = .2 \log 4 = .2 \log 2^2 = .4 \log 2 = .4(.3010) = .1204$

(b) $\log \sqrt[4]{27} = \frac{1}{4} \log 3^3 = \frac{3}{4} \log 3 = \frac{3}{4}(.4771) = .357825$

(c) $\log \sqrt[7]{.75} = \frac{1}{7} \log\left(\frac{3}{4}\right) = \frac{1}{7}(\log 3 - \log 2^2) = \frac{1}{7}(.4771 - 2(.3010)) = -.0178428$

5. (a) $\log_3 2 = \dfrac{\log 2}{\log 3} = \dfrac{.3010}{.4771} = .6309$

(b) $\log_4 27 = \dfrac{\log 27}{\log 4} = \dfrac{\log 3^3}{\log 2^2} = \dfrac{3 \log 3}{2 \log 2} = \dfrac{3(.4771)}{2(.3010)} = 2.3776$

(c) $\log_9 64 = \dfrac{\log 64}{\log 9} = \dfrac{\log 2^6}{\log 3^2} = \dfrac{6 \log 2}{2 \log 3} = \dfrac{6(.3010)}{2(.4771)} = 1.8927$

7. (a) $\ln(1.5) = \ln\left(\frac{3}{2}\right) = \ln 3 - \ln 2 = 1.0986 - .6931 = .4055$

(b) $\ln .75 = \ln\left(\frac{3}{4}\right) = \ln 3 - \ln 2^2 = \ln 3 - 2 \ln 2 = 1.0986 - 2(.6931) = -.2876$

(c) $\ln\left(\frac{27}{16}\right) = \ln 27 - \ln 16 = \ln 3^3 - \ln 2^4$

$= 3 \ln 3 - 4 \ln 2 = 3(1.0986) - 4(.6931) = .5234$

9. (a) $\ln 5e^3 = \ln 5 + \ln e^3 = \ln 5 + 3 \ln e = \ln 5 + 3 \cdot 1 = 3 + \ln 5$

(b) $\ln \sqrt[3]{e^2} = \ln e^{\frac{2}{3}} = \left(\frac{2}{3}\right)\ln e = \frac{2}{3}(1) = \frac{2}{3}$

(c) $\ln\left(\frac{27}{e}\right) = \ln 27 - \ln e = 3 \ln 3 - 1 = 3(1.0986) - 1 = 2.2958$

11. $\log_3(2x + 3) = 4.$ $2x + 3 > 0$ if $2x > -3$ or $x > -1.5$.

$3^{\log_3(2x+3)} = 3^4$, $2x + 3 = 81$ $2x = 78$

$x = 39$ which is larger than -1.5 The solution set is $\left\{39\right\}$.

13. $\log_5(5 - 3x) = 2.$ $5 - 3x > 0$ if $5 > 3x$ or $x < \frac{5}{3}$. $5^{\log_5(5-3x)} = 5^2$

$5 - 3x = 25;$ $-20 = 3x$ $x = -\frac{20}{3}$ which is less than $\frac{5}{3}$. The solution set

is $\left\{-\frac{20}{3}\right\}$.

15. $\log_2(5 - 6x) = 3 + \log_2(3x - 7)$ $5 - 6x > 0$ if $5 > 6x$ or $x < \frac{5}{6}$

$3x - 7 > 0$ if $3x > 7$ or $x > \frac{7}{3}$ Since no number is greater than $\frac{7}{3}$

and less than $\frac{5}{6}$, there are no solutions to the equation.

17. $\log_{12}(x - 2) = 1 - \log_{12}(x - 3)$ $x - 2 > 0$ if $x > 2$ $x - 3 > 0$ if $x > 3$.

Hence any solution must be larger than 3. $\log_{12}(x - 2)(x - 3) = 1$

$(x - 2)(x - 3) = 12^1,$ $(x - 2)(x - 3) = 12$ $x^2 - 5x - 6 = 0$

$(x - 6)(x + 1) = 0$ if $x = 6$ or $x = -1.$ -1 is not a solution and

the solution set is $\left\{6\right\}$.

19. $\log_2(x + 7) = 2 - \log_2(x + 4)$ $x + 7 > 0$ if $x > -7$ $x + 4 > 0$ if $x > -4$

Hence any solution must be larger than -4. $\log_2(x + 7) + \log_2(x + 4) = 2$

$\log_2(x + 7)(x + 4) = 2$ $(x + 7)(x + 4) = 2^2$

$(x + 7)(x + 4) = 4$ $x^2 + 11x + 28 = 4$ $x^2 + 11x + 24 = 0$

$(x + 3)(x + 8) = 0$ $x = -3$ or $x = -8.$ Hence -3 is the only solution

since $-8 < -4$.

21. $\log|x+4| + \log|x+1| = 1$, $x \neq -4$ and $x \neq -1$.

$\log|x+4\|x+1| = 1$ $|x+4\|x+1| = 10$ $|(x+4)(x+1)| = 10$

$|x^2+5x+4| = 10$ $x^2+5x+4 = 10$ or $x^2+5x+4 = -10$

$x^2+5x-6 = 0$ or $x^2+5x+14 = 0$ $(x+6)(x-1) = 0$ or

$x = \dfrac{-5 \pm \sqrt{25-56}}{2}$ which are not real. $x = -6$ or $x = 1$ are the solutions.

23. $x > 0$, $x \neq 1$, and $x > -12$. $x+12 = x^2$

$x^2-x-12 = 0$ $(x-4)(x+3) = 0$ $x = 4$ or -3 4 is the only

solution as -3 can not be a base for a logarithmic function.

25. $x \neq 1$ and $3+x-3x^2 > 0$,   and $x > 0$. $3+x-3x^2 = x^3$

$x^3+3x^2-x-3 = x^2(x+3) - (x+3) = (x^2-1)(x+3)$

$= (x-1)(x+1)(x+3) = 0$ $x = 1$, $x = -3$ or $x = -1$ However none

of these can serve as bases for logarithms.   There are no solutions.

27. $3^{2x-1} = 2e^{3x+5}$ $\ln 3^{2x-1} = \ln\left(2e^{3x+5}\right)$ $(2x-1)\ln 3 = \ln 2 + (3x+5)\ln e$

$2x \ln 3 - \ln 3 = \ln 2 + 3x + 5$ $2x \ln 3 - 3x = \ln 2 + \ln 3 + 5$

$x(2 \ln 3 - 3) = 5 + \ln 6$ $x = \dfrac{5+\ln 6}{2 \ln 3 - 3} = \dfrac{5+\ln 6}{-3+\ln 9} = -8.4603$

29. $A = P(1+\frac{.08}{4})^t$ where t the number of quarters since the deposit

was made. $3P_0 = P_0(1.02)^t$ $3 = (1.02)^t$ $\ln 3 = \ln(1.02)^t = t \ln(1.02)$

$t = \dfrac{\ln 3}{\ln(1.02)} = 55.48$ quarters. Thus 56 quarters are needed.

31. $3P_0 = P_0(1.005)^t$ where t is the number of months since the deposit

was made. $3 = (1.005)^t$ $\ln 3 = t \ln(1.005)$ $t = \dfrac{\ln 3}{\ln(1.005)} = 220.27$ months.

221 months are needed.

33. $3P_0 = P_0 \, e^{.08t}$ where t is the number of years since the deposit was

made. $3 = e^{.08t}$ $\ln 3 = .08t \ln e$, $t = \dfrac{\ln 3}{.08} = 13.73$ years.

35. $P = P_0(1+.30)^t$ $60000 = 20000(1.3)^t$ $3 = (1.3)^t$ $\dfrac{\ln 3}{\ln (1.3)} = t = 4.187$.

Thus the population will reach 60000 in March of 1991.

37. $W(t) = 110\left(1 - e^{-.2t}\right)$. We want t such that $W(t) = 80 = 110\left(1 - e^{-.2t}\right)$

$\frac{8}{11} = 1 - e^{-.2t}$ $e^{-.2t} = \frac{3}{11}$ $\ln\left(e^{-.2t}\right) = \ln\left(\frac{3}{11}\right)$ $-.2t \ln e = \ln\left(\frac{3}{11}\right)$

$t = \dfrac{\ln\left(\frac{3}{11}\right)}{-.2} = 6.496$ weeks. The student should attend school $6\frac{1}{2}$ weeks.

39. $P(t) = P_0\, b^t$. Since the population grew to 195,000 in 15 years,

$195 = 150 b^{15}$. $\frac{195}{150} = b^{15}$ and $\left(\frac{195}{150}\right)^{\frac{1}{15}} = b$. $P(t) = 150,000\left(\frac{195}{150}\right)^{\frac{t}{15}}$.

We want to find t such that $P(t) = 225000 = 150000\left(\frac{195}{150}\right)^{\frac{t}{15}}$.

$\frac{225}{150} = \left(\frac{195}{150}\right)^{\frac{t}{15}}$. $\ln\left(\frac{225}{150}\right) = \ln\left(\frac{195}{150}\right)^{\frac{t}{15}}$. $\ln\left(\frac{225}{150}\right) = \left(\frac{t}{15}\right)\ln\left(\frac{195}{150}\right)$.

$t = \dfrac{15\ln\left(\frac{225}{150}\right)}{\ln\left(\frac{195}{150}\right)} = \dfrac{6.082}{.262} = 23.18$ years or in the year 1993.

41. $P = P_0\, b^t$ Since the half life is 1690 years, $.5 = 1 \cdot b^{1690}$ $b = (.5)^{\frac{1}{1690}}$.

Since we want $\frac{3}{4}$ of the mass to remain, $.75 = 1(.5)^{\frac{t}{1690}}$

$\ln(.75) = \ln\left(.5^{\frac{t}{1690}}\right) = \left(\frac{t}{1690}\right)\ln(.5)$ $t = \dfrac{1690\ln(.75)}{\ln(.5)} = 701.4$ years

43. Let $z = y\log_b x$, $x > 0$. $b^z = b^{y\log_b x} = \left(b^{\log_b x}\right)^y$ [by property 9 of

exponential functions] $= x^y$ [by property 3 of logarithmic functions]. Hence

$\log_b x^y = \log_b b^z = z$ [by property 9 of logarithmic functions] and $z = y\log_b x = \log_b x^y$.

45. Let $x = \log_a b$, $y = \log_b c$, $z = \log_c d$ and $w = \log_a d$. Then
$a^x = b$, $b^y = c$, $c^z = d$ and $a^w = d$. Since $a^x = b$ and $b^y = c$,
$\left(a^x\right)^y = a^{xy} = c$. $c^z = \left(a^{xy}\right)^z = a^{xyz} = d = a^w$. Equating exponents
$w = xyz$ or $\log_a d = \log_a b \log_b c \log_c d$, for $a,b,c,d > 0$ and a,b and $c \neq 1$.
A generalization is $\log_a f = (\log_a b)(\log_b c)(\log_c d)(\log_d f)$
for $a,b,c,d,f > 0$ and a,b,c and $d \neq 1$.

47. $x = \dfrac{72}{5.5} = 13.1$ years

49. $x = \dfrac{72}{7.3} = 9.9$ years

EXERCISE SET 1.8 THE SIGMA NOTATION

1. $\displaystyle\sum_{x \, \varepsilon \, A} f(x) = f(1) + f(3) + f(7) + f(10)$
$$= 4 + 10 + 22 + 31 = 67$$

3. $\displaystyle\sum_{x \, \varepsilon \, A} f(x) = f(-2) + f(-1) + f(0) + f(3) + f(6)$
$$= 5 + 2 + 1 + 10 + 37 = 55$$

5. $\displaystyle\sum_{x \, \varepsilon \, A} f(x) = f(-4) + f(0) + f(2) + f(4) + f(5) + f(9)$
$$= -64 + 0 + 8 + 64 + 125 + 729 = 862$$

7. $\displaystyle\sum_{x \, \varepsilon \, A} f(x) = \sum_{i \, = \, -4}^{4} i^3 = -64 - 27 - 8 - 1 + 0 + 1 + 8 + 27 + 64 = 0$

9. $\displaystyle\sum_{i \, = \, 2}^{6} (3i + 5) = 11 + 14 + 17 + 20 + 23 = 85$

11. $\displaystyle\sum_{i \, = \, 5}^{8} (i^2 + 3i - 2) = (25 + 15 - 2) + (36 + 18 - 2) + (49 + 21 - 2) + (64 + 24 - 2)$
$$= 38 + 52 + 68 + 86 = 244$$

13. $\displaystyle\sum_{i \, = \, -2}^{2} (i^2 + 1) = 5 + 2 + 1 + 2 + 5 = 15$

15. $\displaystyle\sum_{i \, = \, 1}^{150} i^2 = \frac{150(150 + 1)(2(150) + 1)}{6} = \frac{150(151)(301)}{6} = 1{,}136{,}275$

17. $\displaystyle\sum_{i \, = \, 41}^{160} i = \sum_{i \, = \, 1}^{160} i - \sum_{i \, = \, 1}^{40} i = \frac{(160)(161)}{2} - \frac{40(41)}{2} = 12880 - 820 = 12060$

19. $\displaystyle\sum_{i \, = \, 61}^{170} i^3 = \sum_{i \, = \, 1}^{170} i^3 - \sum_{i \, = \, 1}^{60} i^3 = \frac{(170)^2(171)^2}{4} - \frac{60^2(61)^2}{4} = 207{,}917{,}325$

21. $\displaystyle\sum_{i \, = \, 1}^{120} (2i^3 - 5i^2 + 6i - 7) = 2\sum_{i \, = \, 1}^{120} i^3 - 5\sum_{i \, = \, 1}^{120} i^2 + 6\sum_{i \, = \, 1}^{120} i - \sum_{i \, = \, 1}^{120} 7$

$$= \frac{2(120)^2(121)^2}{4} - \frac{5(120)(121)(241)}{6} + \frac{6(120)(121)}{2} - 7(120) =$$

$$105415200 - 2916100 + 43560 - 840 = 102{,}541{,}820$$

23. $\displaystyle\sum_{i \, = \, 0}^{185} -9 = -9(186) = -1674$

25. Terms of an arithmetic progression are being added with a first term of 1 and common difference of 2. Hence the i th term is

$$1 + 2(i-1) = 2i - 1 \text{ and the sum is } \sum_{i=1}^{9} (2i - 1).$$

27. Terms of an arithmetic progression are being added with a first term of 5 and common difference of 3. Hence the i th term is

$$5 + (i-1)3 = 2 + 3i \text{ and the sum is } \sum_{i=1}^{7} (2 + 3i).$$

29. $\qquad 1^2 + 2^2 + 3^2 + 4^2 + 5^2 + 6^2 + 7^2 + 8^2 = \displaystyle\sum_{i=1}^{8} i^2$

31. $\qquad 2^3 + 3^3 + 4^3 + 5^3 + 6^3 = \displaystyle\sum_{i=2}^{6} i^3$

33. $\qquad (-1)^2 + (-1)^3 + (-1)^4 + (-1)^5 + (-1)^6 + (-1)^7 + (-1)^8$

$$+ (-1)^9 + (-1)^{10} + (-1)^{11} = \sum_{i=2}^{11} (-1)^i$$

35. $\qquad 1^2 - 2^2 + 3^2 - 4^2 + 5^2 - 6^2 + 7^2 - 8^2 + 9^2 = \displaystyle\sum_{i=1}^{9} i^2(-1)^{i+1}$

EXERCISE SET 1.9 CHAPTER REVIEW

1.   (a) false   (f) true
        (b) true   (g) true
        (c) false   (h) false
        (d) false   (i) false
        (e) true

3. $2^5 - 1 = 31$

5.

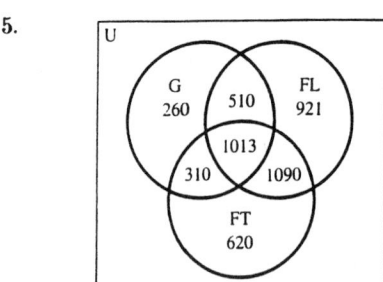

$260 + 510 + 310 + 1013 + 921 + 1090 + 620 = 4724.$
The pollster polled only 4724 voters and should not be paid.

7.   (a) $\sqrt{x+4} = 2x - 7$

$x + 4 = (2x - 7)^2 = 4x^2 - 28x + 49$

$0 = 4x^2 - 29x + 45 = (x - 5)(4x - 9)$

$x = 5 \text{ or } x = 9/4$

$\sqrt{5+4} = 10 - 7 \text{ so } 5 \text{ checks}$

$\sqrt{\frac{9}{4}+4} = \frac{5}{2} \neq 2(\frac{9}{4}) - 7 = \frac{9}{2} - 7$

5 is the only solution.

(b) $\sqrt{7x + 22} = x + 4$

$7x + 22 = x^2 + 8x + 16$

$0 = x^2 + x - 6 = (x + 3)(x - 2)$

$x = -3, \ x = 2 \text{ and both check}$

9.   (a) $3x - 2 < 5x - 10$

$\qquad 8 < 2x$

$\qquad 4 < x$

The solution set is $(4, \infty)$ 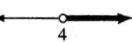

(b) $\frac{x+1}{3} + 1 \geq \frac{2x-3}{7} + 2$

$7x + 7 + 21 \geq 6x - 9 + 42$

$7x + 28 \geq 6x + 33$

$x \geq 5$

The solution set is $[5, \infty)$ 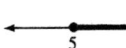

(c) $|x - 2| < 5$

$-5 < x - 2 < 5$

$-3 < x < 7$

The solution set is $(-3, 7)$ 

(d) $|x + 3| \leq 2$

$-2 - \leq x + 3 \leq 2$

$-5 \leq x \leq -1$

The solution set is $[-5, -1]$ 

(e) $|2 - x| \geq 5$

$2 - x \geq 5 \text{ or } 2 - x \leq -5$

$-3 \geq x \text{ or } 7 \leq x$

The solution set is $(-\infty, -3] \cup [7, \infty)$

(f) $|3 + x| + |5 - 2x| < 0$ has no solution as $|3 + x|$ and $|5 - 2x|$ are nonnegative for all real numbers x.

(g) $|2x-3| < 5/4$
$-5/4 < 2x-3 < 5/4$
$7/4 < 2x < /17/4$
$7/8 < x < 17/8$

The solution set is $(7/8, 17/8)$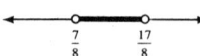

(h) $x^2 + 13x - 14 < 0$
$(x+14)(x-1) < 0$

x			-14				1	
$x+14$	$-$	$-$	0	$+$	$+$	$+$		$+$
$x-1$	$-$	$-$	$-$	$-$	$-$	0		$+$
$x^2 + 13x - 14$	$+$	$+$	0	$-$	$-$	0		$+$

The solution set is $(-14, 1)$

(i) $\dfrac{(x-2)(x+2)}{x+3} \geq 0$

x		-3		-2		2	
$x-2$	$-$	$-$	$-$	$-$	$-$	0	$+$
$x+2$	$-$	$-$	$-$	0	$+$	$+$	$+$
$x+3$	$-$	0	$+$	$+$	$+$	$+$	$+$
$\dfrac{x^2-4}{x+3}$	$-$	U	$+$	0	$-$	0	$+$

The solution set is $(-3,-2] \cup [2,+\infty)$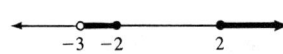

(j) $\dfrac{(x+2)(x-1)}{(3+x)(5-x)} < 0$

x		-3		-2		1		5	
$x+2$	$-$	$-$	$-$	0	$+$	$+$	$+$	$+$	$+$
$x-1$	$-$	$-$	$-$	$-$	$-$	0	$+$	$+$	$+$
$3+x$	$-$	0	$+$	$+$	$+$	$+$	$+$	$+$	$+$
$5-x$	$+$	$+$	$+$	$+$	$+$	$+$	$+$	0	$-$
$\dfrac{x^2+x-2}{15+2x-x^2}$	$-$	U	$+$	0	$-$	0	$+$	U	$-$

The solution is $(-\infty,-3) \cup (-2,1) \cup (5,+\infty)$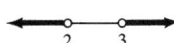

(k) $\dfrac{2x+3-9(x-2)}{x-2} < 0$

$\dfrac{2x+3-9x+18}{x-2} = \dfrac{-7x+21}{x-2} = \dfrac{7(3-x)}{x-2} < 0$

x		2		3	
$3-x$	$+$	$+$	$+$	0	$-$
$x-2$	$-$	0	$+$	$+$	$+$
$\dfrac{21-7x}{x-2}$	$-$	U	$+$	0	$-$

The solution set is $(-\infty,2) \cup (3,+\infty)$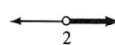

(l) $|2-3x| < 5x-6$

if $5x-6 \ge 0$, $5x \ge 6$ and $x \ge 6/5$ and $|2-3x| < 5x-6$ is equivalent to
$-5x+6 < 2-3x < 5x-6$. Thus,
$-5x+6 < 2-3x$ and $2-3x < 5x-6$.
$4 < 2x$ and $8 < 8x$
$2 < x$ and $1 < x$. Hence, if $x > 2$, x is a solution.
If $5x-6 < 0$, there are no solutions.

The solution set is $(2,+\infty)$

11. $x-2 \ge 0$ if $x \ge 2$. The domain is $\{x | x \ge 2\}$.

13. $f(x) = 2x+3$ is a linear function. The graph is a line with slope 2 and y intercept 3. Two points on the line are

x	0	-1
$f(x)$	3	1

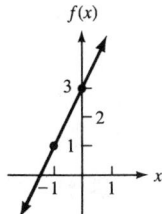

15. $h(x) = -3x^2 + 24x - 33$ is a quadratic function with graph a concave down parabola. The vertex occurs when $x = \dfrac{-24}{-6} = 4$.

Points on the graph include

x	3	4	5
$h(x)$	12	15	12

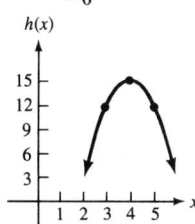

17. (a) The slope is $\dfrac{4-2}{3-(-1)} = \dfrac{2}{4} = \dfrac{1}{2}$. The equation is $(y-2) = \dfrac{1}{2}(x-(-1))$

$y - 2 = \dfrac{1}{2}(x+1)$ $y = \dfrac{1}{2}x + \dfrac{1}{2} + 2$ $y = \dfrac{1}{2}x + \dfrac{5}{2}$

(b) If $2x + 3y = 5$, $3y = 5 - 2x$, $y = \dfrac{5}{3} - \dfrac{2}{3}x$. The slope of the line is $-\dfrac{2}{3}$.
The equation is $y - 5 = -\dfrac{2}{3}(x-2)$.

(c) If $x + 2y = 5$, $2y = 5 - x$, $y = \dfrac{5}{2} - \dfrac{x}{2}$. The slope of the desired line is
$-\dfrac{1}{-\dfrac{1}{2}} = 2$. The equation is $y - 3 = 2(x+1)$ $y = 2x + 5$.

(d) The line is vertical and has equation $x=1$.

(e) The line is horizontal and has equation $y = 4$.

19.   Some points on the graph are

x	-2	-1	0	1	2
$f(x)$	9	3	1	$\dfrac{1}{3}$	$\dfrac{1}{9}$

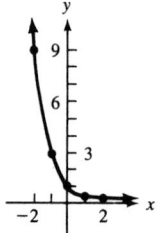

21. $A = 700\left(1 + \dfrac{.06}{12}\right)^{36} = 700(1.005)^{36} = \$837.68$. Hence the interest earned is $137.68.

23. (a) $\log 6 = \log 2 + \log 3 = .3010 + .4771 = .7781$

(b) $\log 8 = \log 2^3 = 3 \log 2 = 3(.3010) = .9030$

(c) $\log \sqrt{3} = \dfrac{1}{2} \log 3 = \dfrac{1}{2}(.4771) = .2385$

(d) $\log \dfrac{4}{27} = \log 4 - \log 27 = 2 \log 2 - 3 \log 3 = 2(.3010) - 3(.4771) = -.8293$

(e) $\log 250 = \log (2 \cdot 125) = \log 2 + \log 5^3 = \log 2 + 3 \log 5 = .3010 + 3 \log \dfrac{10}{2}$

$= .3010 + 3 (\log 10 - \log 2) = .3010 + 3 (1 - .3010) = .3010 + 3 - .9030 = 2.398$

(f) $\log_2 10 = \dfrac{\log 10}{\log 2} = \dfrac{1}{.3010} = 3.3222$

25. (a) $\dfrac{20(21)(41)}{6} = 2870$

(b) $5(30) = 150$

27. $4P = P\left(1 + \dfrac{.06}{12}\right)^t$ where t is the number of months needed. $4 = (1.005)^t$

$\ln 4 = t \ln(1.005)$. $\dfrac{\ln 4}{\ln(1.005)} = t = 277.95$ months ~ 23.2 years.

29.   Let x be the width of the strips added.

new area $= 4800 = (60 + x)(40 + x)$ $4800 = 2400 + 100x + x^2$

$0 = x^2 + 100x - 2400$ $x = \dfrac{-100 \pm \sqrt{10000 + 9600}}{2} = \dfrac{-100 \pm \sqrt{19600}}{2}$

$= \dfrac{-100 \pm 140}{2} = -120\,\text{ft. or } 20\,\text{ft.}$ The width of the strip must be 20 feet.

31.

Quantity sold q	Price p
40	$ 90
80	$ 70

(a)   The slope is $\dfrac{90 - 70}{40 - 80} = -\dfrac{1}{2}$.

The equation is $p - 90 = -\dfrac{1}{2}(q - 40)$ or $p = -\dfrac{1}{2}q + 110$

(b) $R(q) = qP(q) = q\left(-\dfrac{1}{2}q + 110\right)$. $R(q) = -\dfrac{1}{2}q^2 + 110q$

(c)   Profit $= R(q) - C(q) = -\dfrac{1}{2}q^2 + 110q - (60q + 250)$.   Profit $= -\dfrac{1}{2}q^2 + 50q - 250$.

The profit is a maximum when $q = \dfrac{-50}{2(-.5)} = 50$ racquets.

The price per racquet is $-\dfrac{50}{2} + 110 = \$\,85$.

The maximum profit is $-\dfrac{1}{2}(50)^2 + 50(50) - 250 = \$\,1,000$.

33.   Let x be the amount invested at 6%. $12,000 - x$ is invested at 8%.

$$12,000(.065) = x(.06) + (12,000 - x)(.08)$$
$$780 = 960 - .02x$$
$$.02x = 180$$
$$x = 9000$$

$9,000 is invested at 6% and $3,000 at 8%.

35.   Let x be the number of quarts withdrawn.   The amount of winter antifreeze will be

$$x + \dfrac{3}{28}(28 - x) = \dfrac{5}{28}x + 3.$$
$$.25(28) \le \dfrac{25}{28}x + 3 \le .40(28)$$
$$7 \le \dfrac{25}{28}x + 3 \le 11.2$$
$$4 \le \dfrac{25}{28}x \le 8.2$$
$$4.48 \le x \le 9.184$$

CHAPTER 2

SYSTEMS OF LINEAR EQUATIONS AND INEQUALITIES

EXERCISE SET 2.1 SYSTEMS OF TWO LINEAR EQUATIONS IN TWO VARIABLES

1. $\begin{cases} 2x - 3y = 8 \\ 3x + 5y = -7 \end{cases}$ Multiply both sides of the first equation by 5 and both sides of the second equation by 3.

$\begin{cases} 10x - 15y = 40 \\ 9x + 15y = -21 \end{cases}$ Replace the second equation by the sum of the two equations.

$\begin{cases} 10x - 15y = 40 \\ 19x = 19 \end{cases}$ Solve for x.

$\begin{cases} 10x - 15y = 40 \\ x = 1 \end{cases}$ Replace x by 1 in the first equation and solve for y.

$\begin{cases} 10 - 15y = 40 \\ x = 1 \end{cases}$ $\begin{cases} -15y = 30 \\ x = 1 \end{cases}$ $\begin{cases} y = -2 \\ x = 1 \end{cases}$

The solution set is $\{(1, -2)\}$.

3. Multiply the second equation by 2, then replace the second equation by the sum of the equations and solve for y.

$\begin{cases} -2x + 3y = -9 \\ 2x - 8y = 14 \end{cases}$ $\begin{cases} -2x + 3y = -9 \\ -5y = 5 \end{cases}$ $\begin{cases} -2x + 3y = -9 \\ y = -1 \end{cases}$

Substitute -1 for y in the first equation and solve for x.

$\begin{cases} -2x = -6 \\ y = -1 \end{cases}$ $\begin{cases} x = 3 \\ y = -1 \end{cases}$

The solution set is $\{(3, -1)\}$.

5. Multiply the first equation by -3, then replace the second equation by the sum of the equations.

$\begin{cases} -6x + 21y = -51 \\ 6x - 21y = -2 \end{cases}$ $\begin{cases} -6x + 21y = -51 \\ 0 = -53 \end{cases}$

Thus the system is inconsistent, the solution set is $\emptyset$.

7. Multiply the first equation by 2, then replace the second equation by the sum of the equations and solve for x.

$$\begin{cases} 7x - 3y = 37 \\ 2x + 6y = -10 \end{cases} \qquad \begin{cases} 14x - 6y = 74 \\ 2x + 6y = -10 \end{cases} \qquad \begin{cases} 14x - 6y = 74 \\ 16x = 64 \end{cases}$$

$$\begin{cases} 14x - 6y = 74 \\ x = 4 \end{cases} \qquad \text{Substitute } x = 4 \text{ in the first equation and solve for } y.$$

$$\begin{cases} 56 - 6y = 74 \\ x = 4 \end{cases} \qquad \begin{cases} -6y = 18 \\ x = 4 \end{cases} \qquad \begin{cases} y = -3 \\ x = 4 \end{cases}$$

The solution set is $\{(4, -3)\}$.

9. Multiply the first equation by 5 and the second equation by 8, then replace the second equation by the sum of the equations and solve for x.

$$\begin{cases} 25x + 40y = -170 \\ 24x - 40y = 72 \end{cases} \qquad \begin{cases} 25x + 40y = -170 \\ 49x = -98 \end{cases} \qquad \begin{cases} 25x + 40y = -170 \\ x = -2 \end{cases}$$

Substitute -2 for x in the first equation and solve for y.

$$\begin{cases} -50 + 40y = -170 \\ x = -2 \end{cases} \qquad \begin{cases} 40y = -120 \\ x = -2 \end{cases} \qquad \begin{cases} y = -3 \\ x = -2 \end{cases}$$

The solution set is $\{(-2, -3)\}$.

11. Multiply the first equation by -4, then replace the second equation by the sum of the two equations.

$$\begin{cases} -4x - 12y = -20 \\ 4x + 12y = 20 \end{cases} \qquad \begin{cases} -4x - 12y = -20 \\ 0 = 0 \end{cases}$$

Thus the system is consistent and dependent.
If we let $x = c$ in the first equation, $c + 3y = 5$, $3y = 5 - c$, $y = \dfrac{5 - c}{3}$.
The solution set is $\left\{ \left(c, \dfrac{5 - c}{3} \right) \mid c \text{ is a real number} \right\}$.

13. Multiply the first equation by 7 and the second equation by 3, then replace the second equation by the sum of the two equations and solve for x.

$$\begin{cases} 35x - 21y = 266 \\ -12x + 21y = -105 \end{cases} \qquad \begin{cases} 35x - 21y = 266 \\ 23x = 161 \end{cases} \qquad \begin{cases} 35x - 21y = 266 \\ x = 7 \end{cases}$$

Substitute 7 for x in the first equation and solve for y.

$$\begin{cases} 245 - 21y = 266 \\ x = 7 \end{cases} \qquad \begin{cases} -21y = 21 \\ x = 7 \end{cases} \qquad \begin{cases} y = -1 \\ x = 7 \end{cases}$$

The solution set is $\{(7, -1)\}$.

15. Multiply the first equation by 5 and the second equation by 8, then replace the second equation by the sum of the equations and solve for r.

$$\begin{cases} 5r + 8s = 46 \\ 2r - 5s = -37 \end{cases} \qquad \begin{cases} 25r + 40s = 230 \\ 16r - 40s = -296 \end{cases} \qquad \begin{cases} 25r + 40s = 230 \\ 41r = -66 \end{cases}$$

$$\begin{cases} 25r + 40s = 230 \\ r = -66/41 \end{cases}$$
Substitute $-\frac{66}{41}$ for r in the first equation and solve for s.

$$\begin{cases} -1650/41 + 40s = 230 \\ r = -66/41 \end{cases} \qquad \begin{cases} 40s = 11080/41 \\ r = -66/41 \end{cases} \qquad \begin{cases} s = 277/41 \\ r = -66/41 \end{cases}$$

The solution set is $\{(-66/41,\ 277/41)\}$.

17. Multiply the first equation by 12 and the second equation by 60 to eliminate the fractions. Next multiply the first equation by -12, replace the second equation by the sum of the equations and solve for u.

$$\begin{cases} \frac{1}{4}t - \frac{1}{3}u = -\frac{1}{12} \\ \frac{1}{5}t + \frac{1}{6}u = \frac{3}{20} \end{cases} \qquad \begin{cases} 3t - 4u = -1 \\ 12t + 10u = 9 \end{cases} \qquad \begin{cases} -12t + 16u = 4 \\ 12t + 10u = 9 \end{cases}$$

$$\begin{cases} -12t + 16u = 4 \\ 26u = 13 \end{cases} \qquad \begin{cases} -12t + 16u = 4 \\ u = \frac{1}{2} \end{cases}$$

Replace u by $\frac{1}{2}$ in the first equation and solve for t.

$$\begin{cases} -12t + 8 = 4 \\ u = \frac{1}{2} \end{cases} \qquad \begin{cases} -12t = -4 \\ u = \frac{1}{2} \end{cases} \qquad \begin{cases} t = \frac{1}{3} \\ u = \frac{1}{2} \end{cases}$$

The solution set is $\left\{ \left(\frac{1}{3}, \frac{1}{2} \right) \right\}$.

19. $\begin{cases} 3x + y = -1 \\ 2x - 5y = -12 \end{cases}$
Solve for y in the first equation.

$$\begin{cases} y = -1 - 3x \\ 2x - 5y = -12 \end{cases}$$
Substitute $-1 - 3x$ for y in the second equation.

$$\begin{cases} y = -1 - 3x \\ 2x - 5(-1 - 3x) = -12 \end{cases}$$
Simplify.

$$\begin{cases} y = -1 - 3x \\ 17x = -17 \end{cases}$$
Solve for x.

$$\begin{cases} y = -1 - 3x \\ x = -1 \end{cases}$$
Substitute -1 for x in the first equation and solve for y.

$$\begin{cases} y = -1 + 3 \\ x = -1 \end{cases} \qquad \begin{cases} y = 2 \\ x = -1 \end{cases}$$

The solution set is $\{(-1, 2)\}$.

21. Solve the first equation for y.

$$\begin{cases} 5x + 2y = -5 \\ 3x - 7y = -44 \end{cases} \qquad \begin{cases} 2y = -5 - 5x \\ 3x - 7y = -44 \end{cases} \qquad \begin{cases} y = \frac{-5 - 5x}{2} \\ 3x - 7y = -44 \end{cases}$$

Replace y by $(-5 - 5x)/2$ in the second equation and solve the second equation for x.

$$\begin{cases} y = \frac{-5 - 5x}{2} \\ 3x - 7\left(\frac{-5 - 5x}{2}\right) = -44 \end{cases}$$
$$\begin{cases} y = \frac{-5 - 5x}{2} \\ 3x + \frac{35x}{2} = -44 - \frac{35}{2} \end{cases}$$

$$\begin{cases} y = \frac{-5 - 5x}{2} \\ \frac{41x}{2} = \frac{-123}{2} \end{cases}$$
$$\begin{cases} y = \frac{-5 - 5x}{2} \\ x = -3 \end{cases}$$

Replace x by -3 in the first equation and solve for y.

$$\begin{cases} y = \frac{-5 + 15}{2} \\ x = -3 \end{cases}$$
$$\begin{cases} y = 5 \\ x = -3 \end{cases}$$

The solution set is $\{(-3, 5)\}$.

23. Solve the second equation for y.

$$\begin{cases} 3x - 5y = 25 \\ 5x + 6y = 13 \end{cases}$$
$$\begin{cases} 3x - 5y = 25 \\ 6y = 13 - 5x \end{cases}$$
$$\begin{cases} 3x - 5y = 25 \\ y = \frac{13 - 5x}{6} \end{cases}$$

Next replace y by $(13 - 5x)/6$ in the first equation and solve for x.

$$\begin{cases} 3x - \frac{5(13 - 5x)}{6} = 25 \\ y = \frac{13 - 5x}{6} \end{cases}$$
$$\begin{cases} \frac{43x}{6} = \frac{215}{6} \\ y = \frac{13 - 5x}{6} \end{cases}$$
$$\begin{cases} x = 5 \\ y = \frac{13 - 5x}{6} \end{cases}$$

Now replace x by 5 in the second equation and solve for y.

$$\begin{cases} x = 5 \\ y = \frac{13 - 25}{6} \end{cases}$$
$$\begin{cases} x = 5 \\ y = -\frac{12}{6} \end{cases}$$
$$\begin{cases} x = 5 \\ y = -2 \end{cases}$$

The solution set is $\{(5, -2)\}$.

25. Solve the second equation for x.

$$\begin{cases} 8x - 6y = -4 \\ 4x - 3y = -2 \end{cases}$$
$$\begin{cases} 8x - 6y = -4 \\ 4x = 3y - 2 \end{cases}$$
$$\begin{cases} 8x - 6y = -4 \\ x = \frac{3y - 2}{4} \end{cases}$$

Next replace x by $(3y - 2)/4$ in the first equation and simplify.

$$\begin{cases} 8\left(\frac{3y - 2}{4}\right) - 6y = -4 \\ x = \frac{3y - 2}{4} \end{cases}$$
$$\begin{cases} 0 = 0 \\ x = \frac{3y - 2}{4} \end{cases}$$

Thus the system is consistent and dependent. If we let $y = c$ in the second equation, $x = \frac{3c - 2}{4}$ and the solution set is $\left\{\left(\frac{3c - 2}{4}, c\right) \mid c \text{ is an arbitrary real number}\right\}$.

27. Solve the first equation for y.

$$\begin{cases} 3x + 2y = 2 \\ 9x - 8y = -1 \end{cases}$$
$$\begin{cases} 2y = 2 - 3x \\ 9x - 8y = -1 \end{cases}$$
$$\begin{cases} y = 1 - \frac{3}{2}x \\ 9x - 8y = -1 \end{cases}$$

Next replace y by $1 - \frac{3}{2}x$ in the second equation and solve for x.

$$\begin{cases} y = 1 - \frac{3}{2}x \\ 9x - 8\left(1 - \frac{3}{2}x\right) = -1 \end{cases}$$
$$\begin{cases} y = 1 - \frac{3}{2}x \\ 21x = 7 \end{cases}$$
$$\begin{cases} y = 1 - \frac{3}{2}x \\ x = \frac{1}{3} \end{cases}$$

Finally replace x by $\frac{1}{3}$ in the first equation and solve for x.

$$\begin{cases} y = 1 - \frac{3}{2} \cdot \frac{1}{3} \\ x = \frac{1}{3} \end{cases} \qquad\qquad \begin{cases} y = \frac{1}{2} \\ x = \frac{1}{3} \end{cases}$$

The solution set is $\left\{ \left(\frac{1}{3}, \frac{1}{2} \right) \right\}$.

29. Solve the first equation for y.

$$\begin{cases} .5x + .3y = 2.5 \\ 1.2x - 1.4y = -4.6 \end{cases} \qquad \begin{cases} .3y = 2.5 - .5x \\ 1.2x - 1.4y = -4.6 \end{cases} \qquad \begin{cases} y = \frac{25}{3} - \frac{5}{3}x \\ 1.2x - 1.4y = -4.6 \end{cases}$$

Next replace y by $\frac{25}{3} - \frac{5}{3}x$ in the second equation and solve for x.

$$\begin{cases} y = \frac{25}{3} - \frac{5}{3}x \\ 1.2x - 1.4\left(\frac{25}{3} - \frac{5}{3}x \right) = -4.6 \end{cases} \qquad \begin{cases} y = \frac{25}{3} - \frac{5}{3}x \\ 12x - 14\left(\frac{25}{3} - \frac{5}{3}x \right) = -46 \end{cases}$$

$$\begin{cases} y = \frac{25}{3} - \frac{5}{3}x \\ 36x - 14(25 - 5x) = -138 \end{cases} \qquad \begin{cases} y = \frac{25}{3} - \frac{5}{3}x \\ 106x = 212 \end{cases} \qquad \begin{cases} y = \frac{25}{3} - \frac{5}{3}x \\ x = 2 \end{cases}$$

Next substitute 2 for x in the first equation and solve for y.

$$\begin{cases} y = \frac{25}{3} - \frac{10}{3} \\ x = 2 \end{cases} \qquad \begin{cases} y = 5 \\ x = 2 \end{cases}$$

The solution set is $\{(2, 5)\}$.

31. Solve the second equation for y.

$$\begin{cases} 1.3x - 2.5y = 6 \\ 2.6x - 5y = -3 \end{cases} \qquad \begin{cases} 1.3x - 2.5y = 6 \\ 5y = 2.6x + 3 \end{cases} \qquad \begin{cases} 1.3x - 2.5y = 6 \\ y = \frac{2.6x}{5} + .6 \end{cases}$$

$$\begin{cases} 1.3x - 2.5y = 6 \\ y = .52x + .6 \end{cases}$$

Next replace y by $.52x + .6$ in the first equation and simplify.

$$\begin{cases} 1.3x - 2.5(.52x + .6) = 6 \\ y = .52x + .6 \end{cases} \qquad \begin{cases} 1.3x - 1.3x - 1.5 = 6 \\ y = .52x + .6 \end{cases} \qquad \begin{cases} -1.5 = 6 \\ y = .52x + .6 \end{cases}$$

Thus the system is inconsistent and the solution set is $\emptyset$.

33. Solve the first equation for v.

$$\begin{cases} 7u + 2v = 3 \\ -3u + 5v = 28 \end{cases} \qquad \begin{cases} 2v = 3 - 7u \\ -3u + 5v = 28 \end{cases} \qquad \begin{cases} v = \frac{3}{2} - \frac{7}{2}u \\ -3u + 5v = 28 \end{cases}$$

Now replace v by $\frac{3}{2} - \frac{7}{2}u$ in the second equation and solve for u.

$$\begin{cases} v = \frac{3}{2} - \frac{7}{2}u \\ -3u + 5\left(\frac{3}{2} - \frac{7}{2}u \right) = 28 \end{cases} \qquad \begin{cases} v = \frac{3}{2} - \frac{7}{2}u \\ -3u + \frac{15}{2} - \frac{35}{2}u = 28 \end{cases}$$

$$\begin{cases} v = \frac{3}{2} - \frac{7}{2}u \\ -\frac{41}{2}u = \frac{41}{2} \end{cases} \qquad \begin{cases} v = \frac{3}{2} - \frac{7}{2}u \\ u = -1 \end{cases}$$

Finally replace u by -1 in the first equation and solve for v.

$$\begin{cases} v = \frac{3}{2} - \frac{7}{2}(-1) \\ u = -1 \end{cases} \qquad \begin{cases} v = 5 \\ u = -1 \end{cases}$$

The solution set is $\{(-1,5)\}$.

35. $\begin{cases} 2x - 3y = 8 \\ 3x + 5y = -7 \end{cases}$

Let $\qquad C = \begin{vmatrix} 2 & -3 \\ 3 & 5 \end{vmatrix} = 10 - (-9) = 19$

$\qquad A_x = \begin{vmatrix} 8 & -3 \\ -7 & 5 \end{vmatrix} = 40 - 21 = 19$

$\qquad A_y = \begin{vmatrix} 2 & 8 \\ 3 & -7 \end{vmatrix} = -14 - 24 = -38$

Hence the system is equivalent to $\begin{cases} x = 19/19 \\ y = -38/19 \end{cases}$ or $\begin{cases} x = 1 \\ y = -2 \end{cases}$

The solution set is $\{(1, -2)\}$.

37. $\begin{cases} -2x + 3y = -9 \\ x - 4y = 7 \end{cases}$

Let $\qquad C = \begin{vmatrix} -2 & 3 \\ 1 & -4 \end{vmatrix} = 8 - 3 = 5$

$\qquad A_x = \begin{vmatrix} -9 & 3 \\ 7 & -4 \end{vmatrix} = 36 - 21 = 15$

$\qquad A_y = \begin{vmatrix} -2 & -9 \\ 1 & 7 \end{vmatrix} = -14 + 9 = -5$

Hence the system is equivalent to $\begin{cases} x = 15/5 = 3 \\ y = -5/5 = -1 \end{cases}$ and the solution set is $\{(3, -1)\}$.

39. $C = \begin{vmatrix} 2 & -7 \\ 6 & -21 \end{vmatrix} = -42 - (-42) = 0$ and the system cannot be solved by this method. See exercise 5 or 22 for an alternate method.

41. $\qquad C = \begin{vmatrix} 7 & -3 \\ 2 & 6 \end{vmatrix} = 42 + 6 = 48$

$\qquad A_x = \begin{vmatrix} 37 & -3 \\ -10 & 6 \end{vmatrix} = 222 - 30 = 192$

$\qquad A_y = \begin{vmatrix} 7 & 37 \\ 2 & -10 \end{vmatrix} = -70 - 74 = -144$

The system is equivalent to $\begin{cases} x = 192/48 = 4 \\ y = -144/48 = -3 \end{cases}$

The solution set is $\{(4, -3)\}$.

43.

$$C = \begin{vmatrix} 5 & 8 \\ 3 & -5 \end{vmatrix} = -25 - 24 = -49$$

$$A_x = \begin{vmatrix} -34 & 8 \\ 9 & -5 \end{vmatrix} = 170 - 72 = 98$$

$$A_y = \begin{vmatrix} 5 & -34 \\ 3 & 9 \end{vmatrix} = 45 + 102 = 147$$

The system is equivalent to $\begin{cases} x = 98/(-49) = -2 \\ y = 147/(-49) = -3 \end{cases}$

The solution set is $\{(-2, -3)\}$.

45. $C = \begin{vmatrix} 1 & 3 \\ 4 & 12 \end{vmatrix} = 12 - 12 = 0.$ Thus Cramer's Rule does not apply. See Exercise 11 or 28 for an alternate method.

47.

$$C = \begin{vmatrix} 5 & -3 \\ -4 & 7 \end{vmatrix} = 35 - 12 = 23$$

$$A_x = \begin{vmatrix} 38 & -3 \\ -35 & 7 \end{vmatrix} = 266 - 105 = 151$$

$$A_y = \begin{vmatrix} 5 & 38 \\ -4 & -35 \end{vmatrix} = -175 + 152 = -23$$

The system is equivalent to $\begin{cases} x = 151/23 = 7 \\ y = -23/23 = -1 \end{cases}$

The solution set is $\{(7, -1)\}$.

49.

$$C = \begin{vmatrix} 5 & 8 \\ 2 & -5 \end{vmatrix} = -25 - 16 = -41$$

$$A_x = \begin{vmatrix} 46 & 8 \\ -37 & -5 \end{vmatrix} = -230 + 296 = 66$$

$$A_y = \begin{vmatrix} 5 & 46 \\ 2 & -37 \end{vmatrix} = -185 - 92 = -277$$

The system is equivalent to $\begin{cases} x = -66/41 \\ y = 277/41 \end{cases}$

The solution set is $\{(-66/41,\ 277/41)\}$.

51.

$$C = \begin{vmatrix} \frac{1}{4} & -\frac{1}{3} \\ \frac{1}{5} & \frac{1}{6} \end{vmatrix} = \frac{1}{24} + \frac{1}{15} = \frac{13}{120}$$

$$A_t = \begin{vmatrix} -\frac{1}{12} & -\frac{1}{3} \\ \frac{3}{20} & \frac{1}{6} \end{vmatrix} = -\frac{1}{72} + \frac{1}{20} = \frac{-20 + 72}{1440} = \frac{52}{1440} = \frac{13}{360}$$

$$A_u = \begin{vmatrix} \frac{1}{4} & -\frac{1}{12} \\ \frac{1}{5} & \frac{3}{20} \end{vmatrix} = \frac{3}{80} + \frac{1}{60} = \frac{13}{240}.$$

The system is equivalent to $\begin{cases} t = (13/360) \div (13/120) = 120/360 = 1/3 \\ u = (13/240) \div (13/120) = 120/240 = 1/2 \end{cases}$

The solution set is $\left\{\left(\frac{1}{3}, \frac{1}{2}\right)\right\}$.

53. Let $u = x^2$ and $v = y^2$.

$\begin{cases} 2u - v = -17 \\ 3u + 2v = 62 \end{cases}$ Multiply the first equation by 2, add the two equations, and solve the second equation for u.

$\begin{cases} 4u - 2v = -34 \\ 3u + 2v = 62 \end{cases}$ $\begin{cases} 4u - 2v = -34 \\ 7u = 28 \end{cases}$ $\begin{cases} 4u - 2v = -34 \\ u = 4 \end{cases}$

Replace u by 4 in the first equation and solve for v.

$\begin{cases} 16 - 2v = -34 \\ u = 4 \end{cases}$ $\begin{cases} 2v = 50 \\ u = 4 \end{cases}$ $\begin{cases} v = 25 \\ u = 4 \end{cases}$

Replace u and v by x^2 and y^2 respectively and solve for x and y.

$\begin{cases} y^2 = 25 \\ x^2 = 4 \end{cases}$ $\begin{cases} y = \pm 5 \\ x = \pm 2 \end{cases}$

The solution set is $\{(2,5),\ (2,-5),\ (-2,5),\ (-2,-5)\}$.

55. $\begin{cases} \frac{5}{x} + \frac{8}{y} = -34 \\ \frac{3}{x} - \frac{5}{y} = 9 \end{cases}$ Let $\frac{1}{x} = u$ and $\frac{1}{y} = v$ and solve for u and v.

$\begin{cases} 5u + 8v = -34 \\ 3u - 5v = 9 \end{cases}$ $\begin{cases} 25u + 40v = -170 \\ 24u - 40v = 72 \end{cases}$ $\begin{cases} 25u + 40v = -170 \\ 49u = -98 \end{cases}$

$$\begin{cases} 25u + 40v = -170 \\ u = -2 \end{cases} \qquad \begin{cases} -50 + 40v = -170 \\ u = -2 \end{cases} \qquad \begin{cases} 40v = -120 \\ u = -2 \end{cases}$$

$$\begin{cases} v = -3 \\ u = -2 \end{cases}$$

Replace v by $\frac{1}{y}$ and u by $\frac{1}{x}$, then solve for x and y.

$$\begin{cases} \frac{1}{y} = -3 \\ \frac{1}{x} = -2 \end{cases} \qquad \begin{cases} y = -\frac{1}{3} \\ x = -\frac{1}{2} \end{cases}$$

The solution set is $\left\{ \left(-\frac{1}{2}, -\frac{1}{3} \right) \right\}$.

57. Let $u = \frac{1}{x}$ and $v = \frac{1}{y}$.

$$\begin{cases} -2u + 3v = -9 \\ u - 4v = 7 \end{cases}$$

Multiply the second equation by 2 and let the sum of the two equations replace the first equation, and solve the first equation for v.

$$\begin{cases} -2u + 3v = -9 \\ 2u - 8v = 14 \end{cases} \qquad \begin{cases} -5v = 5 \\ 2u - 8v = 14 \end{cases} \qquad \begin{cases} v = -1 \\ 2u - 8v = 14 \end{cases}$$

Replace v by -1 in the second equation, solve for u, replace u by $\frac{1}{x}$, v by $\frac{1}{y}$ and solve for x and y.

$$\begin{cases} v = -1 \\ 2u + 8 = 14 \end{cases} \qquad \begin{cases} v = -1 \\ 2u = 6 \end{cases} \qquad \begin{cases} v = -1 \\ u = 3 \end{cases}$$

$$\begin{cases} \frac{1}{y} = -1 \\ \frac{1}{x} = 3 \end{cases} \qquad \begin{cases} y = -1 \\ x = \frac{1}{3} \end{cases}$$

The solution set is $\left\{ \left(\frac{1}{3}, -1 \right) \right\}$.

59. Let x be the dollars bet with the Dallas fan and y be the dollars bet with the L.A. fan. If Dallas wins her profit will be $7y - 5x$. If L.A. wins her profit will be $8x - 4y$.

To win \$360,

$$\begin{cases} -5x + 7y = 360 \\ 8x - 4y = 360 \end{cases} \qquad \begin{cases} -40x + 56y = 2880 \\ 40x - 20y = 1800 \end{cases}$$

$$\begin{cases} -5x + 7y = 360 \\ 36y = 4680 \end{cases} \qquad \begin{cases} -5x + 7y = 360 \\ y = 130 \end{cases} \qquad \begin{cases} -5x + 910 = 360 \\ y = 130 \end{cases}$$

$$\begin{cases} -5x = -550 \\ y = 130 \end{cases} \qquad \begin{cases} x = 110 \\ y = 130 \end{cases}$$

She must bet \$110 with the Dallas fan and \$130 with the L.A. fan.

61. Let x be the number of pounds of Brand A and y be the number of pounds of Brand B, where $x + y = 100$. The dollar value of Brand A is $x(4.25)$ and the dollar value of B is is $y(3.75)$. Hence we must solve

$$\begin{cases} x + y = 100 \\ 4.25x + 3.75y = 390 \end{cases} \qquad \begin{cases} -3.75x - 3.75y = -375 \\ 4.25x + 3.75y = 390 \end{cases} \qquad \begin{cases} x + y = 100 \\ .5x = 15 \end{cases}$$

$$\begin{cases} x+y=100 \\ x=30 \end{cases} \qquad \begin{cases} y=70 \\ x=30 \end{cases}$$

Mix 30 pounds of Brand A with 70 pounds of Brand B.

63. Let x be the amount invested at 6.5% and y be the amount invested at 9.5%, where $x+y=50,000$. The annual income from the 6.5% investment is $.065x$ and the annual income from the 9.5% investment is $.095y$. Thus

$$\begin{cases} x+y=50,000 \\ .065x+.095y=3790 \end{cases} \qquad \begin{cases} -.065x-.065y=-3250 \\ .065x+.095y=3790 \end{cases}$$

$$\begin{cases} x+y=50,000 \\ .03y=540 \end{cases} \qquad \begin{cases} x+y=50,000 \\ y=18,000 \end{cases} \qquad \begin{cases} x=32,000 \\ y=18,000 \end{cases}$$

$32,000 must be invested at 6.5% and $18,000 at 9.5%.

65.
$$\begin{cases} a_1x+b_1y=c_1 \\ a_2x+b_2y=c_2 \end{cases} \qquad \begin{cases} a_1a_2x+a_2b_1y=a_2c_1 \\ -a_1a_2x-a_1b_2y=-a_1c_2 \end{cases}$$

$$\Rightarrow \qquad a_2b_1y-a_1b_2y=a_2c_1-a_1c_2$$
$$y(a_2b_1-a_1b_2)=a_2c_1-a_1c_2.$$

If $a_2b_1-a_1b_2 \neq 0$, $y=\dfrac{a_2c_1-a_1c_2}{a_2b_1-a_1b_2}=\dfrac{a_1c_2-a_2c_1}{a_2b_1-a_1b_2}$

$$=\dfrac{\begin{vmatrix} a_1 & c_1 \\ a_2 & c_2 \end{vmatrix}}{\begin{vmatrix} a_1 & b_1 \\ a_2 & b_2 \end{vmatrix}}=\dfrac{A_y}{C}. \text{ Similarly, } x=\dfrac{A_x}{C}.$$

EXERCISE SET 2.2 SYSTEMS OF THREE LINEAR EQUATIONS IN THREE VARIABLES

1.
$$\begin{cases} 2x & +3y & -5z & =-1 \\ 3x & -2y & +4z & =-3 \\ 6x & +4y & -3z & =-1 \end{cases} \qquad E_1+E_2 \to E_2$$

$$\begin{cases} 2x & +3y & -5z & =-1 \\ 5x & +y & -z & =-4 \\ 6x & +4y & -3z & =-1 \end{cases} \qquad \begin{array}{l} -3E_2+E_1 \to E_1 \\ -4E_2+E_3 \to E_3 \end{array}$$

$$\begin{cases} -13x & & -2z & =11 \\ 5x & +y & -z & =-4 \\ -14x & & +z & =15 \end{cases} \qquad -E_3+E_1 \to E_1$$

$$\begin{cases} x & & -3z & =-4 \\ 5x & +y & -z & =-4 \\ -14x & & +z & =15 \end{cases} \qquad \begin{array}{l} -5E_1+E_2 \to E_2 \\ 14E_1+E_3 \to E_3 \end{array}$$

$$\begin{cases} x & & -3z & = -4 \\ & y & +14z & = 16 \\ & & -41z & = -41 \end{cases} \qquad -\tfrac{1}{41}E_3 \to E_3$$

$$\begin{cases} x & & -3z & = -4 \\ & y & +14z & = 16 \\ & & z & = 1 \end{cases} \qquad \begin{aligned} & 3E_3 + E_1 \to E_1 \\ & -14E_3 + E_2 \to E_2 \end{aligned}$$

$$\begin{cases} x & & & = -1 \\ & y & & = 2 \\ & & z & = 1 \end{cases}$$

The solution set is $\{(-1, 2, 1)\}$.

3. $\begin{cases} 2x & +3y & +2z & = 8 \\ -x & +4y & -3z & = 1 \\ 3x & -2y & +7z & = 11 \end{cases}$ $\qquad \begin{aligned} & 2E_2 + E_1 \to E_1 \\ & 3E_2 + E_3 \to E_3 \end{aligned}$

$\begin{cases} 2x & +3y & +2z & = 8 \\ & 11y & -4z & = 10 \\ & 10y & -2z & = 14 \end{cases}$ $\qquad \tfrac{1}{2}E_3 \to E_3$

$\begin{cases} 2x & +3y & +2z & = 8 \\ & 11y & -4z & = 10 \\ & 5y & -z & = 7 \end{cases}$ $\qquad -4E_3 + E_2 \to E_3$

$\begin{cases} 2x & +3y & +2z & = 8 \\ & 11y & -4z & = 10 \\ & -9y & & = -18 \end{cases}$ $\qquad -\tfrac{1}{9}E_3 \to E_3$

$\begin{cases} 2x & +3y & +2z & = 8 \\ & 11y & -4z & = 10 \\ & y & & = 2 \end{cases}$ $\qquad -11E_3 + E_2 \to E_2$

$\begin{cases} 2x & +3y & +2z & = 8 \\ & & -4z & = -12 \\ & y & & = 2 \end{cases}$ $\qquad -\tfrac{1}{4}E_2 \to E_2$

$\begin{cases} 2x & +3y & +2z & = 8 \\ & & z & = 3 \\ & y & & = 2 \end{cases}$ $\qquad -3E_3 - 2E_2 + E_1 \to E_1$

$\begin{cases} 2x & & & = -4 \\ & & z & = 3 \\ & y & & = 2 \end{cases}$ $\qquad \tfrac{1}{2}E_1 \to E_1$

$\begin{cases} x & & & = -2 \\ & & z & = 3 \\ & y & & = 2 \end{cases}$

The solution set is $\{(-2, 2, 3)\}$

5. $\begin{cases} x & +y & +3z & = 2 \\ 3x & +2y & -z & = -3 \\ 11x & +8y & +3z & = 6 \end{cases}$ $\qquad \begin{aligned} & 3E_2 + E_3 \to E_3 \\ & 3E_2 + E_1 \to E_2 \end{aligned}$

$$\begin{cases} x & + y & + 3z & = 2 \\ 10x & + 7y & & = -7 \\ 20x & + 14y & & = -3 \end{cases} \qquad -2E_2 + E_3 \rightarrow E_3$$

$$\begin{cases} x & + y & + 3z & = 2 \\ 10x & + 7y & & = -7 \\ & 0 & & = -11 \end{cases}$$

The system is inconsistent, the solution set is $\emptyset$.

7.
$$\begin{cases} -2x & + 3y & + z & = 4 \\ 3x & - 2y & + 2z & = 14 \\ 4x & + 3y & - z & = 6 \end{cases} \qquad \begin{array}{l} 2E_3 + E_2 \rightarrow E_2 \\ E_1 + E_3 \rightarrow E_3 \end{array}$$

$$\begin{cases} -2x & + 3y & + z & = 4 \\ 11x & + 4y & & = 26 \\ 2x & + 6y & & = 10 \end{cases} \qquad \tfrac{1}{2}E_3 \rightarrow E_3$$

$$\begin{cases} -2x & + 3y & + z & = 4 \\ 11x & + 4y & & = 26 \\ x & + 3y & & = 5 \end{cases} \qquad -11E_3 + E_2 \rightarrow E_3$$

$$\begin{cases} -2x & + 3y & + z & = 4 \\ 11x & + 4y & & = 26 \\ & - 29y & & = -29 \end{cases} \qquad -\tfrac{1}{29}E_3 \rightarrow E_3$$

$$\begin{cases} -2x & + 3y & + z & = 4 \\ 11x & + 4y & & = 26 \\ & y & & = 1 \end{cases} \qquad -4E_3 + E_2 \rightarrow E_2$$

$$\begin{cases} -2x & + 3y & + z & = 4 \\ 11x & & & = 22 \\ & y & & = 1 \end{cases} \qquad \tfrac{1}{11}E_2 \rightarrow E_2$$

$$\begin{cases} -2x & + 3y & + z & = 4 \\ x & & & = 2 \\ & y & & = 1 \end{cases} \qquad 2E_2 - 3E_3 + E_1 \rightarrow E_1$$

$$\begin{cases} & z & = 5 \\ x & & = 2 \\ & y & = 1 \end{cases} \qquad \text{The solution set is } \{(2,1,5)\}.$$

9.
$$\begin{cases} 3x & + 2y & - 5z & = -2 \\ 2x & + 3y & - 8z & = 5 \\ -x & + 2y & - 3z & = 8 \end{cases} \qquad \begin{array}{l} 2E_3 + E_2 \rightarrow E_2 \\ 3E_3 + E_1 \rightarrow E_1 \end{array}$$

$$\begin{cases} & 8y & - 14z & = 22 \\ & 7y & - 14z & = 21 \\ -x & + 2y & - 3z & = 8 \end{cases} \qquad \begin{array}{l} E_1 - E_2 \rightarrow E_2 \\ \\ \tfrac{1}{2}E_1 \rightarrow E_1 \end{array}$$

$$\begin{cases} & 4y & - 7z & = 11 \\ & y & & = 1 \\ -x & + 2y & - 3z & = 8 \end{cases} \qquad -4E_2 + E_1 \rightarrow E_1$$

$$\begin{cases} & & - 7z & = 7 \\ & y & & = 1 \\ -x & + 2y & - 3z & = 8 \end{cases} \qquad -\tfrac{1}{7}E_1 \rightarrow E_1$$

$$\begin{cases} & & z & = -1 \\ & y & & = 1 \\ -x & +2y & -3z & = 8 \end{cases} \qquad -E_3 + 2E_2 - 3E_1 \to E_3$$

$$\begin{cases} & & z & = -1 \\ & y & & = 1 \\ x & & & = -3 \end{cases}$$

The solution set is $\{(-3, 1, -1)\}$.

11.
$$\begin{cases} x & +y & +z & = 6 \\ 2x & +y & +3z & = 12 \\ x & -y & +5z & = 10 \end{cases} \qquad \begin{array}{l} E_1 + E_3 \to E_3 \\ E_2 + E_3 \to E_2 \end{array}$$

$$\begin{cases} x & +y & +z & = 6 \\ 3x & & +8z & = 22 \\ 2x & & +6z & = 16 \end{cases} \qquad 2E_2 - 3E_3 \to E_3$$

$$\begin{cases} x & +y & +z & = 6 \\ 3x & & +8z & = 22 \\ & & -2z & = -4 \end{cases} \qquad -\tfrac{1}{2}E_3 \to E_3$$

$$\begin{cases} x & +y & +z & = 6 \\ 3x & & +8z & = 22 \\ & & z & = 2 \end{cases} \qquad E_2 - 8E_3 \to E_2$$

$$\begin{cases} x & +y & +z & = 6 \\ 3x & & & = 6 \\ & & z & = 2 \end{cases} \qquad \tfrac{1}{3}E_2 \to E_2$$

$$\begin{cases} x & +y & +z & = 6 \\ x & & & = 2 \\ & & z & = 2 \end{cases} \qquad E_1 - E_2 - E_3 \to E_1$$

$$\begin{cases} & y & & = 2 \\ x & & & = 2 \\ & & z & = 2 \end{cases}$$

The solution set is $\{(2, 2, 2)\}$.

13.
$$\begin{cases} x & +y & & = -4 \\ x & & +z & = 1 \\ 3x & -y & +2z & = 4 \end{cases} \qquad E_1 + E_3 \to E_3$$

$$\begin{cases} x & +y & & = -4 \\ x & & +z & = 1 \\ 4x & & +2z & = 0 \end{cases} \qquad -4E_2 + E_3 \to E_3$$

$$\begin{cases} x & +y & & = -4 \\ x & & +z & = 1 \\ & & -2z & = -4 \end{cases} \qquad -\tfrac{1}{2}E_3 \to E_3$$

$$\begin{cases} x & +y & & = -4 \\ x & & +z & = 1 \\ & & z & = 2 \end{cases} \qquad E_2 - E_3 \to E_2$$

$$\begin{cases} x & +y & & = -4 \\ x & & & = -1 \\ & & z & = 2 \end{cases} \qquad E_1 - E_2 \to E_1$$

$$\begin{cases} & y & & = -3 \\ x & & & = -1 \\ & & z & = 2 \end{cases}$$

The solution set is $\{(-1, -3, 2)\}$.

15.
$$\begin{cases} 3x & +y & +z & = 0 \\ x & +y & & = 1 \\ 7x & +3y & +3z & = 2 \end{cases} \qquad -3E_1 + E_3 \to E_3$$

$$\begin{cases} 3x & +y & +z & = 0 \\ x & +y & & = 1 \\ -2x & & & = 2 \end{cases} \qquad -\tfrac{1}{2}E_3 \to E_3$$

$$\begin{cases} 3x & +y & +z & = 0 \\ x & +y & & = 1 \\ x & & & = -1 \end{cases} \qquad \begin{array}{l} -E_3 + E_2 \to E_2 \\ -3E_3 + E_1 \to E_1 \end{array}$$

$$\begin{cases} & y & +z & = 3 \\ & y & & = 2 \\ x & & & = -1 \end{cases} \qquad -E_2 + E_1 \to E_1$$

$$\begin{cases} & & z & = 1 \\ & y & & = 2 \\ x & & & = -1 \end{cases} \qquad \text{The solution set is } \{(-1, 2, 1)\}$$

17.
$$\begin{cases} 2x & +3y & +2z & = 8 \\ -x & +4y & -3z & = 1 \\ 3x & -2y & +7z & = 11 \end{cases}$$

Solve the second equation for x to obtain $x = 4y - 3z - 1$.
Replace x by $4y - 3z - 1$ in the 1st and 3rd equations and simplify.

$$\begin{cases} 2(4y - 3z - 1) + 3y + 2z = 8 \\ 3(4y - 3z - 1) - 2y + 7z = 11 \end{cases} \qquad \begin{cases} 11y - 4z = 10 \\ 10y - 2z = 14 \end{cases}$$

Solve the 2nd equation for z.

$$\begin{cases} 11y - 4z = 10 \\ z = 5y - 7 \end{cases}$$

Replace z by $5y - 7$ in the upper equation and solve for y and then z.

$$\begin{cases} 11y - 4(5y - 7) = 10 \\ z = 5y - 7 \end{cases} \qquad \begin{cases} -9y = -18 \\ z = 5y - 7 \end{cases} \qquad \begin{cases} y = 2 \\ z = 5y - 7 \end{cases}$$

$$\begin{cases} y = 2 \\ z = 10 - 7 \end{cases} \qquad \begin{cases} y = 2 \\ z = 3 \end{cases}$$

Since $x = 4y - 3z - 1 = 8 - 9 - 1 = -2$, the solution set is $\{(-2, 2, 3)\}$.

19.
$$\begin{cases} 3x & & +2z & = -9 \\ & 2y & +5z & = 4 \\ 3x & +y & +3z & = -7 \end{cases}$$

From E_3 we obtain $y = -7 - 3x - 3z$. Substituting in E_2, we obtain

$$\begin{cases} 3x + 2z = -9 \\ 2(-7 - 3x - 3z) + 5z = 4 \end{cases} \quad \text{or} \quad \begin{cases} 3x + 2z = -9 \\ -6x - z = 18 \end{cases}$$

We find $x = -3$, $z = 0$ and thus that $y = 2$. The solution set is $\{(-3, 2, 0)\}$.

21.
$$\begin{cases} x & + 2y & - 5z & = 3 \\ 2x & - 3y & + 3z & = -2 \\ 5x & + 3y & - 12z & = 7 \end{cases}$$

Solve the first equation for x, substitute the result in the second and last equations, then simplify.

$$\begin{cases} x = 3 - 2y + 5z \\ 2(3 - 2y + 5z) - 3y + 3z = -2 \\ 5(3 - 2y + 5z) + 3y - 12z = 7 \end{cases} \qquad \begin{cases} x = 3 - 2y + 5z \\ -7y + 13z = -8 \\ -7y + 13z = -8 \end{cases}$$

Since the last two equations are identical, the system has an infinite number of solutions. Let $z = c$. Then the system may be written as

$$\begin{cases} x = 3 - 2y + 5c \\ -7y + 13c = -8 \end{cases} \qquad \begin{cases} x = 3 - 2y + 5c \\ y = \frac{13}{7}c + \frac{8}{7} \end{cases}$$

$$\begin{cases} x = 3 - \frac{26}{7}c - \frac{16}{7} + 5c = \frac{5}{7} + \frac{9}{7}c \\ y = \frac{13}{7}c + \frac{8}{7} \end{cases}$$

The solution set is $\left\{ \left(\frac{5}{7} + \frac{9}{7}c, \frac{13}{7}c + \frac{8}{7}, c \right) \mid c \text{ a real number} \right\}$.

23.
$$\begin{cases} x & + y & + z & = 6 \\ 2x & + y & + 3z & = 12 \\ x & - y & + 5z & = 10 \end{cases}$$

Solve the first equation for x, substitute the result in the last two equations and simplify.

$$\begin{cases} x = 6 - y - z \\ 2(6 - y - z) + y + 3z = 12 \\ 6 - y - z - y + 5z = 10 \end{cases}$$

$$\begin{cases} x = 6 - y - z \\ -y + z = 0 \\ -2y + 4z = 4 \end{cases} \qquad -\tfrac{1}{2}E_3 + E_2 \to E_3$$

$$\begin{cases} x = 6 - y - z \\ -y + z = 0 \\ -z = -2 \end{cases} \qquad \begin{cases} x = 6 - y - z \\ -y + z = 0 \\ z = 2 \end{cases}$$

$$\begin{cases} x = 6 - y - z \\ -y + 2 = 0 \\ z = 2 \end{cases} \qquad \begin{cases} x = 6 - y - z \\ y = 2 \\ z = 2 \end{cases}$$

$$\begin{cases} x = 6 - 2 - 2 = 2 \\ y = 2 \\ z = 2 \end{cases}$$

The solution set is $\{(2, 2, 2)\}$.

25.
$$\begin{cases} x & + y & & = -4 \\ x & & + z & = 1 \\ 3x & - y & + 2z & = 4 \end{cases}$$

Solve the first equation for y, substitute the result in the last equation, and simplify.

$$\begin{cases} y = -4 - x \\ x + z = 1 \\ 3x + 4 + x + 2z = 4 \end{cases}$$

$$\begin{cases} y = -4 - x \\ x + z = 1 \\ 4x + 2z = 0 \end{cases} \qquad\qquad -4E_2 + E_3 \to E_3$$

$$\begin{cases} y = -4 - x \\ x + z = 1 \\ -2z = -4 \end{cases} \qquad\qquad \begin{cases} y = -4 - x \\ x + z = 1 \\ z = 2 \end{cases}$$

$$\begin{cases} y = -4 - x \\ x + 2 = 1 \\ z = 2 \end{cases} \qquad\qquad \begin{cases} y = -4 - x \\ x = -1 \\ z = 2 \end{cases}$$

$$\begin{cases} y = -4 + 1 = -3 \\ x = -1 \\ z = 2 \end{cases}$$

The solution set is $\{(-1, -3, 2)\}$.

27.
$$\begin{cases} 3x & + y & + z & = 0 \\ x & + y & & = 1 \\ 7x & + 3y & + 3z & = 2 \end{cases}$$

Solve the second equation for y, substitute the result in the first and third equations, and simplify.

$$\begin{cases} 3x & + y & + z & = 0 \\ & y & & = 1 - x \\ 7x & + 3y & + 3z & = 2 \end{cases} \qquad \begin{cases} 3x + 1 - x + z & = 0 \\ y & = 1 - x \\ 7x + 3(1 - x) + 3z & = 2 \end{cases}$$

$$\begin{cases} 2x & + z & = -1 \\ y & & = 1 - x \\ 4x & + 3z & = -1 \end{cases}$$

Solve the first equation for z, substitute the result in the third equation and simplify.

$$\begin{cases} z & = -1 - 2x \\ y & = 1 - x \\ 4x + 3(-1 - 2x) & = -1 \end{cases} \qquad \begin{cases} z & = -1 - 2x \\ y & = 1 - x \\ -2x & = 2 \end{cases}$$

$$\begin{cases} z = -1 - 2x \\ y = 1 - x \\ x = -1 \end{cases} \qquad \begin{cases} z = -1 - 2(-1) \\ y = 1 - (-1) \\ x = -1 \end{cases} \qquad \begin{cases} z = 1 \\ y = 2 \\ x = -1 \end{cases}$$

The solution set is $\{(-1, 2, 1)\}$

29. Let x be the number of pounds of A, y the number of pounds of B and z the number of pounds of C where $x + y + z = 400$. The total value of the candy is

$4.20x + 5.25y + 6z = 400(4.50)$ or $420x + 525y + 600z = 180{,}000$.

Furthermore $x = 3(y + z)$. The corresponding system of equations is

$$\begin{cases} x & + y & + z & = 400 \\ 420x & + 525y & + 600z & = 180{,}000 \\ x & - 3y & - 3z & = 0 \end{cases} \qquad 3E_1 + E_3 \to E_3$$

$$\begin{cases} x & +y & +z & = 400 \\ 420x & +525y & +600z & = 180{,}000 \\ 4x & & & = 1200 \end{cases}$$

$$\begin{cases} x & +y & +z & = 400 \\ 420x & +525y & +600z & = 180{,}000 \\ x & & & = 300 \end{cases}$$

$$\begin{cases} & y & +z & = 100 \\ 420(300) & +525y & +600z & = 180{,}000 \\ x & & & = 300 \end{cases}$$

$$\begin{cases} y+z = 100 \\ 525y+600z = 54{,}000 \\ x = 300 \end{cases} \qquad -525E_1 + E_2 \to E_2$$

$$\begin{cases} y+z = 100 \\ 75z = 1500 \\ x = 300 \end{cases} \qquad\qquad \begin{cases} y+z = 100 \\ z = 20 \\ x = 300 \end{cases}$$

$$\begin{cases} y = 80 \\ z = 20 \\ x = 300 \end{cases}$$

Thus mix 300 pounds of kind A, 80 pounds of B, and 20 pounds of C.

31. Let x dollars be invested in bond A, y dollars in bond B, and z dollars in bond C where $x+y+z = 50{,}000$. $x = y+z$ and the yearly income from the investments is

$.062x + .09y + .11z = 4{,}000$ or $62x + 90y + 110z = 4{,}000{,}000$.

The corresponding system of equations is

$$\begin{cases} x & +y & +z & = 50{,}000 \\ x & -y & -z & = 0 \\ 62x & +90y & +110z & = 4{,}000{,}000 \end{cases} \qquad E_1 + E_2 \to E_2$$

$$\begin{cases} x & +y & +z & = 50{,}000 \\ 2x & & & = 50{,}000 \\ 62x & +90y & +110z & = 4{,}000{,}000 \end{cases} \qquad \tfrac{1}{2}E_2 \to E_2$$

$$\begin{cases} x & +y & +z & = 50{,}000 \\ x & & & = 25{,}000 \\ 62x & +90y & +110z & = 4{,}000{,}000 \end{cases} \qquad \begin{aligned} & -E_2 + E_1 \to E_1 \\ & -62E_2 + E_3 \to E_3 \end{aligned}$$

$$\begin{cases} y+z = 25{,}000 \\ x = 25{,}000 \\ 28y+48y = 900{,}000 \end{cases} \qquad -28E_1 + E_3 \to E_3$$

$$\begin{cases} y+z = 25{,}000 \\ x = 25{,}000 \\ 20z = 200{,}000 \end{cases} \qquad\qquad \begin{cases} y+z = 25{,}000 \\ x = 25{,}000 \\ z = 10{,}000 \end{cases}$$

$$\begin{cases} y = 15{,}000 \\ x = 25{,}000 \\ z = 10{,}000 \end{cases}$$

She will invest $25,000 in bond A, $15,000 in bond B and $10,000 in bond C.

33. Let x be the cost of one unit of X, y the cost of a unit of Y and z the cost of a unit of Z.

Ann's cost was $\qquad 2x + 3y + z = 57.75$
Greg's cost was $\qquad x + 2y + 3z = 67.50$
George's cost was $\qquad 2x + 3y + 3z = 82.25$

$$
\begin{cases}
2x & +3y & +z & = 57.75 \\
x & +2y & +3z & = 67.50 \\
2x & +3y & +3z & = 85.25
\end{cases}
\qquad
\begin{array}{l}
2E_2 - E_1 \to E_1 \\[6pt]
-E_1 + E_3 \to E_3
\end{array}
$$

$$
\begin{cases}
 y & +5z & = 77.25 \\
x & +2y & +3z & = 67.50 \\
 & 2z & = 27.50
\end{cases}
\qquad
\tfrac{1}{2}E_3 \to E_3
$$

$$
\begin{cases}
 y & +5z & = 77.25 \\
x & +2y & +3z & = 67.50 \\
 & z & = 13.75
\end{cases}
\qquad
\begin{cases}
y + 5(13.75) = 77.25 \\
x + 2y + 3z = 67.50 \\
z = 13.75
\end{cases}
$$

$$
\begin{cases}
y = 8.50 \\
x + 17 + 41.25 = 67.50 \\
z = 13.75
\end{cases}
\qquad
\begin{cases}
y = 8.50 \\
x = 9.25 \\
z = 13.75
\end{cases}
$$

X costs \$9.25 per unit, Y \$8.50 and Z \$13.75.

35. Let x, y and z be the number of houses of type A, B and C respectively. Since 51 units of roofing were used $3x + 2y + 5z = 51$. Since 67 units of concrete were used $5x + 6y + 4z = 67$. Since 77 units of lumber were used $4x + 7y + 6z = 77$. The corresponding system of equations is

$$
\begin{cases}
3x & +2y & +5z & = 51 \\
5x & +6y & +4z & = 67 \\
4x & +7y & +6z & = 77
\end{cases}
\qquad
\begin{array}{l}
5E_1 - 3E_2 \to E_2 \\
4E_1 - 3E_3 \to E_3
\end{array}
$$

$$
\begin{cases}
3x & +2y & +5z & = 51 \\
 & -8y & +13z & = 54 \\
 & -13y & +2z & = -27
\end{cases}
\qquad
2E_2 - 13E_3 \to E_3
$$

$$
\begin{cases}
3x & +2y & +5z & = 51 \\
 & -8y & +13z & = 54 \\
 & 153y & & = 459
\end{cases}
\qquad
\tfrac{1}{153}E_3 \to E_3
$$

$$
\begin{cases}
3x & +2y & +5z & = 51 \\
 & -8y & +13z & = 54 \\
 & y & & = 3
\end{cases}
\qquad
8E_3 + E_2 \to E_2
$$

$$
\begin{cases}
3x & +2y & +5z & = 51 \\
 & & 13z & = 78 \\
 & y & & = 3
\end{cases}
\qquad
\tfrac{1}{13}E_2 \to E_2
$$

$$
\begin{cases}
3x & +2y & +5z & = 51 \\
 & & z & = 6 \\
 & y & & = 3
\end{cases}
\qquad
-2E_3 - 5E_2 + E_1 \to E_1
$$

$$
\begin{cases}
3x = 15 \\
z = 6 \\
y = 3
\end{cases}
$$

Hence 5 houses of type A were built, 3 of type B and 6 of type C.

37. Let x be the wife's age, y the professor's age and z the son's age. Since the professor is four years older that his wife $x + 4 = y$. Since the wife was 24 when the son was born $z + 24 = x$. Twenty years from now, the wife's age will be $x + 20$, the professor's $y + 20$ and the son's $z + 20$. Hence $x + 20 + z + 20 = 1.5(y + 20)$. The corresponding system of equations is

$$\begin{cases} x & -y & & = -4 \\ -x & & +z & = -24 \\ x & -1.5y & +z & = -10 \end{cases} \qquad \begin{array}{l} E_1 + E_2 \to E_1 \\ E_3 + E_2 \to E_3 \end{array}$$

$$\begin{cases} & -y & +z & = -28 \\ -x & & +z & = -24 \\ & -1.5y & +2z & = -34 \end{cases} \qquad -2E_1 + E_3 \to E_3$$

$$\begin{cases} & -y & +z & = -28 \\ -x & & +z & = -24 \\ & .5y & & = 22 \end{cases} \qquad 2E_3 \to E_3$$

$$\begin{cases} & -y & +z & = -28 \\ -x & & +z & = -24 \\ & y & & = 44 \end{cases} \qquad E_3 + E_1 \to E_1$$

$$\begin{cases} & & z & = 16 \\ -x & & +z & = -24 \\ & y & & = 44 \end{cases} \qquad -E_1 + E_2 \to E_2$$

$$\begin{cases} z = 16 \\ -x = -40 \\ y = 44 \end{cases}$$

Thus the wife's age is 40, the professor's 44 and the son's 16.

39. Let x, y and z be the number of pounds of items A, B and C respectively, where $x + y + z = 200$. The total units of vitamins is $150x + 250y + 420z = 290$. The total number of calories is $600x + 500y + 330z$, or the number of calories per pound is $\frac{600x + 500y + 330z}{200}$ and the number of units of vitamins per pound is $\frac{150x + 250y + 420z}{200}$.

The corresponding system of equations is

$$\begin{cases} x & +y & +z & = 200 \\ 150x & +250y & +420z & = 290(200) \\ 600x & +500y & +330z & = 460(200) \end{cases} \qquad \begin{array}{l} -150E_1 + E_2 \to E_2 \\ -600E_1 + E_3 \to E_3 \end{array}$$

$$\begin{cases} x & +y & +z & = 200 \\ & 100y & +270z & = 28,000 \\ & -100y & -270z & = -28,000 \end{cases} \qquad \begin{cases} x + y + z = 200 \\ y + 2.7z = 280 \end{cases}$$

Let $z = a$, $a \geq 0$, then $y = 280 - 2.7a \geq 0$, so that $a \leq \frac{280}{2.7} = \frac{2800}{27}$

Also $x + (280 - 2.7a) + a = 200$, so $x = -80 + 1.7a \geq 0$, $a \geq \frac{800}{17}$.

The solution set is $\left\{ (1.7a - 80, \, 280 - 2.7a, \, a) \mid \frac{800}{17} \leq a \leq \frac{2800}{27} \right\}$. The cost will be

$C = 1.25x + 1.35y + 1.6z = .08a + 278$. Thus the cost is minimum when a is a minimum, that is when $a = \frac{800}{17}$. The cost will be \$281.76 using no item A, $\frac{2600}{17}$ lbs of B and $\frac{800}{17}$ lbs of C.

1. The graph of $y = x^2 + 6x - 3$ is a concave upward parabola with vertex when
    $x = -6/2 = -3$. Some points on the graph are

x	-4	-3	-2	0
y	-11	-12	-11	-3

The graph of $y = 2x + 2$ is a line passing through $(-4, -6)$ and $(-8, -14)$.

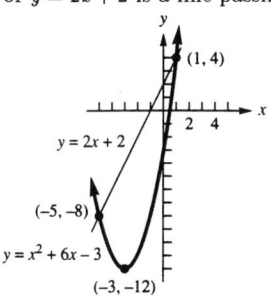

The graphs are shown.
The graphs appear to intersect when
$x = 1$ and $x = -5$, and the points
$(1, 4)$ and $(-5, -8)$ satisfy both
equations. The solution set is
$\{(1, 4), (-5, -8)\}$.

3. The graph of $y = -x^2 + 2x + 5$ is a concave downward parabola with vertex when $x = 1$.
    Some points on the graph are

x	0	1	2
y	5	6	5

The graph of $y = 2x + 1$ is a line passing through $(0, 1)$ and $(2, 5)$.

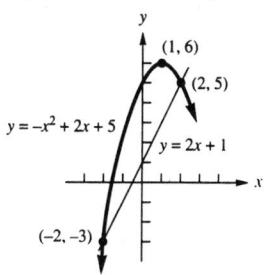

The graphs are shown.
The graphs appear to intersect when
$x = 2$ and $x = -2$, and the points
$(2, 5)$ and $(-2, -3)$ lie on both
graphs. The solution set is
$\{(2, 5), (-2, -3)\}$.

5. The graph of $y = x^2 + 3x + 1$ is a parabola with vertex when $x = -\frac{3}{2}$. Some points on the
    graph are

x	-2	$-\frac{3}{2}$	-1	0
y	-1	$-\frac{5}{4}$	-1	1

The graph of $y = 2x^2 + 4x - 1$ is a parabola with vertex when $x = -1$. Some points on
the graph are

x	0	-3	-2	-1
y	-1	5	-1	-3

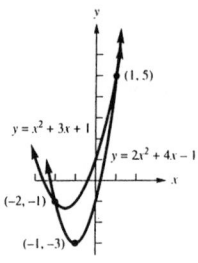

The graphs are shown.
The graphs appear to intersect when
$x = -2$ and $x = 1$, and the points
$(-2, -1)$ and $(1, 5)$ satisfy both
equations. The solution set is
$\{(-2, -1), (1, 5)\}$.

7. The graph of $y = -x^2 + 4x + 6$ is a parabola with vertex when $x = 2$. Some points on the graph are

x	0	1	2	3
y	6	9	10	9

The graph of $y = x^2 + 6x + 2$ is a parabola with vertex when $x = -3$. Some points on the graph are

x	-4	-3	-2	0
y	-6	-7	-6	2

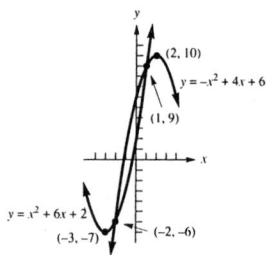

The graphs are shown.
The graphs appear to intersect when
$x = 1$ and $x = -2$, and the points
$(1, 9)$ and $(-2, -6)$ satisfy both
equations. The solution set is
$\{(1, 9), (-2, -6)\}$

9. Some points on the graphs are

x	0	1	2	-1
$y = 3^{-x} + 2$	3	$2.\overline{3}$	$2.\overline{1}$	5
$y = -x^2 + \frac{1}{27}x + \frac{163}{27}$	6.037	5.074	$2.\overline{1}$	5

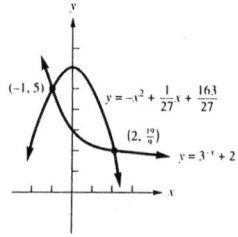

The above points show that the
graphs intersect when $x = 2$ and
$x = -1$, and the graphs show there
are only two points of intersection.
The solution set is $\{(-1, 5),$
$(2, 2.1\overline{1})\}$.

11. Some points on the graphs are

x	0	1	2	3	3.5	4
$y = 2^x$	1	2	4	8	11.3	16
$y = -x^2 + 7x - 4$	-4	2	6	8	8.25	8

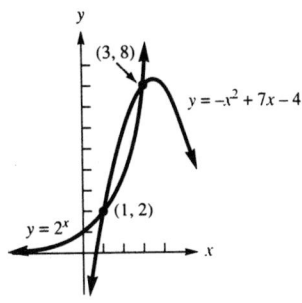

The graphs are shown.
The above points show that $(1, 2)$ and $(3, 8)$ are solutions of the system and the graphs show they are the only solutions. The solution set is $\{(1, 2), (3, 8)\}$.

13. Some points on the graphs are

x	0	1	2	-1
$y = 4^{-x}$	1	.25	.0625	4
$y = \frac{1}{6}(3^{2x})$	$.1\overline{6}$	1.5	20.25	.0185

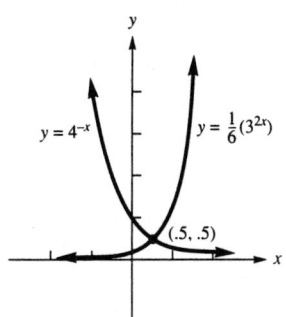

There is one point of intersection when $x = x_0$ where $0 < x_0 < 1$. Adding the point where $x = .5$, we see the solution set is $\{(.5, .5)\}$.

x	.5
$y = 4^{-x}$	.5
$y = \frac{1}{6}(3^{2x})$	.5

15. Some points on the graphs are

x	0	1	2	-1
$y = 2\ln(x+3)$	2.197	2.772	3.219	1.386
$y = 2^{-x/2}$	1	.707	.5	1.414

The graphs are shown.

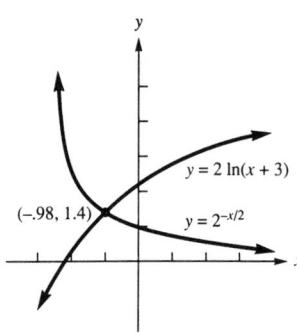

The graphs intersect when $x = x_0$ where $-1 < x_0 < 0$. Looking at some more points:

x	$-.5$	$-.7$	$-.9$	$-.95$	$-.99$	$-.98$	$-.985$
$y = 2\ln(x+3)$	1.832	1.6658	1.4839	1.4357	1.3963	1.4062	1.4012
$y = 2^{-x/2}$	1.189	1.2746	1.366	1.3899	1.4093	1.4044	1.4069

$-1 < x_0 < -.5$
$-1 < x_0 < -.7$
$-1 < x_0 < -.9$
$-1 < x_0 < -.95$
$-.99 < x_0 < -.95$
$-.99 < x_0 < -.98$
$-.985 < x_0 < -.98$

To two decimal places the point of intersection occurs when $x = -.98$ and is approximately $(-.98, 1.40)$.

$-$ 60 $-$

1. $x + y > 1$ if $\qquad y > 1 - x$

$\qquad\qquad$ Also, $\quad x < 2$.

Testing the point $(0, 0)$,

$0 > 1 - 0$ is false. The graph lies above the line

$y = 1 - x$ and to the left of $x = 2$.

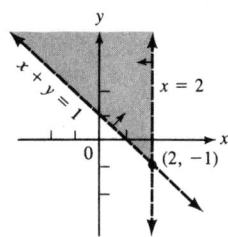

3. $2x + y < 6$ if $\qquad y < 6 - 2x$

$\qquad\qquad$ Also, $\quad x - y > -3$.

Testing the point $(0, 0)$,

$0 < 6$ is true and $0 - 0 > -3$ is true. Hence the

graph lies below the line $y = 6 - 2x$ and below

the line $x - y = -3$.

5. $x + 2y \leq 4$ if $\qquad y \leq 2 - \frac{x}{2}$,

$\qquad\qquad$ Also, $\quad x > -1$ and

$\qquad\qquad\qquad y > -2$.

Testing $(0, 2)$,

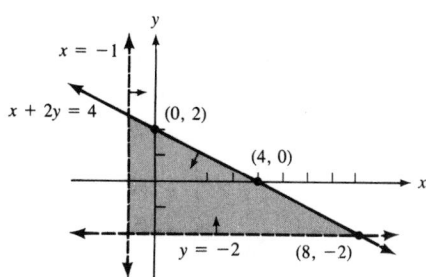

$0 \leq 4$ is true. The graph lies on and below the

line $y = 2 - \frac{x}{2}$, to the right of $x = -1$ and above

$y = -2$.

7. $x - 3y < -3$ if $x + 3 < 3y$ or $\frac{x}{3} + 1 < y$. $x < 1$

and $y > -1$.

Testing $(0, 0)$,

$0 + 1 < 0$ is false. Thus the graph lies above the

line $y = 1 + \frac{x}{3}$, left of the line $x = 1$ and above

the line $y = -1$.

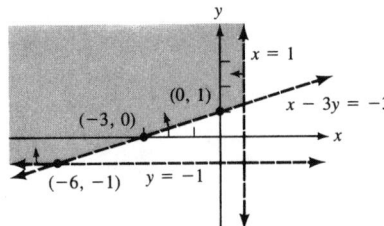

9. The lines $x + y = 1$ and $2x - y = -4$ intersect

according to the solution of the system

$\begin{cases} x + y = 1 \\ 2x - y = -4 \end{cases}$ which is $\{(-1, 2)\}$.

Arrows indicate the result of testing $(0, 0)$. Also

$y \geq -2$.

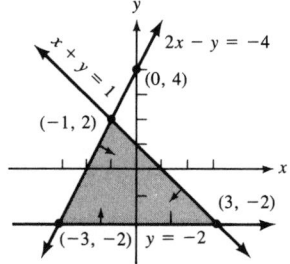

11. The graphs of the lines $2x + y = 6$, $x - y = -3$,

and $2x + 3y = -6$ are shown.

Testing $(0, 0)$,

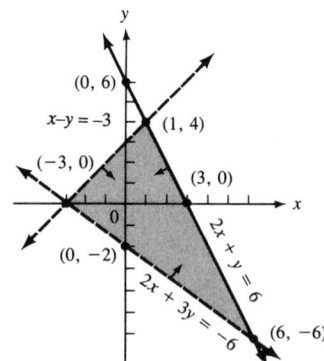

$0 + 0 \leq 6$, so the graph is on or below $2x + y = 6$.

Also $0 - 0 > -3$ is true so the graph lies below

the line $x - y = -3$, and $0 + 0 > -6$ so the

graph is above the line $2x + 3y = -6$.

13. The graphs of $x + y = 1$, $2x - y = -4$ and

$x - 2y = 4$ are shown.

Testing $(0, 0)$,

$0 + 0 \leq 1$, so the graph lies on or below

$x + y = 1$. $0 - 0 \geq -4$ so the graph lies on or

below $2x - y = -4$. $0 - 0 \leq 4$ so the graph lies

on or above $x - 2y = 4$.

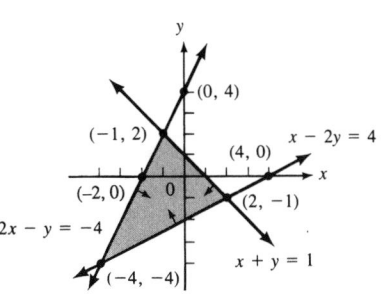

15. The graphs of the lines $x + y = 1$, $2x - y = -4$ and $7x - 8y = 22$ are shown.

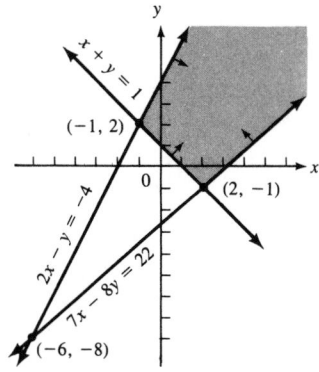

Testing $(0, 0)$, $0 + 0 \geq 1$ is false so the graph lies on or above $x + y = 1$. $0 - 0 \geq -4$ is true so
the graph lies on or below $2x - y = -4$. $0 - 0 \leq 22$ is true so the graph lies on or above
$7x - 8y = 22$.

Solving the systems below we find the points of intersection that are shown.

$$\begin{cases} 2x - y = -4 \\ x + y = 1 \end{cases} \quad \text{or} \quad \begin{cases} x = -1 \\ y = 2 \end{cases}$$

$$\begin{cases} x + y = 1 \\ 7x - 8y = 22 \end{cases} \quad \text{or} \quad \begin{cases} 8x + 8y = 8 \\ 7x - 8y = 22 \end{cases} \quad \text{or} \quad \begin{cases} 15x = 30 \\ 7x - 8y = 22 \end{cases} \quad \text{or} \quad \begin{cases} x = 2 \\ y = -1 \end{cases}$$

17. The graphs of the lines $3x - 4y = -19$, $5x + 3y = 7$, $2x - y = 5$ and $4x + y = 19$ are shown.

Testing the point $(0, 0)$, $0 - 0 > -19$ is true, so

the graph lies below $3x - 4y = -19$. $0 + 0 > 7$ is

false, so the graph lies above $5x + 3y = 7$. $0 < 5$

is true so the graph lies above $2x - y = 5$. $0 < 19$

is true so the graph lies below $4x + y = 19$.

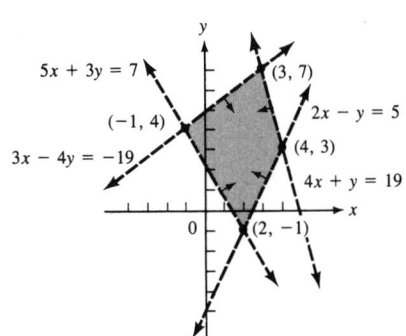

Points of intersection are solutions to the following four systems of equations:

$$\begin{cases} 3x-4y=-19 \\ 5x+3y=7 \end{cases} \rightarrow \begin{cases} 9x-12y=-57 \\ 20x+12y=28 \end{cases} \rightarrow \begin{cases} 29x=-29 \\ 12y=48 \end{cases} \rightarrow \begin{cases} x=-1 \\ y=4 \end{cases}$$

$$\begin{cases} 3x-4y=-19 \\ 4x+y=19 \end{cases} \rightarrow \begin{cases} 3x-4y=-19 \\ 16x+4y=76 \end{cases} \rightarrow \begin{cases} 3x-4y=-19 \\ 19x=57 \end{cases} \rightarrow \begin{cases} y=7 \\ x=3 \end{cases}$$

$$\begin{cases} 4x+y=19 \\ 2x-y=5 \end{cases} \rightarrow \begin{cases} 6x=24 \\ 2x-y=5 \end{cases} \rightarrow \begin{cases} x=4 \\ y=3 \end{cases}$$

$$\begin{cases} 2x-y=5 \\ 5x+3y=7 \end{cases} \rightarrow \begin{cases} 6x-3y=15 \\ 5x+3y=7 \end{cases} \rightarrow \begin{cases} 6x-3y=15 \\ 11x=22 \end{cases} \rightarrow \begin{cases} y=-1 \\ x=2 \end{cases}$$

19.

$$x+y \leq 800$$
$$x \geq 150$$
$$y \geq 100$$
$$x \leq 3y$$

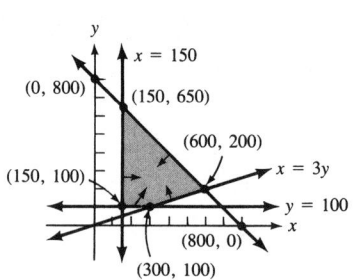

21.

$$3x+2y \leq 38000$$
$$\frac{x}{6}+\frac{y}{10} \leq 2000$$
$$x \geq 2500$$
$$y \geq 3600$$

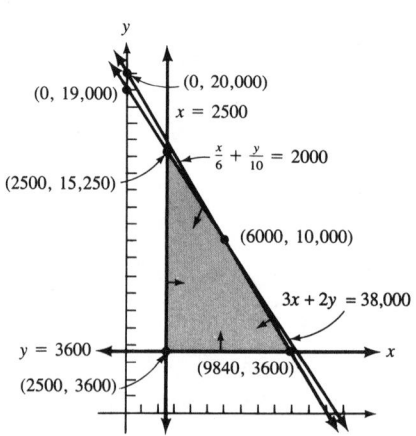

23. x and y are the number of shares of stocks A and B, respectively, where $x + y \leq 300$, $x \geq 50$, and $0 \leq y \leq 200$. Since the number of shares of A does not exceed twice that of B, $x \leq 2y$.

We have the system,

$x + y \leq 300$

$x \geq 50$

$0 \leq y \leq 200$

$x \leq 2y$

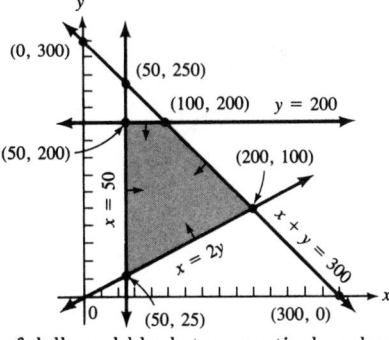

25. x and y are the numbers of dolls and blankets respectively, where $x \geq 0$ and $y \geq 0$.

Since Shirley has 120 hours a month for use, dolls take 3 hours, and blankets take 5 hours, $3x + 5y \leq 120$.

Since the cost for a doll is $4 and for a blanket is $7, and $166 is available, $4x + 7y \leq 166$.

The system is,

$x \geq 0$

$y \geq 0$

$3x + 5y \leq 120$

$4x + 7y \leq 166$

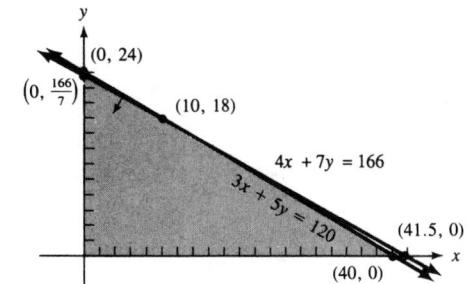

27. $4x + 3y \leq 36$

$2x + 5y \leq 32$

$x \geq 3$

$y \geq 2$

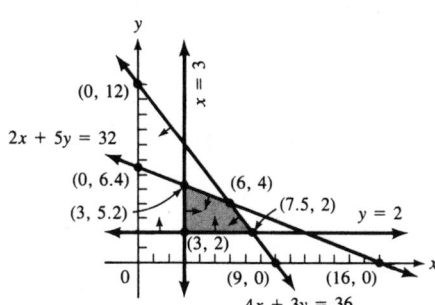

29. $x + y \leq 120$

$.6x + .3y \leq 45$

$x \geq 10$

$0 \leq y \leq 100$

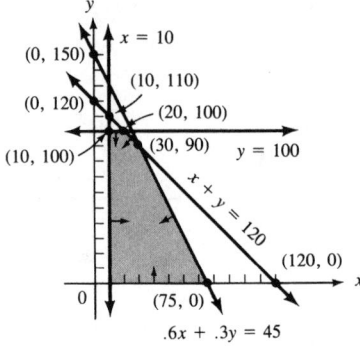

EXERCISE SET 2.5 APPLICATIONS TO BUSINESS AND ECONOMICS

1. a) By definition, the fixed cost is $C(0) = \$25,000$.

b) $R(q) = 5q$. The business breaks even when $R(q) = C(q)$. That is

$$5q = 3q + 25,000$$
$$2q = 25,000$$
$$q = 12,500$$

3. a) $C(q) = R(q)$ when the company breaks even.

$$.25q^2 + 2000q + 525,000 = -2.25q^2 + 4500q$$
$$25q^2 - 25,000q + 5,250,000 = 0$$
$$q^2 - 1000q + 210,000 = 0$$
$$(q - 300)(q - 700) = 0. \quad q = 300 \text{ or } 700.$$

b) The company makes a profit when $R(q) - C(q) > 0$. That is

$$-2.25q^2 + 4500 > .25q^2 + 2000q + 525,000$$
$$0 > (q - 300)(q - 700)$$

q		300		700	
$q - 300$	$-$	0	$+$	$+$	$+$
$q - 700$	$-$	$-$	$-$	0	$+$
$(q - 300)(q - 700) -$	0	$-$	0	$+$

The solution is $300 < q < 700$.

5. a) The company breaks even when $C(q) = R(q)$. That is
$$.5q^2 + 215q + 18,000 = -4q^2 + 800q$$
$$4.5q^2 - 585q + 18,000 = 0$$
$$q = \frac{585 \pm \sqrt{342,225 - 324,000}}{9} = \frac{585 \pm 135}{9}, \quad q = 80 \text{ or } 50.$$

b) A profit occurs when $R(q) - C(q) > 0$. That is
$$4.5q^2 - 585q + 18,000 < 0$$
$$q^2 - 130q + 4000 < 0$$
$$(q - 80)(q - 50) < 0.$$
The foregoing inequality is true when $50 < q < 80$.

7. The equilibrium price occurs when $D(p) = S(p)$. That is
$$75,000 - 120p = -3750 + 105p$$
$$78,750 = 225p$$
$$p = \$350.$$
The equilibrium quantity is $D(350) = 75,000 - 120(350) = 33,000$ units.

9. a) When $I = 30$, $D_1(p) = -60p + 2640 + 360 = -60p + 3000$.
$$D_1(p) = S(p) \text{ if } \quad -60p + 3000 = 42p - 60$$
$$3060 = 102p.$$
The equilibrium price is $30 and the equilibrium quantity is $S(30) = 1200$ thousand units.

b) $D_2(p) = -60p + 2640 + 972 = -60p + 3612$.
$3612 - 60p = 42p - 60$ if $102p = 3672$ or $p = \$36$.
$36 is the equilibrium price and $S(36) = 1452$ thousand units is the equilibrium quantity.

c)

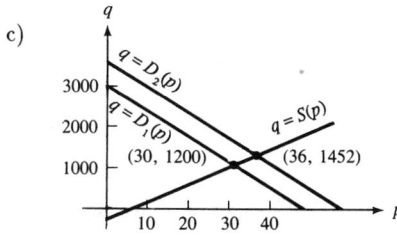

11. **a)** $D_1(p) = -p^2 - 20p + 4000 + 2(850) = -p^2 - 20p + 5700$

$\qquad D_1(p) = S(p)$ if $-p^2 - 20p + 5700 = p^2 + 28p - 510.$

$\qquad 2p^2 + 48p - 6210 = 0,$

$\qquad p^2 + 24p - 3105 = 0,$

$\qquad p = \dfrac{-24 \pm \sqrt{12{,}996}}{2} = \dfrac{-24 \pm 114}{2}.$

Since p cannot be negative $p = \$45$ is the equilibrium price.

The equilibrium quantity is $45^2 + 28(45) - 510 = 2775$ thousand units.

b) $D_2(p) = -p^2 - 20p + 4000 + (1201)2 = -p^2 - 20p + 6402.$

$D_2(p) = S(p)$ if $-p^2 - 20p + 6402 = p^2 + 28p - 510.$

$2p^2 + 48p - 6912 = 0,$

$p = \dfrac{-48 \pm \sqrt{57{,}600}}{4} = \dfrac{-48 \pm 240}{4} = -12 \pm 60.$

Since p cannot be negative, the equilibrium price is $\$48$ and the equilibrium quantity is

$S(48) = 48^2 + 28(48) - 510 = 3138$ thousand units.

c)

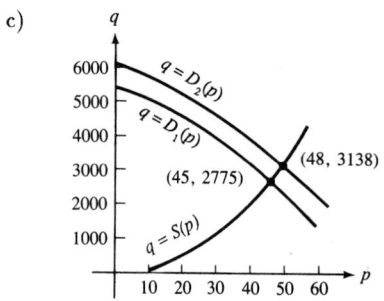

13. **a)** $D(p) = S(p)$ when $-5p + 159 = 3p - 9$, $168 = 8p$, $p = \$21$ is the equilibrium price and 54 thousand units is the equilibrium quantity.

b)   The price received by the supplier is reduced by \$8. Hence

$\qquad -5p + 159 = 3(p - 8) - 9$

$\qquad -5p + 159 = 3p - 33$

$\qquad 192 = 8p$

$p = \$24$ is the equilibrium price and $-5(24) + 159 = 39$ thousand units is the equilibrium quantity.

15. a) $D(p) = mp + b$

$D(12) = 12m + b = 1478$ and $D(40) = 40m + b = 582$

$$\begin{cases} 12m + b = 1478 \\ 40m + b = 582 \end{cases} \quad \begin{cases} 12m + b = 1478 \\ 28m = -896 \end{cases} \quad \begin{cases} 12m + b = 1478 \\ m = -32 \end{cases}$$

$$\begin{cases} b = 1862 \\ m = -32 \end{cases} \qquad D(p) = -32p + 1862.$$

b) $S(p) = mp + b$

$S(12) = 12m + b = 190$ and $S(40) = 40m + b = 862$

$$\begin{cases} 12m + b = 190 \\ 40m + b = 862 \end{cases} \quad \begin{cases} 12m + b = 190 \\ 28m = 672 \end{cases} \quad \begin{cases} 12(24) + b = 190 \\ m = 24 \end{cases}$$

$$\begin{cases} b = -98 \\ m = 24 \end{cases} \qquad S(p) = 24p - 98.$$

c) $D(p) = S(p)$ when $-32p + 1862 = 24p - 98$, $1960 = 56p$, $p = \$35$ is the equilibrium price and $24(35) - 98 = 742$ units is the equilibrium quantity.

d)
$$-32p + 1862 = 24(p - 7) - 98$$
$$-32p + 1862 = 24p - 266$$
$$2128 = 56p$$

$p = \$38$ is the new equilibrium price and $-32(38) + 1862 = 646$ is the new equilibrium quantity.

17. a)
$$D(p) = mp + b$$
$$D(5) = 5m + b = 510$$
$$D(20) = 20m + b = 360$$

$$\begin{cases} 5m + b = 510 \\ 20m + b = 360 \end{cases} \quad \begin{cases} 5m + b = 510 \\ 15m = -150 \end{cases} \quad \begin{cases} -50 + b = 510 \\ m = -10 \end{cases}$$

$$\begin{cases} b = 560 \\ m = -10 \end{cases} \qquad D(p) = -10p + 560.$$

b)
$$S(p) = mp + b$$
$$S(5) = 5m + b = 35$$
$$S(20) = 20m + b = 260$$

$$\begin{cases} 5m + b = 35 \\ 20m + b = 260 \end{cases} \quad \begin{cases} 5m + b = 35 \\ 15m = 225 \end{cases} \quad \begin{cases} 5(15) + b = 35 \\ m = 15 \end{cases}$$

$$\begin{cases} b = -40 \\ m = 15 \end{cases} \qquad S(p) = 15p - 40.$$

c) $D(p) = S(p)$ when $-10p + 560 = 15p - 40$, $25p = 600$, $p = \$24$ is the equilibrium price and $S(24) = 15(24) - 40 = 320$ units is the equilibrium quantity.

d) $S(p) = 15(p+5) - 40 = 15p + 35$.
$S(p) = D(p)$ when $15p + 35 = -10p + 560$, $25p = 525$ and the new equilibrium price is $p = \$21$. The new equilibrium quantity is $S(21) = 15(26) - 40 = 350$ units.

19. a) $S(p) = mp + b$
$S(4) = 4m + b = 34$
$S(6) = 6m + b = 59$

$$\begin{cases} 4m + b = 34 \\ 6m + b = 59 \end{cases} \qquad \begin{cases} 4m + b = 34 \\ 2m = 25 \end{cases} \qquad \begin{cases} 4(12.5) + b = 34 \\ m = 12.5 \end{cases}$$

$$\begin{cases} b = -16 \\ m = 12.5 \end{cases} \qquad S(p) = 12.5p - 16.$$

b) $S(p) = D(p)$ if $12.5p - 16 = -p^2 - p + 156$.
$$p^2 + 13.5p - 172 = 0, \ p = \frac{-13.5 \pm \sqrt{870.25}}{2} = \frac{-13.5 \pm 29.5}{2}.$$
Since p is not negative, the equilibrium price is 8 francs/liter and the equilibrium quantity is $S(8) = 84$ million liters.

c) For the equilibrium quantity to be 114, $D(p) = S(p) = 114$, $D(p) = -p^2 - p + 156 = 114$ if $p^2 + p - 42 = 0$ or $(p+7)(p-6) = 0$.
$p = 6$ francs/liter since p cannot be negative. For the supply to be 114 million liters,
$114 = 12.5(6 + s) - 16$ where s is the amount of the subsidy.
$130 = 75 + 12.5s$, $s = 4.4$ francs/liter.

21. a) $S(10) = 231$ and $S(5) = 81$. The slope of the supply curve is $\frac{231 - 81}{10 - 5} = 30$.
Using the point-slope equation of a line $q - 81 = 30(p - 5)$, $q = 30p - 69 = S(p)$.

b) $D(p) = S(p)$ when
$$-p^2 - 2p + 575 = 30p - 69$$
$$p^2 + 32p - 644 = 0$$
$$(p - 14)(p + 46) = 0$$

$p = \$14/\text{unit}$ is the equilibrium price and $S(14) = 30(14) - 69 = 351$ thousand units is the equilibrium quantity.

c) The new demand will be
$$351 + 81 = -p^2 - 2p + 575$$
$$p^2 + 2p - 143 = 0$$
$$(p + 13)(p - 11) = 0$$

$p = \$11/\text{unit}$ is the new equilibrium price. The supply will be $351 + 81 = 30(11 + s) - 69$ where s is the amount of the subsidy. $432 = 330 + 30s - 69$, $s = \$5.70/\text{unit}$.

23. a) $D_x = S_x$ and $D_y = S_y$ lead to the system of equations
$$\begin{cases} 16 - 2p_x - p_y = -10 + 3p_x + 3p_y \\ 30 - 2p_x - 3p_y = -17 + 8p_x + 3p_y \end{cases}$$ Simplifying we obtain

$$\begin{cases} 5p_x + 4p_y = 26 \\ 10p_x + 6p_y = 47 \end{cases} \qquad E_2 - 2E_1 \rightarrow E_2$$

$$\begin{cases} 5p_x + 4p_y = 26 \\ -2p_y = -5 \end{cases} \qquad -\tfrac{1}{2}E_2 \rightarrow E_2$$

$$\begin{cases} 5p_x + 4p_y = 26 \\ p_y = 2.5 \end{cases} \qquad E_1 - 4E_2 \rightarrow E_1$$

$$\begin{cases} 5p_x = 16 \\ p_y = 2.5 \end{cases}$$

Hence $p_x = \$3.20$ and $p_y = \$2.50$ are the equilibrium prices per gallon of ice cream and container of topping respectively.

b) The equilibrium quantities are $S_x = -10 + 3(3.20) + 3(2.50) = 7.1$ thousand units and $S_y = -17 + 8(3.20) + 3(2.50) = 16.1$ thousand units.

25. To find the equilibrium prices the system must be solved
$$\begin{cases} D_x = S_x \\ D_y = S_y \\ D_z = S_z \end{cases} \qquad \begin{cases} 520 - 30x + 92y - 25z = -120 + 70x - 8y + 75z \\ 140 + 32x - 52y + 18z = -7 - 8x + 98y - 12z \\ 478 - 20x + 28y - 5z = -21 + 40x - 12y + 65z \end{cases}$$

$$\begin{cases} 100x - 100y + 100z = 640 \\ 40x - 150y + 30z = -147 \\ 60x - 40y + 70z = 499 \end{cases} \qquad \tfrac{1}{10}E_1 \rightarrow E_1$$

$$\begin{cases} 10x - 10y + 10z = 64 \\ 40x - 150y + 30z = -147 \\ 60x - 40y + 70z = 499 \end{cases} \qquad \begin{array}{l} 4E_1 - E_2 \rightarrow E_2 \\ 6E_1 - E_3 \rightarrow E_3 \end{array}$$

$$\begin{cases} 10x - 10y + 10z = 64 \\ 110y + 10z = 403 \\ -20y - 10z = -115 \end{cases} \qquad E_3 + E_2 \to E_2$$

$$\begin{cases} 10x - 10y + 10z = 64 \\ 90y = 288 \\ -20y - 10z = -115 \end{cases}$$

$y = 3.20$

$-64 - 10z = -115 \qquad\qquad 10x - 32 + 51 = 64$

$10z = 115 - 64 = 51 \qquad\qquad 10x = 45$

$z = 5.1 \qquad\qquad x = 4.5$

The equilibrium prices for commodities X, Y and Z are \$4.50, \$3.20 and \$5.10 per unit, respectively. At these prices, $D_x = 551.9$, $D_y = 209.4$ and $D_z = 452.1$. Hence the equilibrium quantities for commodities X, Y and Z are 551.9, 209.4 and 452.1 thousand units respectively.

27. The system is

$$\begin{cases} 800 - 23x + 95y - 18z = -35 + 77x - 5y + 82z \\ 320 + 50x - 12y + 35z = -68 - 10x + 118y - 5z \\ 1310 - 10x + 42y - 20z = -112 + 80x - 8y + 100z \end{cases}$$

$$\begin{cases} 100x - 100y + 100z = 835 \\ 60x - 130y + 40z = -388 \\ 90x - 50y + 120z = 1482 \end{cases} \qquad \begin{array}{l} 3E_2 - E_3 \to E_3 \\ \\ 2E_1 - 5E_2 \to E_2 \end{array}$$

$$\begin{cases} 100x & -100y & +100z & = 835 \\ -100x & +450y & & = 3610 \\ 90x & -340y & & = -2586 \end{cases} \qquad \tfrac{1}{10}E_2 \to E_2$$

$$\begin{cases} 100x & -100y & +100z & = 835 \\ -10x & +45y & & = 361 \\ 90x & -340y & & = -2586 \end{cases} \qquad 9E_2 + E_3 \to E_3$$

$$\begin{cases} 100x & -100y & +100z & = 835 \\ -10x & +45y & & = 361 \\ & 65y & & = 663 \end{cases}$$

$y = 10.2$

$10x = 45(10.2) - 361$

$x = 9.8$

$980 - 1020 + 100z = 835$

$z = 8.75$

Hence the equilibrium prices for commodities X, Y and Z are \$9.80, \$10.20 and \$8.75 per unit, respectively. At these prices $S_x = 1386.1$, $S_y = 993.85$ and $S_z = 1465.4$, so the equilibrium quantities for the three commodities are 1386.1, 993.85 and 1465.4 thousand units, respectively.

1.
$$\begin{cases} 2x + 3y = -5 \\ 3x - 5y = 21 \end{cases}$$

Points on the graph of the line $2x + 3y = -5$ include $(0, -5/3)$ and $(-5/2, 0)$.

$(0, -21/5)$ and $(7, 0)$ are on the graph of $3x - 5y = 21$.

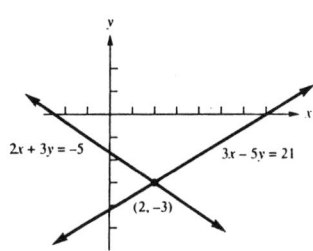

The graphs are shown.

The graphs appear to intersect at the point $(2, -3)$ and $(2, -3)$ satisfies both equations. Hence the solution set is

$$\{(2, -3)\}.$$

3.
$$\begin{cases} 5x + 2y = 0 \\ 4x - 3y = 23 \end{cases}$$

Solving the first equation for y we obtain $y = -\frac{5}{2}x$

and $-\frac{5}{2}x$ replaces y in the second equation to obtain

$$\begin{cases} 5x + 2y = 0 \\ 4x + \frac{15}{2}x = 23 \end{cases} \qquad \begin{cases} 5x + 2y = 0 \\ \frac{23}{2}x = 23 \end{cases} \qquad \begin{cases} 5x + 2y = 0 \\ x = 2 \end{cases}$$

Replacing x by 2 in the first equation we obtain

$$\begin{cases} 10 + 2y = 0 \\ x = 2 \end{cases} \qquad \text{or} \qquad \begin{cases} y = -5 \\ x = 2 \end{cases} \qquad \text{The solution set is } \{(2, -5)\}.$$

5.
$$\begin{cases} 2x - 3y = 7 \\ 3x + 9y = -3 \end{cases} \qquad 3E_1 \rightarrow E_1$$

$$\begin{cases} 6x - 9y = 21 \\ 3x + 9y = -3 \end{cases} \qquad E_1 + E_2 \rightarrow E_2$$

$$\begin{cases} 6x - 9y = 21 \\ 9x = 18 \end{cases} \qquad \frac{1}{2}E_2 \rightarrow E_2$$

$$\begin{cases} 6x - 9y = 21 \\ x = 2 \end{cases} \qquad -6E_2 + E_1 \rightarrow E_1$$

$$\begin{cases} -9y = 9 \\ x = 2 \end{cases} \qquad -\frac{1}{9}E_1 \rightarrow E_1$$

$$\begin{cases} y = -1 \\ x = 2 \end{cases} \qquad \text{The solution set is } \{(2, -1)\}.$$

7.

$$C = \begin{vmatrix} 5 & 1 \\ 3 & -2 \end{vmatrix} = -10 - 3 = -13.$$

$$A_x = \begin{vmatrix} -13 & 1 \\ -13 & -2 \end{vmatrix} = 26 + 13 = 39.$$

$$A_y = \begin{vmatrix} 5 & -13 \\ 3 & -13 \end{vmatrix} = -65 + 39 = -26.$$

$$x = \frac{A_x}{C} = \frac{39}{-13} = -3, \quad y = \frac{A_y}{C} = \frac{-26}{-13} = 2.$$

The solution set is $\{(-3, 2)\}$.

9.

$$\begin{cases} 2x^2 + y^2 = 17 \\ 5x^2 - y^2 = 11 \end{cases} \qquad \text{Let } u = y^2 \text{ and } v = x^2.$$

$$\begin{cases} 2v + u = 17 \\ 5v - u = 11 \end{cases} \qquad E_1 + E_2 \rightarrow E_2$$

$$\begin{cases} 2v + u = 17 \\ 7v = 28 \end{cases} \qquad \tfrac{1}{7}E_2 \rightarrow E_2$$

$$\begin{cases} 2v + u = 17 \\ v = 4 \end{cases} \qquad -2E_2 + E_1 \rightarrow E_1$$

$$\begin{cases} u = 9 \\ v = 4 \end{cases} \qquad \begin{cases} y^2 = 9 \\ x^2 = 4 \end{cases} \qquad \begin{cases} y = \pm 3 \\ x = \pm 2 \end{cases}$$

The solution set is $\{(2,3), (2,-3), (-2,3), (-2,-3)\}$.

11.

$$\begin{cases} \frac{5}{x} + \frac{1}{y} = -13 \\ \frac{3}{x} - \frac{2}{y} = -13 \end{cases} \qquad \text{Let } u = \frac{1}{x} \text{ and } v = \frac{1}{y}.$$

$$\begin{cases} 5u + v = -13 \\ 3u - 2v = -13 \end{cases} \qquad 2E_1 + E_2 \rightarrow E_2$$

$$\begin{cases} 5u + v = -13 \\ 13u = -39 \end{cases} \qquad \begin{cases} 5u + v = -13 \\ u = -3 \end{cases} \qquad \begin{cases} -15 + v = -13 \\ u = -3 \end{cases}$$

$$\begin{cases} v = 2 \\ u = -3 \end{cases} \qquad \begin{cases} \frac{1}{y} = 2 \\ \frac{1}{x} = -3 \end{cases} \qquad \begin{cases} y = \frac{1}{2} \\ x = -\frac{1}{3} \end{cases}$$

The solution set is $\left\{ \left(-\frac{1}{3}, \frac{1}{2} \right) \right\}$.

13. The graph of $y = -2x^2 + 8x + 13$ is a parabola with vertex when $x = 2$. Some points on the graph are

x	0	1	2	3
y	13	19	21	19

The graph of $6x - y = -9$ is a line through the points $(0, 9)$ and $(-\frac{3}{2}, 0)$.

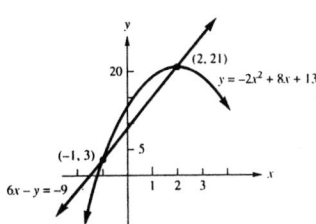

The graphs are shown.
The graphs appear to intersect when $x = 2$ and $x = -1$ and the points $(2, 21)$ and $(-1, 3)$ lie on both graphs. The solution set is $\{(2, 21), (-1, 3)\}$.

15. The graph of $y = x^2 + 5x + 6$ is a parabola with vertex when $x = -\frac{5}{2}$. Some points on the graph are

x	-3	$-\frac{5}{2}$	-2	0
y	0	$-\frac{1}{4}$	0	6

The graph of $y = -x^2 + 3x + 10$ is a parabola with vertex when $x = \frac{3}{2}$.

Some points on the graph are

x	0	1	$\frac{3}{2}$	2
y	10	12	$\frac{49}{4}$	12

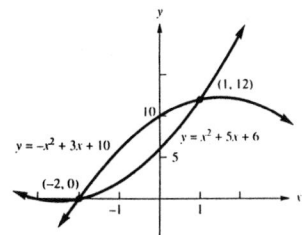

The graphs are shown.
The graphs appear to intersect when $x = 1$ and $x = -2$. The points $(1, 12)$ and $(-2, 0)$ satisfy both equations. The solution set is $\{(1, 12), (-2, 0)\}$.

17.

$$\begin{cases} 12x & -12y & -z & = 0 \\ 6x & +6y & +2z & = 90 \\ 2x & +3y & +z & = 36 \end{cases}$$

$2E_1 + E_2 \rightarrow E_2$
$E_1 + E_3 \rightarrow E_3$

$$\begin{cases} 12x & -12y & -z & = 0 \\ 30x & -18y & & = 90 \\ 14x & -9y & & = 36 \end{cases}$$

$E_2 - 2E_3 \rightarrow E_3$

$\frac{1}{6}E_2 \rightarrow E_2$

$$\begin{cases} 12x & -12y & -z & = 0 \\ 5x & -3y & & = 15 \\ 2x & & & = 18 \end{cases} \qquad \tfrac{1}{2}E_3 \to E_3$$

$$\begin{cases} 12x & -12y & -z & = 0 \\ 5x & -3y & & = 15 \\ x & & & = 9 \end{cases} \qquad E_2 - 5E_3 \to E_2$$

$$\begin{cases} 12x & -12y & -z & = 0 \\ & -3y & & = -30 \\ x & & & = 9 \end{cases} \qquad -\tfrac{1}{3}E_2 \to E_2$$

$$\begin{cases} 12x & -12y & -z & = 0 \\ & y & & = 10 \\ x & & & = 9 \end{cases} \qquad -E_1 + 12E_3 - 12E_2 \to E_1$$

$$\begin{cases} z = -12 \\ y = 10 \\ x = 9 \end{cases}$$

The solution set is $\{(9, 10, -12)\}$.

19.
$$\begin{cases} x & +2y & -7z & = 75 \\ 6x & -2y & +3z & = 75 \\ 2x & +y & +z & = 0 \end{cases} \qquad \begin{matrix} -6E_1 + E_2 \to E_2 \\ -2E_1 + E_3 \to E_3 \end{matrix}$$

$$\begin{cases} x & +2y & -7z & = 75 \\ & -14y & +45z & = -375 \\ & -3y & +15z & = -150 \end{cases} \qquad \begin{matrix} -3E_3 + E_2 \to E_2 \\ \\ -\tfrac{1}{3}E_3 \to E_3 \end{matrix}$$

$$\begin{cases} x & +2y & -7z & = 75 \\ & -5y & & = 75 \\ & y & -5z & = 50 \end{cases} \qquad -\tfrac{1}{5}E_2 \to E_2$$

$$\begin{cases} x & +2y & -7z & = 75 \\ & y & & = -15 \\ & y & -5z & = 50 \end{cases} \qquad E_2 - E_3 \to E_3$$

$$\begin{cases} x & +2y & -7z & = 75 \\ & y & & = -15 \\ & & 5z & = -65 \end{cases} \qquad \tfrac{1}{5}E_3 \to E_3$$

$$\begin{cases} x & +2y & -7z & = 75 \\ & y & & = -15 \\ & & z & = -13 \end{cases} \qquad E_1 - 2E_2 + 7E_3 \to E_3$$

$$\begin{cases} x = 14 \\ y = -15 \\ z = -13 \end{cases}$$

The solution set is $\{(14, -15, -13)\}$.

21.

$$\begin{cases} 2x & +3y & -z & = 2 \\ 5x & -2y & +z & = 3 \\ 9x & +4y & -z & = 9 \end{cases} \qquad \begin{array}{l} E_2 + E_1 \to E_1 \\ E_2 + E_3 \to E_3 \end{array}$$

$$\begin{cases} 7x & +y & & = 5 \\ 5x & -2y & +z & = 3 \\ 14x & +2y & & = 12 \end{cases} \qquad 2E_1 - E_3 \to E_3$$

$$\begin{cases} 7x & +y & & = 5 \\ 5x & -2y & +z & = 3 \\ & 0 & & = -2 \end{cases}$$

Since $0 \neq -2$, the system is inconsistent and the solution set is $\emptyset$.

23.

$$\begin{cases} 2x & -2y & +3z & = 1 \\ x & -3y & -2z & = -9 \\ x & +y & +z & = 6 \end{cases}$$

Solve the third equation for z to obtain $z = 6 - x - y$, replace z by $6 - x - y$ in the first two equations and simplify.

$$\begin{cases} 2x - 2y + 3(6 - x - y) = 1 \\ x - 3y - 2(6 - x - y) = -9 \\ z = 6 - x - y \end{cases}$$

$$\begin{cases} -x - 5y = -17 \\ 3x - y = 3 \\ z = 6 - x - y \end{cases} \qquad 3E_1 + E_2 \to E_2$$

$$\begin{cases} -x - 5y = -17 \\ -16y = -48 \\ z = 6 - x - y \end{cases} \qquad \begin{cases} -x - 5y = -17 \\ y = 3 \\ z = 3 - x \end{cases}$$

$$\begin{cases} -x - 15 = -17 \\ y = 3 \\ z = 3 - x \end{cases} \qquad \begin{cases} x = 2 \\ y = 3 \\ z = 1 \end{cases}$$

The solution set is $\{(2, 3, 1)\}$.

25.

$$\begin{cases} 6x & -2y & -z & = 6 \\ x & -3y & -4z & = 5 \\ 3x & +y & +4z & = 0 \end{cases}$$

Solve the last equation for y to obtain $y = -3x - 4z$. Replace y by $-3x - 4z$ in the first two equations and simplify.

$$\begin{cases} 6x - 2(-3x - 4z) - z = 6 \\ x - 3(-3x - 4z) - 4z = 5 \\ y = -3x - 4z \end{cases}$$

$$\begin{cases} 12x + 7z = 6 \\ 10x - 8z = 5 \\ y = -3x - 4z \end{cases} \qquad 8E_1 + 7E_2 \to E_1$$

$$\begin{cases} 166x = 83 \\ 10x - 8z = 5 \\ y = -3x - 4z \end{cases} \qquad \begin{cases} x = 1/2 \\ 5 - 8z = 5 \\ y = -3/2 - 4z \end{cases}$$

$$\begin{cases} x = 1/2 \\ z = 0 \\ y = -3/2 \end{cases}$$

The solution set is $\{(1/2, -3/2, 0)\}$.

27.
$$\begin{cases} 4x & -y & +z & = 11 \\ 3x & -2y & -z & = 6 \\ 7x & -3y & & = 17 \end{cases}$$

Solve the second equation for z to obtain $z = 3x - 2y - 6$. Replace z by $3x - 2y - 6$ in the first equation and simplify.

$$\begin{cases} 4x - y + 3x - 2y - 6 = 11 \\ z = 3x - 2y - 6 \\ 7x - 3y = 17 \end{cases}$$

$$\begin{cases} 7x - 3y = 17 \\ z = 3x - 2y - 6 \\ 7x - 3y = 17 \end{cases} \qquad E_1 - E_3 \to E_3$$

$$\begin{cases} 7x - 3y = 17 \\ z = 3x - 2y - 6 \\ 0 = 0 \end{cases}$$

There are an infinite number of solutions. Let $x = c$, where c is any real number.

$-3y = 17 - 7c$ or $y = -\frac{17}{3} + \frac{7}{3}c$ and $z = 3c + \frac{34}{3} - \frac{14}{3}c - 6 = \frac{-5c}{3} + \frac{16}{3}$.

The solution set is $\left\{ \left(c, \frac{-17}{3} + \frac{7}{3}c, \frac{-5c}{3} + \frac{16}{3} \right) \mid c \text{ is a real number} \right\}$.

29.
$$\begin{cases} 5x & & -z & = 16 \\ 3x & -y & -2z & = 9 \\ x & +2y & +3z & = 4 \end{cases}$$

Solve the first equation for z to obtain $z = 5x - 16$. Replace z by $5x - 16$ in the last two equations and simplify.

$$\begin{cases} z = 5x - 16 \\ 3x - y - 2(5x - 16) = 9 \\ x + 2y + 3(5x - 16) = 4 \end{cases}$$

$$\begin{cases} z = 5x - 16 \\ -7x - y = -23 \\ 16 + 2y = 52 \end{cases} \qquad 2E_2 + E_3 \to E_3$$

$$\begin{cases} z = 5x - 16 \\ -7x - y = -23 \\ 2x = 6 \end{cases} \qquad \begin{cases} z = 5x - 16 \\ -7x - y = -23 \\ x = 3 \end{cases}$$

$$\begin{cases} z = 15 - 16 \\ -21 - y = -23 \\ x = 3 \end{cases} \qquad \begin{cases} z = -1 \\ y = 2 \\ x = 3 \end{cases}$$

The solution set is $\{(3, 2, -1)\}$.

31.
$$\begin{cases} x > 2 \\ 2x - y \le 9 \\ x + y \le 12 \end{cases}$$

The graphs of $x = 2$, $2x - y = 9$ and $x + y = 12$ are shown. Testing $(0,0)$, $0 - 0 \le 9$ so the graph is above or on the line $2x - y = 9$, and $0 \le 12$ so the graph is on or below the line $x + y = 12$.

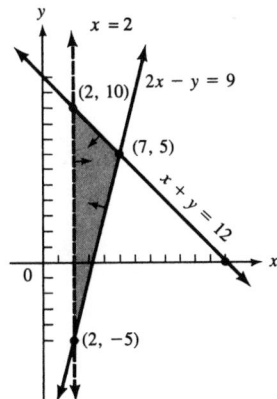

33.
$$\begin{cases} x \le 7 \\ y \le 8 \\ x \ge 0 \\ y \ge 0 \\ x + y \ge 5 \end{cases}$$

Testing $(0,0)$, $0 < 5$ hence the graph is on or above the line $x + y = 5$.

35. Let x be the value of the first store and y the value of the second.
For 1987 the profit was $.10x - .05y = 47,500$.
For 1988 the profit was $.08x + .06y = 93,000$.
The system of equations is

$$\begin{cases} .10x - .05y = 47,500 \\ .08x + .06y = 93,000 \end{cases}$$
$6E_1 + 5E_2 \rightarrow E_2$

$$\begin{cases} .10x - .05y = 47,500 \\ x = 750,000 \end{cases}$$
$-.1E_2 + E_1 \rightarrow E_1$

$$\begin{cases} -.05y = -27,500 \\ x = 750,000 \end{cases}$$
$-20E_1 \rightarrow E_1$

$$\begin{cases} y = 550,000 \\ x = 750,000 \end{cases}$$

The value of the first store is \$750,000 and that of the second store is \$550,000.

37. Let x be the daily wage of the adults and y the wage of the teenagers. The total daily wages are $15x + 8y = 1336$. Since 2 adults together earn \$48 more that 3 teens $3y + 48 = 2x$. The sytem is

$$\begin{cases} 15x + 8y = 1336 \\ 2x - 3y = 48 \end{cases} \qquad\qquad 3E_1 + 8E_2 \rightarrow E_2$$

$$\begin{cases} 15x + 8y = 1336 \\ 61x = 4392 \end{cases} \qquad\qquad \frac{1}{61}E_2 \rightarrow E_2$$

$$\begin{cases} 15x + 8y = 1336 \\ x = 72 \end{cases} \qquad\qquad -15E_2 + E_1 \rightarrow E_1$$

$$\begin{cases} 8y = 256 \\ x = 72 \end{cases} \qquad\qquad \begin{cases} y = 32 \\ x = 72 \end{cases}$$

The adults' wage is \$72, and the teens' is \$32.

39. Let x be the number of days in which Kris could do the job by himself, and y the number of days for Pete. Kris can do $\frac{1}{x}$ of the job in one day and Pete can do $\frac{1}{y}$ of the job in one day. If Kris works 2 days and Pete works 5 days, we have $2(\frac{1}{x}) + 5(\frac{1}{y}) = \frac{3}{8}$. Furthermore, $8(\frac{1}{x}) + 10(\frac{1}{y}) = 1$. The system is

$$\begin{cases} 2/x + 5/y = 3/8 \\ 8/x + 10/y = 1 \end{cases} \qquad \text{Let } u = \frac{1}{x} \text{ and } v = \frac{1}{y}$$

$$\begin{cases} 2u + 5v = 3/8 \\ 8u + 10v = 1 \end{cases} \qquad\qquad -4E_1 + E_2 \rightarrow E_2$$

$$\begin{cases} 2u + 5v = 3/8 \\ -10v = -1/2 \end{cases} \qquad\qquad -.1E_2 \rightarrow E_2$$

$$\begin{cases} 2u + 5v = 3/8 \\ v = 1/20 \end{cases} \qquad\qquad -5E_2 + E_1 \rightarrow E_1$$

$$\begin{cases} 2u = 1/8 \\ v = 1/20 \end{cases} \qquad\qquad \frac{1}{2}E_1 \rightarrow E_1$$

$$\begin{cases} u = 1/16 \\ v = 1/20 \end{cases}$$

Replacing u by $\frac{1}{x}$ and v by $\frac{1}{y}$

$\frac{1}{x} = \frac{1}{16}$ and $x = 16$

$\frac{1}{y} = \frac{1}{20}$ and $y = 20$.

Kris can do the job in 16 days and Pete can do it in 20 days.

41. Since x and y denote the daily production of deluxe and standard models respectively and since it takes 4 hours to assemble a deluxe model and 3 hours to assemble a standard model $4x + 3y \leq 78(8) = 624$. Since it takes 2 hours to paint a deluxe model and 1.2 hours to paint a standard model $2x + 1.2y \leq 36(8) = 288$.

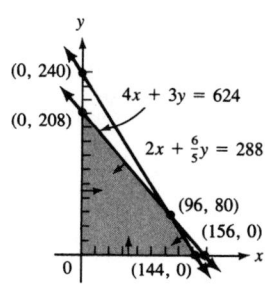

The system is
$$\begin{cases} x \geq 0, \quad y \geq 0 \\ 4x + 3y \leq 624 \\ 2x + 1.2y \leq 288 \end{cases}$$

43.   At equilibrium we have the system

$$\begin{cases} 387{,}500 - 50{,}000\,p_x + 25{,}000\,p_y = -137{,}500 + 75{,}000\,p_x - 12{,}500\,p_y \\ 556{,}250 + 20{,}000\,p_x - 62{,}500\,p_y = -257{,}500 - 5{,}000\,p_x + 312{,}500\,p_y \end{cases}$$

$$\begin{cases} 125{,}000\,p_x - 37{,}500\,p_y = 525{,}000 \\ 25{,}000\,p_x - 375{,}000\,p_y = -813{,}750 \end{cases} \qquad \begin{aligned} -10\mathrm{E}_1 + \mathrm{E}_2 &\rightarrow \mathrm{E}_2 \\ .01\mathrm{E}_1 &\rightarrow \mathrm{E}_1 \end{aligned}$$

$$\begin{cases} 1250\,p_x - 375\,p_y = 5250 \\ -1{,}225{,}000\,p_x = -6{,}063{,}750 \end{cases}$$

$$\begin{cases} 1250(4.95) - 375\,p_y = 5250 \\ p_x = 4.95 \end{cases} \qquad \begin{cases} p_y = 2.50 \\ p_x = 4.95 \end{cases}$$

a) The equilibrium price for gourmet chocolate is \$4.95 per pound and for regular is \$2.50 per pound.

b) At these prices $S_x = 202{,}500$ and $S_y = 499{,}000$. The equilibrium demands for gourmet and regular chocolate are 202,500 pounds and 499,000 pounds, respectively.

CHAPTER 3

MATRICES

EXERCISE SET 3.1 MATRIX NOTATION AND MATRIX ARITHMETIC

1. Since A has two rows and three columns, its dimension is 2×3.
   Since B has three rows and two columns, its dimension is 3×2.
   Since C has three rows and four columns, its dimension is 3×4.

3.
$$[a_{ij}] = \begin{bmatrix} 3(1)+2(1) & 3(1)+2(2) & 3(1)+2(3) \\ 3(2)+2(1) & 3(2)+2(2) & 3(2)+2(3) \end{bmatrix} = \begin{bmatrix} 5 & 7 & 9 \\ 8 & 10 & 12 \end{bmatrix}$$

5.
$$[c_{ij}] = \begin{bmatrix} 1^2+2(1) & 1^2+2(2) & 1^2+2(3) \\ 2^2+2(1) & 2^2+2(2) & 2^2+2(3) \\ 3^2+2(1) & 3^2+2(2) & 3^2+2(3) \end{bmatrix} = \begin{bmatrix} 3 & 5 & 7 \\ 6 & 8 & 10 \\ 11 & 13 & 15 \end{bmatrix}$$

7.
$$[a_{ij}] = \begin{bmatrix} 0 & 2(1)+2 & 2(1)+3 & 2(1)+4 \\ 2(2)+1 & 0 & 2(2)+3 & 2(2)+4 \\ 2(3)+1 & 2(3)+2 & 0 & 2(3)+4 \\ 2(4)+1 & 2(4)+2 & 2(4)+3 & 0 \end{bmatrix} = \begin{bmatrix} 0 & 4 & 5 & 6 \\ 5 & 0 & 7 & 8 \\ 7 & 8 & 0 & 10 \\ 9 & 10 & 11 & 0 \end{bmatrix}$$

9.
$$[c_{ij}] = \begin{bmatrix} 1 & 3(1)+2 & 3(1)+3 & 3(1)+4 \\ 3(2)+1 & 2^2 & 3(2)+3 & 3(2)+4 \\ 3(3)+1 & 3(3)+2 & 3^2 & 3(3)+1 \end{bmatrix} = \begin{bmatrix} 1 & 5 & 6 & 7 \\ 7 & 4 & 9 & 10 \\ 10 & 11 & 9 & 13 \end{bmatrix}$$

11.

$$[a_{ij}] = \begin{bmatrix} -1 & 2(1) & 2(1) & 2(1) \\ 2(2) & 1 & 2(2) & 2(2) \\ 2(3) & 2(3) & -1 & 2(3) \\ 2(4) & 2(4) & 2(4) & 1 \end{bmatrix} = \begin{bmatrix} -1 & 2 & 2 & 2 \\ 4 & 1 & 4 & 4 \\ 6 & 6 & -1 & 6 \\ 8 & 8 & 8 & 1 \end{bmatrix}$$

13.

$$\begin{bmatrix} -3(3) & -3(4) & -3(-2) \\ -3(2) & -3(-4) & -3(0) \\ -3(7) & -3(6) & -3(12) \end{bmatrix} = \begin{bmatrix} -9 & -12 & 6 \\ -6 & 12 & 0 \\ -21 & -18 & -36 \end{bmatrix}$$

15. It is not possible to add the two matrices since they don't have the same dimension.

17.

$$\begin{bmatrix} 2(3) & 2(2) \\ 2(9) & 2(0) \\ 2(4) & 2(2) \\ 2(6) & 2(5) \end{bmatrix} + \begin{bmatrix} 3(8) & 3(5) \\ 3(-1) & 3(3) \\ 3(3) & 3(-2) \\ 3(-3) & 3(0) \end{bmatrix} = \begin{bmatrix} 6 & 4 \\ 18 & 0 \\ 8 & 4 \\ 12 & 10 \end{bmatrix} + \begin{bmatrix} 24 & 15 \\ -3 & 9 \\ 9 & -6 \\ -9 & 0 \end{bmatrix} = \begin{bmatrix} 30 & 19 \\ 15 & 9 \\ 17 & -2 \\ 3 & 10 \end{bmatrix}$$

19.

$$\begin{bmatrix} 15 & 40 & 5 \\ 25 & 5 & -10 \\ 45 & 10 & 15 \end{bmatrix} - \begin{bmatrix} 4 & -18 & 8 \\ 0 & 6 & 14 \\ 10 & 0 & 4 \end{bmatrix} = \begin{bmatrix} 11 & 58 & -3 \\ 25 & -1 & -24 \\ 35 & 10 & 11 \end{bmatrix}$$

21. Equating corresponding elements of the two matrices,
$$y = 3$$
$$x = 5$$
$$z = 7$$

23. Equating corresponding elements of the matrices,
$$2 - x = 3x - 2, \ 3 + y = 2y + 1, \ 2 + z = 3z - 6, \ 4w = 2w + 6$$
which lead to
$$x = 1, \ y = 2, \ z = 4, \ w = 3$$

25. Adding the two matrices on the left we obtain $\begin{bmatrix} 2x + 3y & 3 \\ 7 & 3x - 2y \end{bmatrix} = \begin{bmatrix} 5 & 3 \\ 7 & -12 \end{bmatrix}$

Equating corresponding elements leads to the system

$$\begin{cases} 2x + 3y = 5 \\ 3x - 2y = -12 \end{cases} \quad \begin{cases} 4x + 6y = 10 \\ 9x - 6y = -36 \end{cases}$$

$$\begin{cases} 13x = -26 \\ 3x - 2y = -12 \end{cases} \quad \begin{cases} x = -2 \\ -6 - 2y = -12 \end{cases} \quad \begin{cases} x = -2 \\ y = 3 \end{cases}$$

27. Adding the matrices on the left we obtain

$$
\begin{bmatrix}
2x+y+z & 5 & 0 \\
7 & 3x-2y & 5 \\
6 & 4y & 3y-2z
\end{bmatrix}
=
\begin{bmatrix}
7-2z & 5 & 0 \\
7 & -z & 5 \\
6 & 4y & 3+x
\end{bmatrix}
$$

Equating corresponding elements leads to the system

$$
\begin{cases} 2x+y+z = 7-2z \\ 3x-2y = -z \\ 3y-2z = 3+x \end{cases}
\quad
\begin{cases} 2x+y+3z = 7 \\ 3x-2y+z = 0 \\ x-3y+2z = -3 \end{cases}
\quad
\begin{matrix} 2E_1+E_2 \to E_2 \\ 3E_1+E_3 \to E_3 \end{matrix}
\quad
\begin{cases} 2x+y+3z = 7 \\ 7x+7z = 14 \\ 7x+11z = 18 \end{cases}
$$

$$
\begin{matrix} -E_2+E_3 \to E_3 \\ (1/7)E_2 \to E_2 \end{matrix}
\quad
\begin{cases} 2x+y+3z = 7 \\ x+z = 2 \\ 4z = 4 \end{cases}
\quad
\begin{cases} 2x+y+3z = 7 \\ x+z = 2 \\ z = 1 \end{cases}
\quad
\begin{cases} 2x+y+3z = 7 \\ x = 1 \\ z = 1 \end{cases}
\quad
\begin{cases} y = 2 \\ x = 1 \\ z = 1 \end{cases}
$$

29.

$$
\begin{bmatrix}
1 & 0 & 4 & 6 \\
3 & 2 & 1 & 4 \\
4 & 0 & 2 & 4 \\
0 & 3 & 1 & 5 \\
2 & 4 & 1 & 0
\end{bmatrix}
+
\begin{bmatrix}
0 & 2 & 5 & 3 \\
5 & 0 & 4 & 2 \\
2 & 5 & 0 & 7 \\
0 & 3 & 5 & 2 \\
5 & 3 & 1 & 0
\end{bmatrix}
=
\begin{bmatrix}
1 & 2 & 9 & 9 \\
8 & 2 & 5 & 6 \\
6 & 5 & 2 & 11 \\
0 & 6 & 6 & 7 \\
7 & 7 & 2 & 0
\end{bmatrix}
$$

31. a) If $i \neq j$, $a_{ij} + a_{ji}$ is the number of games won by player i when playing player j plus the number of games won by j against i. Since i and j play 3 games, $a_{ij} + a_{ji} = 3$.

b) $a_{ii} = 0$ since a player cannot defeat itself

c)

$$
\begin{bmatrix}
0 & 2 & 2 & 1 & 3 \\
1 & 0 & 3 & 1 & 2 \\
1 & 0 & 0 & 2 & 0 \\
2 & 2 & 1 & 0 & 2 \\
0 & 1 & 3 & 1 & 0
\end{bmatrix}
+
\begin{bmatrix}
0 & 3 & 1 & 2 & 1 \\
0 & 0 & 2 & 1 & 3 \\
2 & 1 & 0 & 3 & 0 \\
1 & 2 & 0 & 0 & 1 \\
2 & 0 & 3 & 2 & 0
\end{bmatrix}
=
\begin{bmatrix}
0 & 5 & 3 & 3 & 4 \\
1 & 0 & 5 & 2 & 5 \\
3 & 1 & 0 & 5 & 0 \\
3 & 4 & 1 & 0 & 3 \\
2 & 1 & 6 & 3 & 0
\end{bmatrix}
$$

d) Player 1 won $0+5+3+3+4 = 15$ games
Player 2 won $1+0+5+2+5 = 13$ games
Player 3 won 9 games
Player 4 won 11 games
Player 5 won 12 games

Hence Player 1 won the most games.

33. (a) The total production matrix is given by

$$\begin{bmatrix} 40 & 24 & 32 \\ 36 & 16 & 20 \end{bmatrix} + \begin{bmatrix} 20 & 15 & 25 \\ 35 & 30 & 20 \end{bmatrix} = \begin{bmatrix} 60 & 39 & 57 \\ 71 & 46 & 40 \end{bmatrix}$$

(b) The new production matrix for the Montreal plant is $\begin{bmatrix} 50 & 30 & 40 \\ 45 & 20 & 25 \end{bmatrix}$ and

the new matrix for the Chicago plant is $\begin{bmatrix} 24 & 18 & 30 \\ 42 & 36 & 24 \end{bmatrix}$.

The total production matrix,
which is the sum of these two matrices, is $\begin{bmatrix} 74 & 48 & 70 \\ 87 & 56 & 49 \end{bmatrix}$.

EXERCISE SET 3.2 MATRIX MULTIPLICATION

1.
$$AB = \begin{bmatrix} 1 & 3 & -2 \\ 0 & 2 & 5 \\ 2 & 4 & 0 \end{bmatrix} \cdot \begin{bmatrix} 2 & 7 \\ 3 & 5 \\ 7 & 2 \end{bmatrix} = \begin{bmatrix} 2+9-14 & 7+15-4 \\ 0+6+35 & 0+10+10 \\ 4+12+0 & 14+20+0 \end{bmatrix} = \begin{bmatrix} -3 & 18 \\ 41 & 20 \\ 16 & 34 \end{bmatrix}$$

BA is undefined since B has 2 columns and A has 3 rows.

AC is undefined since A has 3 columns and C has 2 rows

$$CA = \begin{bmatrix} -1 & 2 & 1 \\ 9 & 0 & 3 \end{bmatrix} \cdot \begin{bmatrix} 1 & 3 & -2 \\ 0 & 2 & 5 \\ 2 & 4 & 0 \end{bmatrix} = \begin{bmatrix} -1+0+2 & -3+4+4 & 2+10+0 \\ 9+0+6 & 27+0+12 & -18+0+0 \end{bmatrix} = \begin{bmatrix} 1 & 5 & 12 \\ 15 & 39 & -18 \end{bmatrix}$$

$$BC = \begin{bmatrix} 2 & 7 \\ 3 & 5 \\ 7 & 2 \end{bmatrix} \cdot \begin{bmatrix} -1 & 2 & 1 \\ 9 & 0 & 3 \end{bmatrix} = \begin{bmatrix} -2+63 & 4+0 & 2+21 \\ -3+45 & 6+0 & 3+15 \\ -7+18 & 14+0 & 7+6 \end{bmatrix} = \begin{bmatrix} 61 & 4 & 23 \\ 42 & 6 & 18 \\ 11 & 14 & 13 \end{bmatrix}$$

$$CB = \begin{bmatrix} -1 & 2 & 1 \\ 9 & 0 & 3 \end{bmatrix} \cdot \begin{bmatrix} 2 & 7 \\ 3 & 5 \\ 7 & 2 \end{bmatrix} = \begin{bmatrix} -2+6+7 & -7+10+2 \\ 18+0+21 & 63+0+6 \end{bmatrix} = \begin{bmatrix} 11 & 5 \\ 39 & 69 \end{bmatrix}$$

3. Since A has 2 columns and B has 3 rows, AB is undefined.
Since B has 3 columns and A has 2 rows, BA is undefined.
Since A has 2 columns and C has 1 row, AC is undefined.
Since C has 3 columns and A has 2 rows, CA is undefined.
Since B has 3 columns and C has 1 row, BC is undefined.

$$CB = \begin{bmatrix} 2 & 5 & 1 \end{bmatrix} \cdot \begin{bmatrix} 3 & 5 & -2 \\ 9 & 3 & 0 \\ 0 & -1 & 3 \end{bmatrix} = \begin{bmatrix} 51 & 24 & -1 \end{bmatrix}$$

5. Since A has 4 columns and B has 2 rows, AB is undefined.

$$BA = \begin{bmatrix} 9 & 5 & -2 \\ 0 & 3 & 2 \end{bmatrix} \begin{bmatrix} 5 & 0 & 8 & 1 \\ 5 & 1 & 0 & 2 \\ 0 & 2 & 5 & 7 \end{bmatrix} = \begin{bmatrix} 70 & 1 & 62 & 5 \\ 15 & 7 & 10 & 20 \end{bmatrix}$$

Since A has 4 columns and C has 3 rows, AC is undefined.
Since C has 1 column and A has 3 rows, CA is undefined.

$$BC = \begin{bmatrix} 9 & 5 & -2 \\ 0 & 3 & 2 \end{bmatrix} \cdot \begin{bmatrix} 5 \\ 3 \\ 6 \end{bmatrix} = \begin{bmatrix} 48 \\ 21 \end{bmatrix}$$

Since C has 1 column and B has 2 rows, CB is undefined.

7. $4 \begin{bmatrix} 3 & 1 & 7 \\ 9 & 4 & -2 \end{bmatrix} \cdot \left(\begin{bmatrix} 2 & 4 & -6 \\ 0 & -2 & 10 \\ 14 & 0 & 4 \end{bmatrix} + \begin{bmatrix} 15 & 30 & 0 \\ 15 & 0 & 5 \\ 25 & 20 & 45 \end{bmatrix} \right)$

$$= 4 \begin{bmatrix} 3 & 1 & 7 \\ 0 & 4 & -2 \end{bmatrix} \cdot \begin{bmatrix} 17 & 34 & -6 \\ 15 & -2 & 15 \\ 39 & 20 & 49 \end{bmatrix}$$

$$= 4 \begin{bmatrix} 339 & 240 & 340 \\ -18 & -48 & -38 \end{bmatrix} = \begin{bmatrix} 1356 & 960 & 1360 \\ -72 & -192 & -152 \end{bmatrix}$$

9. $\begin{bmatrix} 5 & 9 \\ 10 & 13 \\ 5 & -1 \end{bmatrix} \cdot \begin{bmatrix} 2 & 5 & 7 \\ 4 & 2 & 3 \end{bmatrix} = \begin{bmatrix} 46 & 43 & 62 \\ 72 & 76 & 109 \\ 6 & 23 & 32 \end{bmatrix}$

11.

$$AB = \begin{bmatrix} 2 & 5 & -1 \\ 4 & 0 & 7 \end{bmatrix} \cdot \begin{bmatrix} 3 & -2 \\ 0 & 4 \\ 5 & 1 \end{bmatrix} = \begin{bmatrix} 1 & 15 \\ 47 & -1 \end{bmatrix}$$

$$BA = \begin{bmatrix} 3 & -2 \\ 0 & 4 \\ 5 & 1 \end{bmatrix} \cdot \begin{bmatrix} 2 & 5 & -1 \\ 4 & 0 & 7 \end{bmatrix} = \begin{bmatrix} -2 & 15 & -17 \\ 16 & 0 & 28 \\ 14 & 25 & 2 \end{bmatrix}$$

13.

$$AB = \begin{bmatrix} 3 & 1 & 7 & -9 \\ 0 & 4 & 2 & 3 \end{bmatrix} \cdot \begin{bmatrix} 3 & 5 \\ 0 & -3 \\ 3 & 7 \\ 6 & 1 \end{bmatrix} = \begin{bmatrix} -24 & 52 \\ 24 & 5 \end{bmatrix}$$

$$BA = \begin{bmatrix} 3 & 5 \\ 0 & -3 \\ 3 & 7 \\ 6 & 1 \end{bmatrix} \cdot \begin{bmatrix} 3 & 1 & 7 & -9 \\ 0 & 4 & 2 & 3 \end{bmatrix} = \begin{bmatrix} 9 & 23 & 31 & -12 \\ 0 & -12 & -6 & -9 \\ 9 & 31 & 35 & -6 \\ 18 & 10 & 44 & -51 \end{bmatrix}$$

15.

$$AB = \begin{bmatrix} 5 & 2 \\ 7 & 3 \end{bmatrix} \cdot \begin{bmatrix} 3 & -2 \\ -7 & 5 \end{bmatrix} = \begin{bmatrix} 15-14 & -10+10 \\ 21-21 & -14+15 \end{bmatrix} = \begin{bmatrix} 1 & 0 \\ 0 & 1 \end{bmatrix}$$

$$BA = \begin{bmatrix} 3 & -2 \\ -7 & 5 \end{bmatrix} \cdot \begin{bmatrix} 5 & 2 \\ 7 & 3 \end{bmatrix} = \begin{bmatrix} 15-14 & 6-6 \\ -35+35 & -14+15 \end{bmatrix} = \begin{bmatrix} 1 & 0 \\ 0 & 1 \end{bmatrix}$$

17.

$$AB = \begin{bmatrix} 1 & 2 & 3 \\ 4 & 5 & 6 \\ 7 & 8 & 9 \end{bmatrix} \cdot \begin{bmatrix} 9 & 8 & 7 \\ 6 & 5 & 4 \\ 3 & 2 & 1 \end{bmatrix} = \begin{bmatrix} 9+12+9 & 8+10+6 & 7+8+3 \\ 36+30+18 & 32+25+12 & 28+20+6 \\ 63+48+27 & 56+40+18 & 49+32+9 \end{bmatrix}$$

$$= \begin{bmatrix} 30 & 24 & 18 \\ 84 & 69 & 54 \\ 138 & 114 & 90 \end{bmatrix}$$

$$BA = \begin{bmatrix} 9 & 8 & 7 \\ 6 & 5 & 4 \\ 3 & 2 & 1 \end{bmatrix} \cdot \begin{bmatrix} 1 & 2 & 3 \\ 4 & 5 & 6 \\ 7 & 8 & 9 \end{bmatrix} = \begin{bmatrix} 9+32+49 & 18+40+56 & 27+48+63 \\ 6+20+28 & 12+25+32 & 18+30+36 \\ 3+8+7 & 6+10+8 & 9+12+9 \end{bmatrix}$$

$$= \begin{bmatrix} 90 & 114 & 138 \\ 54 & 69 & 84 \\ 18 & 24 & 30 \end{bmatrix}$$

19.

$$AB = \begin{bmatrix} 3 & 1 & 2 \\ 0 & 7 & -2 \\ 3 & 1 & 3 \end{bmatrix} \cdot \begin{bmatrix} 0 & 2 & 0 \\ 3 & 6 & 1 \\ 2 & 1 & -6 \end{bmatrix} = \begin{bmatrix} 0+3+4 & 6+6+2 & 0+1-12 \\ 0+21-4 & 0+42-2 & 0+7+12 \\ 0+3+6 & 6+6+3 & 0+1-18 \end{bmatrix}$$

$$= \begin{bmatrix} 7 & 14 & -11 \\ 17 & 40 & 19 \\ 9 & 15 & -17 \end{bmatrix}$$

$$BA = \begin{bmatrix} 0 & 2 & 0 \\ 3 & 6 & 1 \\ 2 & 1 & -6 \end{bmatrix} \cdot \begin{bmatrix} 3 & 1 & 2 \\ 0 & 7 & -2 \\ 3 & 1 & 3 \end{bmatrix} = \begin{bmatrix} 0+0+0 & 0+14+0 & 0-4+0 \\ 9+0+3 & 3+42+1 & 6-12+3 \\ 6+0-18 & 2+7-6 & 4-2-18 \end{bmatrix}$$

$$= \begin{bmatrix} 0 & 14 & -4 \\ 12 & 46 & -3 \\ -12 & 3 & -16 \end{bmatrix}$$

21.

Let $A = \begin{bmatrix} 1 & 2 \\ 3 & 4 \end{bmatrix}$ and $B = \begin{bmatrix} 1 & 0 \\ 0 & 1 \end{bmatrix}$

$$AB = \begin{bmatrix} 1 & 2 \\ 3 & 4 \end{bmatrix} = BA$$

23.

$$\begin{bmatrix} a & c \\ b & d \end{bmatrix} \begin{bmatrix} 6 & 10 \\ 3 & 5 \end{bmatrix} = 0$$

$$6a + 3c = 0 \Rightarrow 2a + c = 0$$
$$10a + 5c = 0 \Rightarrow 2a + c = 0$$
$$6b + 3d = 0 \Rightarrow 2b + d = 0$$
$$10b + 5d = 0 \Rightarrow 2b + c = 0$$

Let $c = -2a$ and $d = -2b$.

If $A = \begin{bmatrix} 1 & -2 \\ 3 & -6 \end{bmatrix}$, $AB = 0$

25.

$$I_3A = \begin{bmatrix} 1 & 0 & 0 \\ 0 & 1 & 0 \\ 0 & 0 & 1 \end{bmatrix} \cdot \begin{bmatrix} 3 & 5 \\ 2 & -1 \\ 4 & 7 \end{bmatrix} = \begin{bmatrix} 3+0+0 & 5+0+0 \\ 0+2+0 & 0-1+0 \\ 0+0+4 & 0+0+7 \end{bmatrix} = \begin{bmatrix} 3 & 5 \\ 2 & -1 \\ 4 & 7 \end{bmatrix}$$

$$AI_2 = \begin{bmatrix} 3 & 5 \\ 2 & -1 \\ 4 & 7 \end{bmatrix} \cdot \begin{bmatrix} 1 & 0 \\ 0 & 1 \end{bmatrix} = \begin{bmatrix} 3+0 & 0+5 \\ 2+0 & 0-1 \\ 4+0 & 0+7 \end{bmatrix} = \begin{bmatrix} 3 & 5 \\ 2 & -1 \\ 4 & 7 \end{bmatrix}$$

27.

$$AB = \begin{bmatrix} 7 & 5 \\ 4 & 3 \end{bmatrix} \cdot \begin{bmatrix} 3 & -5 \\ -4 & 7 \end{bmatrix} = \begin{bmatrix} 21-20 & -35+35 \\ 12-12 & -20+21 \end{bmatrix} = \begin{bmatrix} 1 & 0 \\ 0 & 1 \end{bmatrix}$$

$$BA = \begin{bmatrix} 3 & -5 \\ -4 & 7 \end{bmatrix} \cdot \begin{bmatrix} 7 & 5 \\ 4 & 3 \end{bmatrix} = \begin{bmatrix} 21-20 & 15-15 \\ -28+28 & -20+21 \end{bmatrix} = \begin{bmatrix} 1 & 0 \\ 0 & 1 \end{bmatrix}$$

29.

$$AB = \begin{bmatrix} 3 & -5 & 8 \\ -4 & 7 & 13 \end{bmatrix} \cdot \begin{bmatrix} 7 & 5 \\ 4 & 3 \\ 0 & 0 \end{bmatrix} = \begin{bmatrix} 1 & 0 \\ 0 & 1 \end{bmatrix}.$$

$$BA = \begin{bmatrix} 7 & 5 \\ 4 & 3 \\ 0 & 0 \end{bmatrix} \cdot \begin{bmatrix} 3 & -5 & 8 \\ -4 & 7 & 13 \end{bmatrix} = \begin{bmatrix} 1 & 0 & 121 \\ 0 & 1 & 71 \\ 0 & 0 & 0 \end{bmatrix}$$

31.

$$\begin{bmatrix} 5 & 2 \\ 7 & 3 \end{bmatrix}\begin{bmatrix} x \\ y \end{bmatrix} = \begin{bmatrix} -3 \\ 5 \end{bmatrix}.$$

33.

$$\begin{bmatrix} 11 & 6 \\ 9 & 5 \end{bmatrix} \cdot \begin{bmatrix} x \\ y \end{bmatrix} = \begin{bmatrix} 7 \\ -5 \end{bmatrix}.$$

35.

$$\begin{bmatrix} 1 & -4 & 0 & 1 \\ 5 & 0 & -5 & 4 \\ 0 & 1 & 3 & 1 \\ 2 & 2 & -1 & 2 \end{bmatrix}\begin{bmatrix} x \\ y \\ z \\ w \end{bmatrix} = \begin{bmatrix} -2 \\ 3 \\ 5 \\ -1 \end{bmatrix}$$

37.

$$P = \begin{bmatrix} 8 & 5 & 9 & 3 \\ 7 & 6 & 8 & 2 \\ 9 & 5 & 7 & 2 \end{bmatrix} \qquad R = \begin{bmatrix} 1000 & 1500 & 1300 & 1400 \\ 700 & 950 & 875 & 910 \\ 2200 & 1900 & 2100 & 1875 \\ 800 & 950 & 875 & 650 \end{bmatrix}$$

a)

$$PR = \begin{bmatrix} 33{,}700 & 36{,}700 & 36{,}300 & 34{,}575 \\ 30{,}400 & 33{,}300 & 32{,}900 & 31{,}560 \\ 29{,}500 & 33{,}450 & 32{,}525 & 31{,}575 \end{bmatrix}$$

The number in the i-th row and j-th column is the cost of project j using supplier i.

b)   Use supplier C for project I,
        supplier B for project II,
        supplier C for project III,
    and supplier B for project IV

39.

$$\begin{bmatrix} 2 & 4 & 3 \\ 3 & 2 & 4 \\ 4 & 3 & 2 \end{bmatrix} \cdot \begin{bmatrix} 3 & 2 \\ 4 & 3 \\ 5 & 1 \end{bmatrix} = \begin{bmatrix} 37 & 19 \\ 37 & 16 \\ 34 & 19 \end{bmatrix}$$

From To →	Vancouver	Victoria
Quebec	37	19
Montreal	37	16
Toronto	34	19

41.

$$\begin{bmatrix} .15 & .15 & .2 & .5 \\ .3 & .3 & .3 & .1 \\ .25 & .2 & .25 & .3 \end{bmatrix} \cdot \begin{bmatrix} 87 & 92 & 90 & 78 & 98 & 25 \\ 90 & 95 & 85 & 89 & 79 & 45 \\ 88 & 96 & 95 & 94 & 97 & 32 \\ 79 & 98 & 92 & 93 & 100 & 51 \end{bmatrix}$$

$$\begin{bmatrix} 83.65 & 96.25 & 91.25 & 90.35 & 95.95 & 42.4 \\ 87.4 & 94.70 & 90.2 & 87.6 & 92.2 & 35.7 \\ 85.45 & 95.40 & 90.85 & 88.7 & 94.55 & 38.55 \end{bmatrix}$$

The number in the i-th row and j-th column is the final average for student j taking the course with professor i.

43.

(a) $AB = \begin{bmatrix} 2 & 1 \\ 6 & 3 \end{bmatrix} \cdot \begin{bmatrix} 0 & 2 \\ 12 & 2 \end{bmatrix} = \begin{bmatrix} 12 & 6 \\ 36 & 18 \end{bmatrix}$

$BA = \begin{bmatrix} 0 & 2 \\ 12 & 2 \end{bmatrix} \cdot \begin{bmatrix} 2 & 1 \\ 6 & 3 \end{bmatrix} = \begin{bmatrix} 12 & 6 \\ 36 & 18 \end{bmatrix}.$ $\quad AB = BA$

(b) det $A = 2(3) - 1(6) = 0$ and det $B = 0(2) - 2(12) = -24$.
Suppose $A = C^m$ and $B = C^n$ where C is a 2×2 matrix. By Exercise 42,
det $A = (\det C)^m$; hence det $C = 0$, det $B = (\det C)^n = -24$, and det $C \neq 0$.
Thus we have a contradiction. It follows that either $A \neq C^m$ or $B \neq C^n$.

EXERCISE SET 3.3 SOLUTION OF LINEAR SYSTEMS BY ROW REDUCTION

1. $\begin{bmatrix} 2 & 3 & | & -5 \\ 3 & -5 & | & 17 \end{bmatrix}$

3. The system may be rewritten as

$$\begin{cases} -7x + 2y + 5z = 4 \\ 3x \qquad + 7z = 0 \\ \qquad -5y + 2z = 7 \end{cases}$$

and the augmented matrix is

$$\begin{bmatrix} -7 & 2 & 5 & | & 4 \\ 3 & 0 & 7 & | & 0 \\ 0 & -5 & 2 & | & 7 \end{bmatrix}$$

5. $\begin{cases} 2x \qquad = 3 \\ 5x + 3y = -2 \end{cases}$

7.
$$\begin{cases} 2x & +3z+4w=7 \\ -3x+2y+5z & =2 \\ 2x-y & +4w=-3 \end{cases}$$

9.
$$\begin{bmatrix} 2 & 1 & \bigm| & 5 \\ 6 & 3 & \bigm| & 1 \end{bmatrix} \xrightarrow{-3R_1+R_2\to R_2} \begin{bmatrix} 2 & 1 & \bigm| & 5 \\ 0 & 0 & \bigm| & -14 \end{bmatrix}$$

The system is inconsistent, since the last row of the augmented matrix gives $0x+0y=1$, which has no solution.

11.
$$\begin{bmatrix} 8 & 12 & \bigm| & 16 \\ 6 & 9 & \bigm| & 12 \end{bmatrix} \begin{smallmatrix} \frac{1}{4}R_1\to R_1 \\ \frac{1}{3}R_2\to R_2 \end{smallmatrix} \begin{bmatrix} 2 & 3 & \bigm| & 4 \\ 2 & 3 & \bigm| & 4 \end{bmatrix} \xrightarrow{R_1-R_2\to R_2} \begin{bmatrix} 2 & 3 & \bigm| & 4 \\ 0 & 0 & \bigm| & 0 \end{bmatrix}$$

Since the last row gives no information, the solution set depends only on the equation $2x+3y=4$ or $x=\frac{1}{2}(4-3y)$. The solution set is $\left\{ \left(\frac{1}{2}(4-3c),\ c\right) \mid c \text{ is a real number}\right\}$.

13.
$$\begin{bmatrix} 6 & 15 & \bigm| & 2 \\ 4 & 10 & \bigm| & 1 \end{bmatrix} \xrightarrow{4R_1-6R_2\to R_2} \begin{bmatrix} 6 & 15 & \bigm| & 2 \\ 0 & 0 & \bigm| & 2 \end{bmatrix}$$

The solution set is $\emptyset$ as the last row of the augmented matrix gives $0x+0y=2$.

15.
$$\begin{bmatrix} 5 & -5 & 4 & \bigm| & 8 \\ 0 & 3 & 1 & \bigm| & 5 \\ 2 & -1 & 2 & \bigm| & 5 \end{bmatrix} \xrightarrow{2R_1-5R_3\to R_3} \begin{bmatrix} 5 & -5 & 4 & \bigm| & 8 \\ 0 & 3 & 1 & \bigm| & 5 \\ 0 & -5 & -2 & \bigm| & -9 \end{bmatrix}$$

$$\begin{smallmatrix} R_1-R_3\to R_1 \\ 5R_2+3R_3\to R_3 \end{smallmatrix} \begin{bmatrix} 5 & 0 & 6 & \bigm| & 17 \\ 0 & 3 & 1 & \bigm| & 5 \\ 0 & 0 & -1 & \bigm| & -2 \end{bmatrix} \begin{smallmatrix} R_3+R_2\to R_2 \\ 6R_3+R_1\to R_1 \end{smallmatrix}$$

$$\begin{bmatrix} 5 & 0 & 0 & \bigm| & 5 \\ 0 & 3 & 0 & \bigm| & 3 \\ 0 & 0 & -1 & \bigm| & -2 \end{bmatrix} \begin{smallmatrix} \frac{1}{5}R_1\to R_1 \\ \frac{1}{3}R_2\to R_2 \\ -R_3\to R_3 \end{smallmatrix} \begin{bmatrix} 1 & 0 & 0 & \bigm| & 1 \\ 0 & 1 & 0 & \bigm| & 1 \\ 0 & 0 & 1 & \bigm| & 2 \end{bmatrix}$$

The solution set is $\{(1,\ 1,\ 2)\}$

17.
$$\begin{bmatrix} 1 & 2 & 3 & \bigm| & 2 \\ 3 & 3 & 0 & \bigm| & 4 \\ 2 & 1 & -3 & \bigm| & 3 \end{bmatrix} \xrightarrow{R_2\longleftrightarrow R_3} \begin{bmatrix} 1 & 2 & 3 & \bigm| & 2 \\ 2 & 1 & -3 & \bigm| & 3 \\ 3 & 3 & 0 & \bigm| & 4 \end{bmatrix} \xrightarrow{R_2+R_1\to R_1}$$

$$\begin{bmatrix} 3 & 3 & 0 & | & 5 \\ 2 & 1 & -3 & | & 3 \\ 3 & 3 & 0 & | & 4 \end{bmatrix} \quad R_3 - R_1 \to R_1 \quad \begin{bmatrix} 0 & 0 & 0 & | & -1 \\ 2 & 1 & -3 & | & 3 \\ 3 & 3 & 0 & | & 4 \end{bmatrix}$$

The system is inconsistent as the 1st row of the augmented matrix gives
$0x + 0y + 0z = -1$.

19.

$$\begin{bmatrix} 3 & 2 & -7 & | & 3 \\ 2 & 3 & 5 & | & 15 \\ 2 & 2 & -5 & | & 3 \end{bmatrix} \quad \begin{matrix} R_2 - R_3 \to R_3 \\ 2R_1 - 3R_2 \to R_2 \end{matrix} \quad \begin{bmatrix} 3 & 2 & -7 & | & 3 \\ 0 & -5 & -29 & | & -39 \\ 0 & 1 & 10 & | & 12 \end{bmatrix} \quad R_3 \longleftrightarrow R_2$$

$$\begin{bmatrix} 3 & 2 & -7 & | & -3 \\ 0 & 1 & 10 & | & 12 \\ 0 & -5 & -29 & | & -39 \end{bmatrix} \quad \begin{matrix} 5R_2 + R_3 \to R_3 \\ R_1 - 2R_2 \to R_1 \end{matrix} \quad \begin{bmatrix} 3 & 0 & -27 & | & -21 \\ 0 & 1 & 10 & | & 12 \\ 0 & 0 & 21 & | & 21 \end{bmatrix} \quad \begin{matrix} \frac{1}{21}R_3 \to R_3 \\ \frac{1}{3}R_1 \to R_1 \end{matrix}$$

$$\begin{bmatrix} 1 & 0 & -9 & | & -7 \\ 0 & 1 & 10 & | & 12 \\ 0 & 0 & 1 & | & 1 \end{bmatrix} \quad \begin{matrix} 9R_3 + R_1 \to R_1 \\ -10R_3 + R_2 \to R_2 \end{matrix} \quad \begin{bmatrix} 1 & 0 & 0 & | & 2 \\ 0 & 1 & 0 & | & 2 \\ 0 & 0 & 1 & | & 1 \end{bmatrix}$$

The solution set is $\{(2, 2, 1)\}$

21.

$$\begin{bmatrix} 2 & 1 & -3 & | & 2 \\ 3 & 2 & 4 & | & 1 \\ 8 & 5 & 5 & | & 4 \end{bmatrix} \quad \begin{matrix} 4R_1 - R_3 \to R_3 \\ 3R_1 - 2R_2 \to R_2 \end{matrix} \quad \begin{bmatrix} 2 & 1 & -3 & | & 2 \\ 0 & -1 & -17 & | & 4 \\ 0 & -1 & -17 & | & 4 \end{bmatrix} \quad \begin{matrix} R_2 - R_3 \to R_3 \\ \frac{1}{2}R_1 \to R_1 \\ -R_2 \to R_2 \end{matrix}$$

$$\begin{bmatrix} 1 & \frac{1}{2} & -\frac{3}{2} & | & 1 \\ 0 & 1 & 17 & | & -4 \\ 0 & 0 & 0 & | & 0 \end{bmatrix} \quad \text{The augmented matrix is in row-reduced echelon form.}$$

The first row corresponds to the equation $x + \frac{1}{2}y - \frac{3}{2}z = 1$, the second to $y + 17z = -4$,
and the last gives no information.
If $z = a$, $y = -4 - 17a$ and $x = 1 - \frac{1}{2}(-4 - 17a) + \frac{3}{2}(a) = 3 + 10a$. The solution set is
$\{(3 + 10a, \; -4 - 17a, \; a) \mid a \text{ is a real number}\}$

23.

$$\begin{bmatrix} 3 & -1 & 3 & | & -1 \\ 2 & 2 & 3 & | & 8 \\ 0 & -3 & -1 & | & -10 \end{bmatrix} \quad 2R_1 - 3R_2 \to R_2 \quad \begin{bmatrix} 3 & -1 & 3 & | & -1 \\ 0 & -8 & -3 & | & -26 \\ 0 & -3 & -1 & | & -10 \end{bmatrix} \quad \begin{matrix} 3R_1 - R_3 \to R_3 \\ 3R_2 - 8R_3 \to R_3 \end{matrix}$$

$$\begin{bmatrix} 9 & 0 & 10 & | & 7 \\ 0 & -8 & -3 & | & -26 \\ 0 & 0 & -1 & | & 2 \end{bmatrix} \quad \begin{matrix} 10R_3 + R_1 \to R_1 \\ R_2 + 3R_3 \to R_2 \end{matrix} \quad \begin{bmatrix} 9 & 0 & 0 & | & 27 \\ 0 & 8 & 0 & | & 32 \\ 0 & 0 & -1 & | & 2 \end{bmatrix} \quad \begin{matrix} \frac{1}{9}R_1 \to R_1 \\ \frac{1}{8}R_2 \to R_2 \\ -R_3 \to R_3 \end{matrix}$$

$$\begin{bmatrix} 1 & 0 & 0 & | & 3 \\ 0 & 1 & 0 & | & 4 \\ 0 & 0 & 1 & | & -2 \end{bmatrix}$$ The solution set is $\{(3, 4, -2)\}$

25.
$$\begin{bmatrix} 2 & -1 & 2 & | & 14 \\ 4 & 1 & 5 & | & 37 \\ 11 & -8 & 10 & | & 69 \end{bmatrix} \begin{array}{l} 2R_1-R_2\to R_2 \\ 11R_1-2R_3\to R_3 \end{array} \begin{bmatrix} 2 & -1 & 2 & | & 14 \\ 0 & -3 & -1 & | & -9 \\ 0 & 5 & 2 & | & 16 \end{bmatrix} \begin{array}{l} 5R_1+R_3\to R_1 \\ 5R_2+3R_3\to R_3 \end{array}$$

$$\begin{bmatrix} 10 & 0 & 12 & | & 86 \\ 0 & -3 & -1 & | & -9 \\ 0 & 0 & 1 & | & 3 \end{bmatrix} \begin{array}{l} R_2+R_3\to R_2 \\ -12R_3+R_1\to R_1 \end{array} \begin{bmatrix} 10 & 0 & 0 & | & 50 \\ 0 & -3 & 0 & | & -6 \\ 0 & 0 & 1 & | & 3 \end{bmatrix} \begin{array}{l} \frac{1}{10}R_1\to R_1 \\ -\frac{1}{3}R_2\to R_2 \end{array}$$

$$\begin{bmatrix} 1 & 0 & 0 & | & 5 \\ 0 & 1 & 0 & | & 2 \\ 0 & 0 & 1 & | & 3 \end{bmatrix}$$ The solution set is $\{(5, 2, 3)\}$

27. The system may be rewritten as

$$\begin{bmatrix} 7 & -8 & -5 & 6 & | & 27 \\ 1 & -4 & 0 & 1 & | & 7 \\ 3 & -11 & 3 & 4 & | & 22 \\ 4 & -6 & -1 & 4 & | & 18 \end{bmatrix} \begin{array}{l} 3R_2\to R_3\to R_3 \\ 4R_2-R_4\to R_4 \\ R_1-7R_2\to R_3 \end{array} \begin{bmatrix} 0 & 20 & -5 & -1 & | & -22 \\ 1 & -4 & 0 & 1 & | & 7 \\ 0 & -1 & -3 & -1 & | & -1 \\ 0 & -10 & 1 & 0 & | & 10 \end{bmatrix} \begin{array}{l} R_2\longleftrightarrow R_1 \\ -R_3\to R_3 \end{array}$$

$$\begin{bmatrix} 1 & -4 & 0 & 1 & | & 7 \\ 0 & 20 & -5 & -1 & | & -22 \\ 0 & 1 & 3 & 1 & | & 1 \\ 0 & -10 & 1 & 0 & | & 10 \end{bmatrix} \begin{array}{l} 5R_4+R_2\to R_2 \\ -3R_4+R_3\to R_3 \end{array}$$

$$\begin{bmatrix} 1 & -4 & 0 & 1 & | & 7 \\ 0 & -30 & 0 & -1 & | & 28 \\ 0 & 31 & 0 & 1 & | & -29 \\ 0 & -10 & 1 & 0 & | & 10 \end{bmatrix} \quad R_2+R_3\to R_3$$

$$\begin{bmatrix} 1 & -4 & 0 & 1 & | & 7 \\ 0 & -30 & 0 & -1 & | & 28 \\ 0 & 1 & 0 & 0 & | & -1 \\ 0 & -10 & 1 & 0 & | & 10 \end{bmatrix} \quad R_3\longleftrightarrow R_2 \quad \begin{bmatrix} 1 & -4 & 0 & 1 & | & 7 \\ 0 & 1 & 0 & 0 & | & -1 \\ 0 & -30 & 0 & -1 & | & 28 \\ 0 & -10 & 1 & 0 & | & 10 \end{bmatrix} \begin{array}{l} R_1+R_3\to R_1 \\ R_3\longleftrightarrow R_4 \end{array}$$

$$\left[\begin{array}{cccc|c} 1 & -34 & 0 & 0 & 35 \\ 0 & 1 & 0 & 0 & -1 \\ 0 & -10 & 1 & 0 & 10 \\ 0 & -30 & 0 & -1 & 28 \end{array}\right] \begin{array}{l} 10R_2 + R_3 \to R_3 \\ 34R_2 + R_1 \to R_1 \\ -30R_2 - R_4 \to R_4 \end{array} \left[\begin{array}{cccc|c} 1 & 0 & 0 & 0 & 1 \\ 0 & 1 & 0 & 0 & -1 \\ 0 & 0 & 1 & 0 & 0 \\ 0 & 0 & 0 & 1 & 2 \end{array}\right]$$

Hence $w = 1$, $x = -1$, $y = 0$ and $z = 2$

29.
$$\left[\begin{array}{cccc|c} 0 & 1 & 3 & 1 & 1 \\ 5 & 1 & -2 & 5 & 3 \\ 1 & -2 & 6 & 3 & 3 \\ 2 & 3 & 2 & 3 & 2 \end{array}\right] \begin{array}{l} -5R_3 + R_2 \to R_2 \\ -2R_3 + R_4 \to R_4 \end{array} \left[\begin{array}{cccc|c} 0 & 1 & 3 & 1 & 1 \\ 0 & 11 & -32 & -10 & -12 \\ 1 & -2 & 6 & 3 & 3 \\ 0 & 7 & -10 & -3 & -4 \end{array}\right]$$

$$\begin{array}{l} R_2 - 11R_1 \to R_2 \\ R_3 + 2R_1 \to R_3 \\ R_4 - 7R_1 \to R_4 \end{array} \left[\begin{array}{cccc|c} 0 & 1 & 3 & 1 & 1 \\ 0 & 0 & -65 & -21 & -23 \\ 1 & 0 & 12 & 5 & 5 \\ 0 & 0 & -31 & -10 & -11 \end{array}\right] \begin{array}{l} R_1 \to R_2 \\ R_2 \to R_3 \\ R_3 \to R_1 \end{array} \left[\begin{array}{cccc|c} 1 & 0 & 12 & 5 & 5 \\ 0 & 1 & 3 & 1 & 1 \\ 0 & 0 & -65 & -21 & -23 \\ 0 & 0 & -31 & -10 & -11 \end{array}\right]$$

$$-31R_3 + 65R_4 \to R_4 \left[\begin{array}{cccc|c} 1 & 0 & 12 & 5 & 5 \\ 0 & 1 & 3 & 1 & 1 \\ 0 & 0 & -65 & -21 & -23 \\ 0 & 0 & 0 & -1 & -2 \end{array}\right] \begin{array}{l} R_1 - 5R_4 \to R_1 \\ R_2 - R_4 \to R_2 \\ R_3 + 21R_4 \to R_3 \end{array}$$

$$\left[\begin{array}{cccc|c} 1 & 0 & 12 & 0 & 15 \\ 0 & 1 & 3 & 0 & 3 \\ 0 & 0 & -65 & 0 & -65 \\ 0 & 0 & 0 & 1 & -2 \end{array}\right] -\frac{1}{65}R_3 \to R_3 \left[\begin{array}{cccc|c} 1 & 0 & 12 & 0 & 15 \\ 0 & 1 & 3 & 0 & 3 \\ 0 & 0 & 1 & 0 & 1 \\ 0 & 0 & 0 & 1 & -2 \end{array}\right]$$

$$\begin{array}{l} R_1 - 12R_3 \to R_1 \\ R_2 - 3R_3 \to R_2 \end{array} \left[\begin{array}{cccc|c} 1 & 0 & 0 & 0 & 3 \\ 0 & 1 & 0 & 0 & 0 \\ 0 & 0 & 1 & 0 & 1 \\ 0 & 0 & 0 & 1 & -2 \end{array}\right]$$ The solution set is $\{(3, 0, 1, -2)\}$

31.
$$\left[\begin{array}{ccc|c} 2 & 3 & -1 & 2 \\ 4 & 6 & -2 & 3 \end{array}\right] 2R_1 - R_2 \to R_2 \left[\begin{array}{ccc|c} 2 & 3 & -1 & 2 \\ 0 & 0 & 0 & 1 \end{array}\right]$$

The last row gives $0x + 0y + 0z = 1$. Thus the solution set is $\emptyset$.

33.

$$\begin{bmatrix} 1 & 3 & | & -7 \\ 3 & 1 & | & 3 \\ 2 & -5 & | & 19 \end{bmatrix} \begin{matrix} -2R_1+R_3\to R_3 \\ -3R_1+R_2\to R_2 \end{matrix} \begin{bmatrix} 1 & 0 & | & -7 \\ 0 & -8 & | & 24 \\ 0 & -1 & | & 33 \end{bmatrix} R_2-8R_3\to R_2 \begin{bmatrix} 1 & 3 & | & -7 \\ 0 & 0 & | & -240 \\ 0 & -1 & | & 33 \end{bmatrix}$$

The middle row gives $0x + 0y = -240$. Thus the solution set is $\emptyset$.

35.

$$\begin{bmatrix} 4 & 5 & 2 & -5 & | & -2 \\ 2 & 0 & 1 & 6 & | & 22 \\ 2 & 2 & 1 & -1 & | & 3 \end{bmatrix} \begin{matrix} 2R_2-R_1\to R_2 \\ 2R_3-R_1\to R_3 \end{matrix} \begin{bmatrix} 4 & 5 & 2 & -5 & | & -2 \\ 0 & -5 & 0 & 17 & | & 46 \\ 0 & -1 & 0 & 3 & | & 8 \end{bmatrix} \begin{matrix} R_1+R_2\to R_1 \\ R_2-5R_3\to R_2 \\ -R_3\to R_3 \end{matrix}$$

$$\begin{bmatrix} 4 & 0 & 2 & 12 & | & 44 \\ 0 & 0 & 0 & 2 & | & 6 \\ 0 & 1 & 0 & -3 & | & -8 \end{bmatrix} \begin{matrix} \frac{1}{4}R_1\to R_1 \\ R_3\to R_2 \text{ and} \\ \frac{1}{2}R_2\to R_3 \end{matrix} \begin{bmatrix} 1 & 0 & \frac{1}{2} & 3 & | & 11 \\ 0 & 1 & 0 & -3 & | & -8 \\ 0 & 0 & 0 & 1 & | & 3 \end{bmatrix}$$

Hence $z = 3$, $x - 3z = -8$ or $x = 1$

$w = 11 - \frac{1}{2}y - 3z = 11 - \frac{y}{2} - 9 = 2 - y/2$.

The solution set is $\left\{(2 - \frac{a}{2}, 1, a, 3) \mid a \text{ is a real number}\right\}$

37.

$$\begin{bmatrix} 1 & 1 & -3 & 0 & | & 0 \\ 2 & 0 & 2 & 3 & | & 13 \\ 8 & 3 & -4 & 8 & | & 34 \end{bmatrix} \begin{matrix} 2R_1-R_2\to R_2 \\ 8R_1-R_3\to R_3 \end{matrix} \begin{bmatrix} 1 & 1 & -3 & 0 & | & 0 \\ 0 & 2 & -8 & -3 & | & -13 \\ 0 & 5 & -20 & -8 & | & -34 \end{bmatrix}$$

$$\begin{matrix} 2R_1-R_2\to R_1 \\ 5R_2-2R_3\to R_3 \end{matrix} \begin{bmatrix} 2 & 0 & 2 & 3 & | & 13 \\ 0 & 2 & -8 & -3 & | & -13 \\ 0 & 0 & 0 & 1 & | & 3 \end{bmatrix} \begin{matrix} \frac{1}{2}R_2\to R_2 \\ \frac{1}{2}R_1\to R_1 \end{matrix} \begin{bmatrix} 1 & 0 & 1 & \frac{3}{2} & | & \frac{13}{2} \\ 0 & 1 & -4 & -\frac{3}{2} & | & -\frac{13}{2} \\ 0 & 0 & 0 & 1 & | & 3 \end{bmatrix}$$

From the last row, $z = 3$. From the middle row $x = -\frac{13}{2} + 4y + \frac{3}{2}z = -2 + 4y$.

From the 1st row $w = \frac{13}{2} - y - \frac{3}{2}z = \frac{13}{2} - y - \frac{9}{2} = 2 - y$. Let $y = a$.

The solution set is $\left\{(2 - a, -2 + 4a, a, 3) \mid a \text{ is a real number}\right\}$

39.

$$\begin{bmatrix} 3 & 7 & 2 & -7 & | & 21 \\ 2 & 4 & 2 & -5 & | & 17 \\ 2 & 3 & 3 & 5 & | & -10 \end{bmatrix} \begin{matrix} 2R_1-3R_2\to R_2 \\ 2R_1-3R_3\to R_3 \end{matrix} \begin{bmatrix} 3 & 7 & 2 & -7 & | & 21 \\ 0 & 2 & -2 & 1 & | & -9 \\ 0 & 5 & -5-29 & | & 72 \end{bmatrix} \begin{matrix} 2R_1-7R_2\to R_1 \\ 5R_2-2R_3\to R_3 \end{matrix}$$

$$\begin{bmatrix} 6 & 0 & 18 & -21 & | & 105 \\ 0 & 2 & -2 & 1 & | & -9 \\ 0 & 0 & 0 & 63 & | & -189 \end{bmatrix} \begin{matrix} \frac{1}{63}R_3\to R_3 \\ \frac{1}{3}R_1\to R_1 \end{matrix} \begin{bmatrix} 2 & 0 & 6 & -7 & | & 35 \\ 0 & 2 & -2 & 1 & | & -9 \\ 0 & 0 & 0 & 1 & | & -3 \end{bmatrix} \begin{matrix} -R_3+R_2\to R_2 \\ 7R_3+R_1\to R_1 \end{matrix}$$

$$\begin{bmatrix} 2 & 0 & 6 & 0 & | & 14 \\ 0 & 2 & -2 & 0 & | & -6 \\ 0 & 0 & 0 & 1 & | & -3 \end{bmatrix} \begin{matrix} \frac{1}{2}R_2 \to R_2 \\ \frac{1}{2}R_1 \to R_1 \end{matrix} \begin{bmatrix} 1 & 0 & 3 & 0 & | & 7 \\ 0 & 1 & -1 & 0 & | & -3 \\ 0 & 0 & 0 & 1 & | & -3 \end{bmatrix}$$

From the last row, $z = 3$.

From the middle row, $x = y - 3$.

From the first row, $w = 7 - 3y$.

Let $y = a$, where a is an arbitrary real number.

The solution set is $\{(7 - 3a, \ a - 3, \ a, \ -3) \mid a \text{ is a real number}\}$

41.
$$\begin{bmatrix} 4 & 5 & 2 & -5 & | & 0 \\ 2 & 0 & 1 & 6 & | & 0 \\ 2 & 2 & 1 & -1 & | & 0 \end{bmatrix} \begin{matrix} R_1 - 2R_2 \to R_2 \\ R_2 - R_3 \to R_3 \end{matrix} \begin{bmatrix} 4 & 5 & 2 & -5 & | & 0 \\ 0 & 5 & 0 & -17 & | & 0 \\ 0 & -2 & 0 & 7 & | & 0 \end{bmatrix} \begin{matrix} R_1 - R_2 \to R_1 \\ 2R_2 + 5R_3 \to R_3 \end{matrix}$$

$$\begin{bmatrix} 4 & 0 & 2 & 12 & | & 0 \\ 0 & 5 & 0 & -17 & | & 0 \\ 0 & 0 & 0 & 1 & | & 0 \end{bmatrix} \begin{matrix} -12R_3 + R_1 \to R_1 \\ 17R_3 + R_2 \to R_2 \end{matrix} \begin{bmatrix} 4 & 0 & 2 & 0 & | & 0 \\ 0 & 5 & 0 & 0 & | & 0 \\ 0 & 0 & 0 & 1 & | & 0 \end{bmatrix} \begin{matrix} \frac{1}{4}R_1 \to R_1 \\ \frac{1}{5}R_2 \to R_2 \end{matrix}$$

$$\begin{bmatrix} 1 & 0 & \frac{1}{2} & 0 & | & 0 \\ 0 & 1 & 0 & 0 & | & 0 \\ 0 & 0 & 0 & 1 & | & 0 \end{bmatrix} \quad \text{The last row gives } z = 0, \text{ the middle row gives } x = 0,$$
$$\text{and the first row gives } w = -\tfrac{1}{2}y.$$

Let $y = a$ where a is an arbitrary real number.

The solution set is $\{(-\frac{a}{2}, \ 0, \ a, \ 0) \mid a \text{ is a real number}\}$

43.
$$\begin{bmatrix} 1 & 1 & -3 & 0 & | & 0 \\ 2 & 0 & 2 & 3 & | & 0 \\ 8 & 3 & -4 & 8 & | & 0 \end{bmatrix} \ 3R_1 - R_3 \to R_3 \ \begin{bmatrix} 1 & 1 & -3 & 0 & | & 0 \\ 2 & 0 & 2 & 3 & | & 0 \\ -5 & 0 & -5 & -8 & | & 0 \end{bmatrix} \begin{matrix} R_2 \to R_1 \\ 5R_1 + R_3 \to R_2 \end{matrix}$$

$$\begin{bmatrix} 2 & 0 & 2 & 3 & | & 0 \\ 0 & 5 & -20 & -8 & | & 0 \\ -5 & 0 & -5 & -8 & | & 0 \end{bmatrix} \ 5R_1 + 2R_3 \to R_3 \ \begin{bmatrix} 2 & 0 & 2 & 3 & | & 0 \\ 0 & 5 & -20 & -8 & | & 0 \\ 0 & 0 & 0 & -1 & | & 0 \end{bmatrix} \begin{matrix} 3R_3 + R_1 \to R_1 \\ -8R_3 + R_2 \to R_2 \\ -R_3 \to R_3 \end{matrix}$$

$$\begin{bmatrix} 2 & 0 & 2 & 0 & | & 0 \\ 0 & 5 & -20 & 0 & | & 0 \\ 0 & 0 & 0 & 1 & | & 0 \end{bmatrix} \begin{matrix} \frac{1}{2}R_1 \to R_1 \\ \frac{1}{5}R_2 \to R_2 \end{matrix} \begin{bmatrix} 1 & 0 & 1 & 0 & | & 0 \\ 0 & 1 & -4 & 0 & | & 0 \\ 0 & 0 & 0 & 1 & | & 0 \end{bmatrix}$$

The last row gives $z = 0$. The middle row gives $x = 4y$, and the top row gives $w = -y$.
The solution set is $\{(-a, \ 4a, \ a, \ 0) \mid a \text{ is a real number}\}$.

45.

$$\begin{bmatrix} 3 & 7 & 2 & -7 & | & 0 \\ 2 & 4 & 2 & -5 & | & 0 \\ 2 & 3 & 3 & 5 & | & 0 \end{bmatrix} \begin{array}{c} R_2 - R_3 \to R_3 \\ 2R_1 - 3R_2 \to R_2 \end{array} \begin{bmatrix} 3 & 7 & 2 & -7 & | & 0 \\ 0 & 2 & -2 & 1 & | & 0 \\ 0 & 1 & -1 & -10 & | & 0 \end{bmatrix} \begin{array}{c} -7R_3 + R_1 \to R_1 \\ -2R_3 + R_2 \to R_3 \end{array}$$

$$\begin{bmatrix} 3 & 0 & 9 & 63 & | & 0 \\ 0 & 2 & -2 & 1 & | & 0 \\ 0 & 0 & 0 & 21 & | & 0 \end{bmatrix} \begin{array}{c} -3R_3 + R_1 \to R_1 \\ \frac{1}{21}R_3 \to R_3 \end{array} \begin{bmatrix} 3 & 0 & 9 & 0 & | & 0 \\ 0 & 2 & -2 & 1 & | & 0 \\ 0 & 0 & 0 & 1 & | & 0 \end{bmatrix} \begin{array}{c} \frac{1}{3}R_1 \to R_1 \\ R_2 - R_3 \to R_2 \end{array}$$

$$\begin{bmatrix} 1 & 0 & 3 & 0 & | & 0 \\ 0 & 2 & -2 & 0 & | & 0 \\ 0 & 0 & 0 & 1 & | & 0 \end{bmatrix} \tfrac{1}{2}R_2 \to R_2 \begin{bmatrix} 1 & 0 & 3 & 0 & | & 0 \\ 0 & 1 & -1 & 0 & | & 0 \\ 0 & 0 & 0 & 1 & | & 0 \end{bmatrix}.$$

Row 3 gives $z = 0$. Row 2 gives $x = y$. Row 1 gives $w = -3y$.
The solution set is $\{(-3a,\ a,\ a,\ 0) \mid a \text{ is a real number}\}$.

47.

$$\begin{bmatrix} 3 & -7 & | & 2 & | & -4 & | & 3 \\ -2 & 5 & | & 5 & | & 7 & | & 6 \end{bmatrix} \quad 2R_1 + 3R_2 \to R_2$$

$$\begin{bmatrix} 3 & -7 & | & 2 & | & -4 & | & 3 \\ 0 & 1 & | & 19 & | & 13 & | & 24 \end{bmatrix} \quad 7R_2 + R_1 \to R_1$$

$$\begin{bmatrix} 3 & 0 & | & 135 & | & 87 & | & 171 \\ 0 & 1 & | & 19 & | & 13 & | & 24 \end{bmatrix} \quad \tfrac{1}{3}R_1 \to R_1$$

$$\begin{bmatrix} 1 & 0 & | & 45 & | & 29 & | & 57 \\ 0 & 1 & | & 19 & | & 13 & | & 24 \end{bmatrix}$$

The solution sets are a) $\{(45, 19)\}$, b) $\{(29, 13)\}$, c) $\{(57, 24)\}$

49.

$$\begin{bmatrix} 1 & 0 & -2 & | & 2 & | & 0 & | & 7 \\ 2 & 2 & -5 & | & 0 & | & 3 & | & -2 \\ 4 & 5 & -10 & | & 7 & | & 5 & | & 4 \end{bmatrix} \begin{array}{c} 2R_1 - R_2 \to R_2 \\ 4R_1 - R_3 \to R_3 \end{array}$$

$$\begin{bmatrix} 1 & 0 & -2 & | & 2 & | & 0 & | & 7 \\ 0 & -2 & 1 & | & 4 & | & -3 & | & 16 \\ 0 & -5 & 2 & | & 1 & | & -5 & | & 24 \end{bmatrix} \quad 5R_2 - 2R_3 \to R_3$$

$$\begin{bmatrix} 1 & 0 & -2 & 2 & 0 & 7 \\ 0 & -2 & 1 & 4 & -3 & 16 \\ 0 & 0 & 1 & 18 & -5 & 32 \end{bmatrix} \begin{array}{l} R_2 - R_3 \to R_2 \\ R_1 + 2R_3 \to R_3 \end{array}$$

$$\begin{bmatrix} 1 & 0 & 0 & 38 & -10 & 71 \\ 0 & -2 & 0 & -14 & 2 & -16 \\ 0 & 0 & 1 & 18 & -5 & 32 \end{bmatrix} -\tfrac{1}{2}R_2 \to R_2$$

$$\begin{bmatrix} 1 & 0 & 0 & 38 & -10 & 71 \\ 0 & 1 & 0 & 7 & -1 & 8 \\ 0 & 0 & 1 & 18 & -5 & 32 \end{bmatrix}$$

The solution sets are a) $\{(38,\ 7,\ 18)\}$, b) $\{(-10,\ -1,\ -5)\}$, c) $\{(71,\ 8,\ 32)\}$

51.
$$\begin{bmatrix} 5 & -5 & 4 & 9 & 6 & 2 \\ 0 & 3 & 1 & 5 & 1 & 13 \\ 2 & -1 & 2 & 1 & 7 & -5 \end{bmatrix} 2R_1 - 5R_3 \to R_3$$

$$\begin{bmatrix} 5 & -5 & 4 & 9 & 6 & 2 \\ 0 & 3 & 1 & 5 & 1 & 13 \\ 0 & -5 & -2 & 13 & -23 & 29 \end{bmatrix} \begin{array}{l} R_1 - R_3 \to R_1 \\ 5R_2 + 3R_3 \to R_3 \end{array}$$

$$\begin{bmatrix} 5 & 0 & 6 & -4 & 29 & -27 \\ 0 & 3 & 1 & 5 & 1 & 13 \\ 0 & 0 & -1 & 64 & -64 & 152 \end{bmatrix} \begin{array}{l} R_2 + R_3 \to R_2 \\ -R_3 \to R_3 \\ R_1 + 6R_3 \to R_1 \end{array}$$

$$\begin{bmatrix} 5 & 0 & 0 & 380 & -355 & 885 \\ 0 & 3 & 0 & 69 & -63 & 165 \\ 0 & 0 & 1 & -64 & 64 & -152 \end{bmatrix} \begin{array}{l} \tfrac{1}{5}R_1 \to R_1 \\ \tfrac{1}{3}R_2 \to R_2 \end{array}$$

$$\begin{bmatrix} 1 & 0 & 0 & 76 & -71 & 177 \\ 0 & 1 & 0 & 23 & -21 & 55 \\ 0 & 0 & 1 & -64 & 64 & -152 \end{bmatrix}$$

The solution sets are a) $\{(76,\ 23,\ -64)\}$ b) $\{(-71,\ -21,\ 64)\}$ c) $\{(177,\ 55,\ -152)\}$

53. (a)

Monday	Tuesday	Wednesday

$$\begin{cases} 120x + 50y = 750 \\ 80x + 30y = 490 \end{cases} \qquad \begin{cases} 120x + 50y = 820 \\ 80x + 30y = 540 \end{cases} \qquad \begin{cases} 120x + 50y = 730 \\ 80x + 30y = 470 \end{cases}$$

Thursday	Friday

$$\begin{cases} 120x + 50y = 1020 \\ 80x + 30y = 660 \end{cases} \qquad \begin{cases} 120x + 50y = 1040 \\ 80x + 30y = 680 \end{cases}$$

where the first equation in each system is the units of cargo of type A transported and the second equation is the units of cargo of type B on that day.

(b) We first divide each equation by 10.

$$\left[\begin{array}{cc|c|c|c|c|c} 12 & 5 & 75 & 82 & 73 & 102 & 104 \\ 8 & 3 & 49 & 54 & 47 & 66 & 68 \end{array} \right] \quad 2R_1 - 3R_2 \rightarrow R_2$$

$$\left[\begin{array}{cc|c|c|c|c|c} 12 & 5 & 75 & 82 & 73 & 102 & 104 \\ 0 & 1 & 3 & 2 & 5 & 6 & 4 \end{array} \right] \quad R_1 - 5R_2 \rightarrow R_1$$

$$\left[\begin{array}{cc|c|c|c|c|c} 12 & 0 & 60 & 72 & 48 & 72 & 84 \\ 0 & 1 & 3 & 2 & 5 & 6 & 4 \end{array} \right] \quad \tfrac{1}{12}R_1 \rightarrow R_1$$

$$\left[\begin{array}{cc|c|c|c|c|c} 1 & 0 & 5 & 6 & 4 & 6 & 7 \\ 0 & 1 & 3 & 2 & 5 & 6 & 4 \end{array} \right]$$

The solution sets are $\{(5, 3)\}, \{(6, 2)\}, \{(4, 5)\}, \{(6, 6)\}, \{(7, 4)\}$

55. (a)

January	February	March

$$\begin{cases} 5x + 3y + 2z = 690 \\ 6x + 4y + 3z = 910 \\ 2x + y + z = 280 \end{cases} \qquad \begin{cases} 5x + 3y + 2z = 675 \\ 6x + 4y + 3z = 900 \\ 2x + y + z = 275 \end{cases} \qquad \begin{cases} 5x + 3y + 2z = 785 \\ 6x + 4y + 3z = 930 \\ 2x + y + z = 320 \end{cases}$$

(b)

$$\left[\begin{array}{ccc|c|c|c} 5 & 3 & 2 & 690 & 675 & 785 \\ 6 & 4 & 3 & 910 & 900 & 930 \\ 2 & 1 & 1 & 280 & 275 & 320 \end{array} \right] \quad \begin{array}{l} R_2 - 3R_3 \rightarrow R_2 \\ 2R_1 - 5R_3 \rightarrow R_3 \end{array}$$

$$\left[\begin{array}{ccc|c|c|c} 5 & 3 & 2 & 690 & 675 & 785 \\ 0 & 1 & 0 & 70 & 75 & -30 \\ 0 & 1 & -1 & -20 & -25 & -30 \end{array} \right] \quad \begin{array}{l} R_1 - 3R_2 \rightarrow R_1 \\ R_2 - R_3 \rightarrow R_3 \end{array}$$

$$\left[\begin{array}{ccc|c|c|c} 5 & 0 & 2 & 480 & 450 & 875 \\ 0 & 1 & 0 & 70 & 75 & -30 \\ 0 & 0 & 1 & 90 & 100 & 0 \end{array} \right] \quad R_1 - 2R_3 \rightarrow R_1$$

$$\begin{bmatrix} 5 & 0 & 0 & | & 300 & | & 250 & | & 875 \\ 0 & 1 & 0 & | & 70 & | & 75 & | & -30 \\ 0 & 0 & 1 & | & 90 & | & 100 & | & 0 \end{bmatrix} \quad \tfrac{1}{5}R_1 \to R_1$$

$$\begin{bmatrix} 1 & 0 & 0 & | & 60 & | & 50 & | & 175 \\ 0 & 1 & 0 & | & 70 & | & 75 & | & -30 \\ 0 & 0 & 1 & | & 90 & | & 100 & | & 0 \end{bmatrix}$$

The solution sets are $\{(60, 70, 90)\}$, $\{(50, 75, 100)\}$, $\{(175, -30, 0)\}$. However, the last solution is not within the domain of this problem as x, y and z must be non-negative.

57. At full capacity,

$7x + 5y + 4z = 285$, the total assembly time, and

$2x + 1.5y + z = 78$, the total testing time, where

x, y and z are the number of units of Models A, B and C respectively.

Solving the system,

$$\begin{bmatrix} 7 & 5 & 4 & | & 285 \\ 2 & 1.5 & 1 & | & 78 \end{bmatrix} \quad 2R_1 - 7R_2 \to R_2 \quad \begin{bmatrix} 7 & 5 & 4 & | & 285 \\ 0 & -.5 & 1 & | & 24 \end{bmatrix} \quad 10R_2 + R_1 \to R_1$$

$$\begin{bmatrix} 7 & 0 & 14 & | & 525 \\ 0 & -.5 & 1 & | & 24 \end{bmatrix} \quad \begin{matrix} \tfrac{1}{7}R_1 \to R_1 \\ -2R_2 \to R_2 \end{matrix} \quad \begin{bmatrix} 1 & 0 & 2 & | & 75 \\ 0 & 1 & -2 & | & -48 \end{bmatrix}$$

The last row gives $y = -48 + 2z$ and the first row gives $x = 75 - 2z$.

Solutions take the form

$$x = 75 - 2a, \quad y = 2a - 48, \quad \text{and} \quad z = a$$

where $a \geq 0$, $2a \leq 75$, or $a \leq 37.5$, and $2a - 48 \geq 0$ or $a \geq 24$

The total profit is given by $50x + 60y + 40z = 50(75 - 2a) + 60(2a - 48) + 40a = 870 + 60a$. This profit is largest when a is as large as possible, that is, when $a = 37.5$. The profit is $870 + 60(37.5) = \$3120$ and occurs when $x = 75 - 75 = 0$, $y = 75 - 48 = 27$ and $z = 37.5$.

59. Let x, y and z be the number of standard, regular and deluxe models, respectively. If all the time on the machines is used,

$$20x + 30y + 35z = 35(60) \quad \text{and} \quad 15x + 20y + 30z = 28(60).$$

Solving the system,

$$\begin{bmatrix} 20 & 30 & 35 & | & 2100 \\ 15 & 20 & 30 & | & 1680 \end{bmatrix} \quad \begin{matrix} \tfrac{1}{5}R_1 \to R_1 \\ \tfrac{1}{5}R_2 \to R_2 \end{matrix} \quad \begin{bmatrix} 4 & 6 & 7 & | & 420 \\ 3 & 4 & 6 & | & 336 \end{bmatrix} \quad 3R_1 - 4R_2 \to R_2$$

$$\begin{bmatrix} 4 & 6 & 7 & | & 420 \\ 0 & 2 & -3 & | & -84 \end{bmatrix} \quad R_1 - 3R_2 \to R_1 \quad \begin{bmatrix} 4 & 0 & 16 & | & 672 \\ 0 & 2 & -3 & | & -84 \end{bmatrix} \quad \begin{matrix} \tfrac{1}{4}R_1 \to R_1 \\ \tfrac{1}{2}R_2 \to R_2 \end{matrix}$$

$$\begin{bmatrix} 1 & 0 & 4 & | & 168 \\ 0 & 1 & -\frac{3}{2} & | & -42 \end{bmatrix}.$$ Solutions occur when $y = \frac{3}{2}z - 42$ and $x = 168 - 4z$.

Letting $z = a$, $x = 168 - 4a$, $y = \frac{3}{2}a - 42$ and $z = a$ where $a \geq 0$, $168 - 4a \geq 0$ and $\frac{3}{2}a - 42 \geq 0$ or $28 \leq a \leq 42$, when all the time available on the two machines is utilized. The total profit is $45x + 56y + 60z = 45(168 - 4a) + 56(\frac{3}{2}a - 42) + 60a = 5208 - 36a$. This is largest when a is smallest, and the maximum profit is $5208 - 36(28) = \$4200$. This maximum profit occurs when $x = 168 - 4(28) = 56$, $y = 0$ and $z = 28$.

61. Let w, x, y and z be the numbers of small, medium, large and super large boats produced respectively. At full capacity, $w + 1.5x + 2y + 2.5z = 295$, $1.5w + 2x + 2.5y + 3z = 385$ and $.3w + .4x + .7y + .6z = 85$. Solving the system,

$$\begin{bmatrix} 1.5 & 2 & 2.5 & 3 & | & 385 \\ 1 & 1.5 & 2 & 2.5 & | & 295 \\ .3 & .4 & .7 & .6 & | & 85 \end{bmatrix} \begin{matrix} R_1 - 1.5R_2 \to R_1 \\ .3R_2 - R_3 \to R_3 \end{matrix} \begin{bmatrix} 0 & -.25 & -.5 & -.75 & | & -57.5 \\ 1 & 1.5 & 2 & 2.5 & | & 295 \\ 0 & .05 & -.1 & .15 & | & 3.5 \end{bmatrix}$$

$$\begin{matrix} -100R_1 \to R_1 \\ 10R_2 \to R_2 \\ 100R_3 \to R_3 \end{matrix} \begin{bmatrix} 0 & 25 & 50 & 75 & | & 5750 \\ 10 & 15 & 20 & 25 & | & 2950 \\ 0 & 5 & -10 & 15 & | & 350 \end{bmatrix} \begin{matrix} \frac{1}{25}R_1 \to R_1 \\ \frac{1}{5}R_2 \to R_2 \\ \frac{1}{5}R_3 \to R_3 \end{matrix} \begin{bmatrix} 0 & 1 & 2 & 3 & | & 230 \\ 2 & 3 & 4 & 5 & | & 590 \\ 0 & 1 & -2 & 3 & | & 70 \end{bmatrix}$$

$$\begin{matrix} R_1 - R_3 \to R_1 \\ R_2 - 3R_3 \to R_2 \end{matrix} \begin{bmatrix} 0 & 0 & 4 & 0 & | & 160 \\ 2 & 0 & 10 & -4 & | & 380 \\ 0 & 1 & -2 & 3 & | & 70 \end{bmatrix} \begin{matrix} \frac{1}{2}R_2 \to R_1 \\ \frac{1}{4}R_1 \to R_3 \\ R_3 \to R_2 \end{matrix} \begin{bmatrix} 1 & 0 & 5 & -2 & | & 190 \\ 0 & 1 & -2 & 3 & | & 70 \\ 0 & 0 & 1 & 0 & | & 40 \end{bmatrix}$$

The solutions take the form $y = 40$, $z = a$, $x = 70 + 2(40) - 3a = 150 - 3a$ and $w = 190 - 5(40) + 2a = 2a - 10$, where $a \geq 0$, $2a \geq 10$, and $3a \leq 150$. The last 3 inequalities lead to $5 \leq a \leq 30$.

The profit is $40w + 50x + 75y + 90z = 40(2a - 10) + 50(150 - 3a) + 75(40) + 90a$
$= 80a - 400 + 7500 - 150a + 3000 + 90a = 10,100 + 20a$.

This is a maximum when a is largest. The maximum profit is $10,100 + 20(30) = \$10,700$. This occurs when $w = 50$, $x = 60$, $y = 40$ and $z = 30$.

63. Since at equilibrium, the demands and corresponding supplies are equal,

$$\begin{cases} 520 - 30x + 92y - 25z = -120 + 70x - 8y + 75z \\ 140 + 32x - 52y + 18z = -7 - 8x + 98y - 12z \\ 478 - 20x + 28y - 5z = -21 + 40x - 12y + 65z \end{cases}$$

Simplifying,

$$\begin{cases} 100x - 100y + 100z = 640 \\ 40x - 150y + 30z = -147 \\ 60x - 40y + 70z = 499 \end{cases}$$

Solving,

$$
\begin{bmatrix}
100 & -100 & 100 & \bigg| & 640 \\
40 & -150 & 30 & \bigg| & -147 \\
60 & -40 & 70 & \bigg| & 499
\end{bmatrix}
\quad
\begin{array}{c}
3R_2 - 2R_3 \to R_3 \\
2R_1 - 5R_2 \to R_2
\end{array}
\quad
\begin{bmatrix}
100 & -100 & 100 & \bigg| & 640 \\
0 & 550 & 50 & \bigg| & 2015 \\
0 & -370 & -50 & \bigg| & -1439
\end{bmatrix}
$$

$$
\begin{array}{c}
11R_1 + 2R_2 \to R_1 \\
37R_2 + 55R_3 \to R_3
\end{array}
\quad
\begin{bmatrix}
1100 & 0 & 1200 & \bigg| & 11{,}070 \\
0 & 550 & 50 & \bigg| & 2015 \\
0 & 0 & -900 & \bigg| & -4590
\end{bmatrix}
\quad
\begin{array}{c}
-\frac{1}{90}R_3 \to R_3 \\
\frac{1}{10}R_1 \to R_1 \\
\frac{1}{5}R_2 \to R_2
\end{array}
$$

$$
\begin{bmatrix}
110 & 0 & 120 & \bigg| & 1107 \\
0 & 110 & 10 & \bigg| & 403 \\
0 & 0 & 10 & \bigg| & 51
\end{bmatrix}
\quad
\begin{array}{c}
R_1 - 12R_3 \to R_1 \\
R_2 - R_3 \to R_2
\end{array}
\quad
\begin{bmatrix}
110 & 0 & 0 & \bigg| & 495 \\
0 & 110 & 0 & \bigg| & 352 \\
0 & 0 & 10 & \bigg| & 51
\end{bmatrix}
$$

$$
\begin{array}{c}
\frac{1}{110}R_1 \to R_1 \\
\frac{1}{110}R_2 \to R_2 \\
\frac{1}{10}R_3 \to R_3
\end{array}
\quad
\begin{bmatrix}
1 & 0 & 0 & \bigg| & 4.5 \\
0 & 1 & 0 & \bigg| & 3.2 \\
0 & 0 & 1 & \bigg| & 5.1
\end{bmatrix}
$$

The equilibrium price for commodity X is \$4.50, for Y is \$3.20 and for Z is \$5.10.

The equilibrium quantity for X is $D_X = 520 - 30(4.50) + 92(3.2) - 25(5.1) = 551{,}900$
for Y is $D_Y = 140 + 32(4.5) - 52(3.2) + 18(5.1) = 209{,}400$
and for Z is $D_Z = 478 - 20(4.5) + 28(3.2) - 5(5.1) = 452{,}100$

65. At equilibrium, supply equals demand. Hence

$$
\begin{cases}
800 - 23x + 95y - 18z = -35 + 77x - 5y + 82z \\
320 + 50x - 12y + 35z = -68 - 10x + 118y - 5z \\
1310 - 10x + 42y - 20z = -112 + 80x - 8y + 100z
\end{cases}
$$

Simplifying, we obtain the system

$$
\begin{cases}
100x - 100y + 100z = 835 \\
60x - 130y + 40z = -388 \\
90x - 50y + 120z = 1422
\end{cases}
$$

Solving the system

$$
\begin{bmatrix}
100 & -100 & 100 & \bigg| & 835 \\
60 & -130 & 40 & \bigg| & -388 \\
90 & -50 & 120 & \bigg| & 1422
\end{bmatrix}
\quad
\begin{array}{c}
3R_1 - 5R_2 \to R_2 \\
9R_1 - 10R_3 \to R_3
\end{array}
\quad
\begin{bmatrix}
100 & -100 & 100 & \bigg| & 835 \\
0 & 350 & 100 & \bigg| & 4445 \\
0 & -400 & -300 & \bigg| & -6705
\end{bmatrix}
$$

$$
\begin{array}{c}
R_1 - R_2 \to R_1 \\
\frac{1}{5}R_2 \to R_2 \\
3R_2 + R_3 \to R_3
\end{array}
\quad
\begin{bmatrix}
100 & -450 & 0 & \bigg| & -3610 \\
0 & 70 & 20 & \bigg| & 889 \\
0 & 650 & 0 & \bigg| & 6630
\end{bmatrix}
\quad
\begin{array}{c}
\frac{1}{10}R_1 \to R_1 \\
\frac{1}{10}R_3 \to R_3
\end{array}
\quad
\begin{bmatrix}
10 & -45 & 0 & \bigg| & -361 \\
0 & 70 & 20 & \bigg| & 889 \\
0 & 65 & 0 & \bigg| & 663
\end{bmatrix}
$$

$$\begin{array}{c} 13R_2 - 14R_3 \to R_2 \\ 13R_1 + 9R_3 \to R_1 \end{array} \left[\begin{array}{ccc|c} 130 & 0 & 0 & 1274 \\ 0 & 0 & 260 & 2275 \\ 0 & 65 & 0 & 663 \end{array}\right] \begin{array}{c} R_1/130 \to R_1 \\ R_2/260 \to R_3 \\ R_3/65 \to R_2 \end{array}$$

$$\left[\begin{array}{ccc|c} 1 & 0 & 0 & 9.8 \\ 0 & 1 & 0 & 10.2 \\ 0 & 0 & 1 & 8.75 \end{array}\right]$$

The equilibrium prices for commodities X, Y and Z are \$9.80, \$10.20 and \$8.75, respectively.

The corresponding equilibrium quantities are
$$S_X = (-35 + 77(9.8) - 5(10.2) + 82(8.75))1000 = 1,386,100$$
$$S_Y = 993,850$$
$$S_Z = 1,465,400$$

EXERCISE SET 3.4   MULTIPLICATIVE INVERSE

1. $\left[\begin{array}{cc} 1 & 0 \\ 1 & 0 \end{array}\right] \cdot \left[\begin{array}{cc} a & c \\ b & d \end{array}\right] = \left[\begin{array}{cc} a & c \\ a & c \end{array}\right] \neq \left[\begin{array}{cc} 1 & 0 \\ 0 & 1 \end{array}\right]$ since a cannot equal both zero and one.

3. $\left[\begin{array}{cc} 3 & 2 \\ 9 & 6 \end{array}\right]\left[\begin{array}{cc} a & c \\ b & d \end{array}\right] = \left[\begin{array}{cc} 3a + 2b & 3c + 2d \\ 9a + 6b & 9c + 6d \end{array}\right] = \left[\begin{array}{cc} 3a + 2b & 3c + 2d \\ 3(3a + 2b) & 3(3c + 2d) \end{array}\right] \neq \left[\begin{array}{cc} 1 & 0 \\ 0 & 1 \end{array}\right],$

as $3a + 2b$ cannot be equal to both zero and one.

5. $\left[\begin{array}{cc|cc} 5 & 7 & 1 & 0 \\ 2 & 3 & 0 & 1 \end{array}\right] \quad 2R_1 + 5R_2 \to R_2 \quad \left[\begin{array}{cc|cc} 5 & 7 & 1 & 0 \\ 0 & -1 & 2 & -5 \end{array}\right] \quad 7R_2 + R_1 \to R_1$

$\left[\begin{array}{cc|cc} 5 & 0 & 15 & -35 \\ 0 & -1 & 2 & -5 \end{array}\right] \begin{array}{c} \frac{1}{5}R_1 \to R_1 \\ -R_2 \to R_2 \end{array} \left[\begin{array}{cc|cc} 1 & 0 & 3 & -7 \\ 0 & 1 & -2 & 5 \end{array}\right].$

The inverse matrix is $\left[\begin{array}{cc} 3 & -7 \\ -2 & 5 \end{array}\right].$

7. $\left[\begin{array}{cc|cc} 9 & 13 & 1 & 0 \\ 2 & 3 & 0 & 1 \end{array}\right] \quad 2R_1 - 9R_2 \to R_2 \quad \left[\begin{array}{cc|cc} 9 & 13 & 1 & 0 \\ 0 & -1 & 2 & -9 \end{array}\right] \quad R_1 + 13R_2 \to R_1$

$\left[\begin{array}{cc|cc} 9 & 0 & 27 & -117 \\ 0 & -1 & 2 & -9 \end{array}\right] \begin{array}{c} R_1/9 \to R_1 \\ -R_2 \to R_2 \end{array} \left[\begin{array}{cc|cc} 1 & 0 & 3 & -13 \\ 0 & 1 & -2 & 9 \end{array}\right].$

The inverse matrix is $\begin{bmatrix} 3 & -13 \\ -2 & 9 \end{bmatrix}$.

9. $\left[\begin{array}{cc|cc} 3 & 8 & 1 & 0 \\ 5 & 12 & 0 & 1 \end{array}\right]$ $\quad 5R_1 - 3R_2 \to R_2 \quad$ $\left[\begin{array}{cc|cc} 3 & 8 & 1 & 0 \\ 0 & 4 & 5 & -3 \end{array}\right]$ $\quad R_1 - 2R_2 \to R_1$

$\left[\begin{array}{cc|cc} 3 & 0 & -9 & 6 \\ 0 & 4 & 5 & -3 \end{array}\right]$ $\quad \begin{array}{c} R_1/3 \to R_1 \\ R_2/4 \to R_2 \end{array} \quad$ $\left[\begin{array}{cc|cc} 1 & 0 & -3 & 2 \\ 0 & 1 & \frac{5}{4} & -\frac{3}{4} \end{array}\right]$.

The inverse is $\begin{bmatrix} -3 & 2 \\ \frac{5}{4} & -\frac{3}{4} \end{bmatrix}$.

11. $\left[\begin{array}{ccc|ccc} 5 & -5 & 4 & 1 & 0 & 0 \\ 0 & 3 & 1 & 0 & 1 & 0 \\ 2 & -1 & 2 & 0 & 0 & 1 \end{array}\right]$ $\quad 2R_1 - 5R_3 \to R_3 \quad$ $\left[\begin{array}{ccc|ccc} 5 & -5 & 4 & 1 & 0 & 0 \\ 0 & 3 & 1 & 0 & 1 & 0 \\ 0 & -5 & -2 & 2 & 0 & -5 \end{array}\right]$

$\begin{array}{c} R_1 - R_3 \to R_1 \\ 5R_2 + 3R_3 \to R_3 \end{array}$ $\left[\begin{array}{ccc|ccc} 5 & 0 & 6 & -1 & 0 & 5 \\ 0 & 3 & 1 & 0 & 1 & 0 \\ 0 & 0 & -1 & 6 & 5 & -15 \end{array}\right]$ $\begin{array}{c} R_2 + R_3 \to R_3 \\ R_1 + 6R_3 \to R_1 \end{array}$

$\left[\begin{array}{ccc|ccc} 5 & 0 & 0 & 35 & 30 & -85 \\ 0 & 3 & 0 & 6 & 6 & -15 \\ 0 & 0 & -1 & 6 & 5 & -15 \end{array}\right]$ $\begin{array}{c} R_1/5 \to R_1 \\ R_2/3 \to R_2 \\ -R_3 \to R_3 \end{array}$ $\left[\begin{array}{ccc|ccc} 1 & 0 & 0 & 7 & 6 & -17 \\ 0 & 1 & 0 & 2 & 2 & -5 \\ 0 & 0 & 1 & -6 & -5 & 15 \end{array}\right]$.

The inverse matrix is $\begin{bmatrix} 7 & 6 & -17 \\ 2 & 2 & -5 \\ -6 & -5 & 15 \end{bmatrix}$.

13. $\left[\begin{array}{ccc|ccc} 7 & 6 & -17 & 1 & 0 & 0 \\ 2 & 2 & -5 & 0 & 1 & 0 \\ -6 & -5 & 15 & 0 & 0 & 1 \end{array}\right]$ $\begin{array}{c} 3R_2 + R_3 \to R_3 \\ 2R_1 - 7R_2 \to R_2 \end{array}$ $\left[\begin{array}{ccc|ccc} 7 & 6 & -17 & 1 & 0 & 0 \\ 0 & -2 & 1 & 2 & -7 & 0 \\ 0 & 1 & 0 & 0 & 3 & 1 \end{array}\right]$

$\begin{array}{c} 2R_3 + R_2 \to R_2 \\ R_1 - 6R_3 \to R_1 \end{array}$ $\left[\begin{array}{ccc|ccc} 7 & 0 & -17 & 1 & -18 & -6 \\ 0 & 0 & 1 & 2 & -1 & 2 \\ 0 & 1 & 0 & 0 & 3 & 1 \end{array}\right]$ $\begin{array}{c} R_1 + 17R_2 \to R_1 \\ R_3 \to R_2 \end{array}$

$$\begin{bmatrix} 7 & 0 & 0 & | & 35 & -35 & 28 \\ 0 & 1 & 0 & | & 0 & 3 & 1 \\ 0 & 0 & 1 & | & 2 & -1 & 2 \end{bmatrix} \quad R_1/7 \to R_1 \quad \begin{bmatrix} 1 & 0 & 0 & | & 5 & -5 & 4 \\ 0 & 1 & 0 & | & 0 & 3 & 1 \\ 0 & 0 & 1 & | & 2 & -1 & 2 \end{bmatrix}.$$

The inverse matrix is $\begin{bmatrix} 5 & -5 & 4 \\ 0 & 3 & 1 \\ 2 & -1 & 2 \end{bmatrix}$.

15.
$$\begin{bmatrix} 1 & 0 & -2 & | & 1 & 0 & 0 \\ 2 & 5 & -6 & | & 0 & 1 & 0 \\ 4 & 5 & -10 & | & 0 & 0 & 1 \end{bmatrix} \quad -R_2 + R_3 \to R_3 \quad \begin{bmatrix} 1 & 0 & -2 & | & 1 & 0 & 0 \\ 2 & 5 & -6 & | & 0 & 1 & 0 \\ 2 & 0 & -4 & | & 0 & -1 & 1 \end{bmatrix}$$

$$-2R_1 + R_3 \to R_3 \quad \begin{bmatrix} 1 & 0 & -2 & | & 1 & 0 & 0 \\ 2 & 5 & -6 & | & 0 & 1 & 0 \\ 0 & 0 & 0 & | & -2 & -1 & 1 \end{bmatrix}$$

Because there is a row of zeros to the left of the vertical line, the matrix has no inverse.

17.
$$\begin{bmatrix} -7 & 1 & 1 & | & 1 & 0 & 0 \\ 3 & 2 & 3 & | & 0 & 1 & 0 \\ 5 & 1 & 2 & | & 0 & 0 & 1 \end{bmatrix} \quad \begin{matrix} 3R_1 + 7R_2 \to R_2 \\ 5R_1 + 7R_3 \to R_3 \end{matrix} \quad \begin{bmatrix} -7 & 1 & 1 & | & 1 & 0 & 0 \\ 0 & 17 & 24 & | & 3 & 7 & 0 \\ 0 & 12 & 19 & | & 5 & 0 & 7 \end{bmatrix}$$

$$\begin{matrix} 17R_1 - R_2 \to R_1 \\ 12R_2 - 17R_3 \to R_3 \end{matrix} \quad \begin{bmatrix} -119 & 0 & -7 & | & 14 & -7 & 0 \\ 0 & 17 & 24 & | & 3 & 7 & 0 \\ 0 & 0 & -35 & | & -49 & 84 & -119 \end{bmatrix} \quad \begin{matrix} R_1/7 \to R_1 \\ R_3/7 \to R_3 \end{matrix}$$

$$\begin{bmatrix} -17 & 0 & -1 & | & 2 & -1 & 0 \\ 0 & 17 & 24 & | & 3 & 7 & 0 \\ 0 & 0 & -5 & | & -7 & 12 & -17 \end{bmatrix} \quad \begin{matrix} 5R_1 - R_3 \to R_1 \\ 5R_2 + 24R_3 \to R_2 \end{matrix}$$

$$\begin{bmatrix} -85 & 0 & 0 & | & 17 & -17 & 17 \\ 0 & 85 & 0 & | & -153 & 323 & -408 \\ 0 & 0 & -5 & | & -7 & 12 & -17 \end{bmatrix} \quad \begin{matrix} R_1/-85 \to R_1 \\ R_2/85 \to R_2 \\ R_3/(-5) \to R_3 \end{matrix}$$

(continued)

$$\left[\begin{array}{ccc|ccc} 1 & 0 & 0 & -1/5 & 1/5 & -1/5 \\ 0 & 1 & 0 & -9/5 & 19/5 & -24/5 \\ 0 & 0 & 1 & 7/5 & -12/5 & 17/5 \end{array}\right].$$

The inverse matrix is $\dfrac{1}{5}\left[\begin{array}{ccc} -1 & 1 & -1 \\ -9 & 19 & -24 \\ 7 & -12 & 17 \end{array}\right].$

19. $\left[\begin{array}{ccc|ccc} 2 & -1 & 7 & 1 & 0 & 0 \\ 3 & 9 & 0 & 0 & 1 & 0 \\ 5 & 0 & 2 & 0 & 0 & 1 \end{array}\right]$ $9R_1 + R_2 \rightarrow R_1$ $\left[\begin{array}{ccc|ccc} 21 & 0 & 63 & 9 & 1 & 0 \\ 3 & 9 & 0 & 0 & 1 & 0 \\ 5 & 0 & 2 & 0 & 0 & 1 \end{array}\right]$

$2R_1 - 63R_3 \rightarrow R_1$ $\left[\begin{array}{ccc|ccc} -273 & 0 & 0 & 18 & 2 & -63 \\ 3 & 9 & 0 & 0 & 1 & 0 \\ 5 & 0 & 2 & 0 & 0 & 1 \end{array}\right]$ $\begin{array}{l} 5R_1 + 273R_3 \rightarrow R_3 \\ R_1 + 91R_2 \rightarrow R_2 \end{array}$

$\left[\begin{array}{ccc|ccc} -273 & 0 & 0 & 18 & 2 & -63 \\ 0 & 819 & 0 & 18 & 93 & -63 \\ 0 & 0 & 546 & 90 & 10 & -42 \end{array}\right]$ $\begin{array}{l} -R_1/273 \rightarrow R_1 \\ R_2/819 \rightarrow R_2 \\ R_3/546 \rightarrow R_3 \end{array}$

$\left[\begin{array}{ccc|ccc} 1 & 0 & 0 & -18/273 & -2/273 & 63/273 \\ 0 & 1 & 0 & 6/273 & 31/273 & -21/273 \\ 0 & 0 & 1 & 45/273 & 5/273 & -21/273 \end{array}\right].$

The inverse matrix is $\dfrac{1}{273}\left[\begin{array}{ccc} -18 & -2 & 63 \\ 6 & 31 & -21 \\ 45 & 5 & -21 \end{array}\right].$

21. $\left[\begin{array}{cccc|cccc} 7 & -8 & -5 & 6 & 1 & 0 & 0 & 0 \\ 1 & -4 & 0 & 1 & 0 & 1 & 0 & 0 \\ 3 & -11 & 3 & 4 & 0 & 0 & 1 & 0 \\ 4 & -6 & -1 & 4 & 0 & 0 & 0 & 1 \end{array}\right]$ $\begin{array}{l} R_1 - 5R_4 \rightarrow R_1 \\ R_3 + 3R_4 \rightarrow R_4 \end{array}$

$$\left[\begin{array}{cccc|cccc}
-13 & 22 & 0 & -14 & 1 & 0 & 0 & -5 \\
1 & -4 & 0 & 1 & 0 & 1 & 0 & 0 \\
3 & -11 & 3 & 4 & 0 & 0 & 1 & 0 \\
15 & -29 & 0 & 16 & 0 & 0 & 1 & 3
\end{array}\right] \quad \begin{array}{l} 4R_3 - R_4 \to R_3 \\ R_2 \leftrightarrow R_1 \end{array}$$

$$\left[\begin{array}{cccc|cccc}
1 & -4 & 0 & 1 & 0 & 1 & 0 & 0 \\
-13 & 22 & 0 & -14 & 1 & 0 & 0 & -5 \\
-3 & -15 & 12 & 0 & 0 & 0 & 3 & -3 \\
15 & -29 & 0 & 16 & 0 & 0 & 1 & 3
\end{array}\right] \quad \begin{array}{l} 16R_1 - R_4 \to R_1 \\ 14R_1 + R_2 \to R_2 \\ R_3/3 \to R_3 \end{array}$$

$$\left[\begin{array}{cccc|cccc}
1 & -35 & 0 & 0 & 0 & 16 & -1 & -3 \\
1 & -34 & 0 & 0 & 1 & 14 & 0 & -5 \\
-1 & -5 & 4 & 0 & 0 & 0 & 1 & -1 \\
15 & -29 & 0 & 16 & 0 & 0 & 1 & 3
\end{array}\right] \quad \begin{array}{l} -R_1 + R_2 \to R_2 \\ R_1 + R_3 \to R_3 \\ 15R_1 - R_4 \to R_4 \end{array}$$

$$\left[\begin{array}{cccc|cccc}
1 & -35 & 0 & 0 & 0 & 16 & -1 & -3 \\
0 & 1 & 0 & 0 & 1 & -2 & 1 & -2 \\
0 & -40 & 4 & 0 & 0 & 16 & 0 & -4 \\
0 & -496 & 0 & -16 & 0 & 240 & -16 & -48
\end{array}\right] \quad \begin{array}{l} R_3/4 \to R_3 \\ R_4/16 \to R_4 \end{array}$$

$$\left[\begin{array}{cccc|cccc}
1 & -35 & 0 & 0 & 0 & 16 & -1 & -3 \\
0 & 1 & 0 & 0 & 1 & -2 & 1 & -2 \\
0 & -10 & 1 & 0 & 0 & 4 & 0 & -1 \\
0 & -31 & 0 & -1 & 0 & 15 & -1 & -3
\end{array}\right] \quad \begin{array}{l} R_1 + 35R_2 \to R_1 \\ 10R_2 + R_3 \to R_3 \\ 31R_2 + R_4 \to R_4 \\ -R_4 \to R_4 \end{array}$$

$$\left[\begin{array}{cccc|cccc}
1 & 0 & 0 & 0 & 35 & -54 & 34 & -73 \\
0 & 1 & 0 & 0 & 1 & -2 & 1 & -2 \\
0 & 0 & 1 & 0 & 10 & -16 & 10 & -21 \\
0 & 0 & 0 & 1 & -31 & +47 & -30 & +65
\end{array}\right].$$

$$\text{The inverse matrix is} \quad \left[\begin{array}{cccc}
35 & -54 & 34 & -73 \\
1 & -2 & 1 & -2 \\
10 & -16 & 10 & -21 \\
-31 & 47 & -30 & 65
\end{array}\right].$$

23.

$$\left[\begin{array}{cccc|cccc}
0 & 1 & 3 & 1 & 1 & 0 & 0 & 0 \\
5 & 1 & -2 & 5 & 0 & 1 & 0 & 0 \\
1 & -2 & 6 & 3 & 0 & 0 & 1 & 0 \\
2 & 3 & 2 & 3 & 0 & 0 & 0 & 1
\end{array}\right]$$

$R_2 - 5R_3 \to R_2$
$R_4 - 2R_3 \to R_4$
$R_3 \leftrightarrow R_1$

$$\left[\begin{array}{cccc|cccc}
1 & -2 & 6 & 3 & 0 & 0 & 1 & 0 \\
0 & 11 & -32 & -10 & 0 & 1 & -5 & 0 \\
0 & 1 & 3 & 1 & 1 & 0 & 0 & 0 \\
0 & 7 & -10 & -3 & 0 & 0 & -2 & 1
\end{array}\right]$$

$R_1 + 2R_3 \to R_3$
$R_2 - 11R_3 \to R_2$
$R_4 - 7R_3 \to R_3$

$$\left[\begin{array}{cccc|cccc}
1 & 0 & 12 & 5 & 2 & 0 & 1 & 0 \\
0 & 0 & -65 & -21 & -11 & 1 & -5 & 0 \\
0 & 1 & 3 & 1 & 1 & 0 & 0 & 0 \\
0 & 0 & -31 & -10 & -7 & 0 & -2 & 1
\end{array}\right]$$

$3R_2 + 65R_3 \to R_3$
$65R_1 + 12R_2 \to R_1$
$31R_2 - 65R_4 \to R_4$

$$\left[\begin{array}{cccc|cccc}
65 & 0 & 0 & 73 & -2 & 12 & 5 & 0 \\
0 & 0 & -65 & -21 & -11 & 1 & -5 & 0 \\
0 & 65 & 0 & 2 & 32 & 3 & -15 & 0 \\
0 & 0 & 0 & -1 & 114 & 31 & -25 & -65
\end{array}\right]$$

$R_1 + 73R_4 \to R_1$
$R_2 - 21R_4 \to R_2$
$R_3 + 2R_4 \to R_3$
$-R_4 \to R_4$

$$\left[\begin{array}{cccc|cccc}
65 & 0 & 0 & 0 & 8320 & 2275 & -1820 & -4745 \\
0 & 0 & -65 & 0 & -2405 & -650 & 520 & 1365 \\
0 & 65 & 0 & 0 & 260 & 65 & -65 & -130 \\
0 & 0 & 0 & 1 & -114 & -31 & 25 & 65
\end{array}\right]$$

$R_1/65 \to R_1$
$-R_2/65 \to R_2$
$R_3/65 \to R_3$
$R_2 \leftrightarrow R_3$

$$\left[\begin{array}{cccc|cccc}
1 & 0 & 0 & 0 & 128 & 35 & -28 & -73 \\
0 & 1 & 0 & 0 & 4 & 1 & -1 & -2 \\
0 & 0 & 1 & 0 & 37 & 10 & -8 & -21 \\
0 & 0 & 0 & 1 & -114 & -31 & 25 & 65
\end{array}\right].$$

The inverse matrix is $\begin{bmatrix}
128 & 35 & -28 & -73 \\
4 & 1 & -1 & -2 \\
37 & 10 & -8 & -21 \\
-114 & -31 & 25 & 65
\end{bmatrix}.$

25.
$$\begin{bmatrix} 5 & 7 \\ 2 & 3 \end{bmatrix} \cdot \begin{bmatrix} x \\ y \end{bmatrix} = \begin{bmatrix} 2 \\ 5 \end{bmatrix}$$

$$\left(\begin{bmatrix} 3 & -7 \\ -2 & 5 \end{bmatrix} \cdot \begin{bmatrix} 5 & 7 \\ 2 & 3 \end{bmatrix} \right) \cdot \begin{bmatrix} x \\ y \end{bmatrix} = \begin{bmatrix} 3 & -7 \\ -2 & 5 \end{bmatrix} \cdot \begin{bmatrix} 2 \\ 5 \end{bmatrix}$$

$$\begin{bmatrix} 1 & 0 \\ 0 & 1 \end{bmatrix} \cdot \begin{bmatrix} x \\ y \end{bmatrix} = \begin{bmatrix} x \\ y \end{bmatrix} = \begin{bmatrix} -29 \\ 21 \end{bmatrix}$$

The solution set is $\{(-29, 21)\}$

27.
$$\begin{bmatrix} 9 & 13 \\ 2 & 3 \end{bmatrix} \cdot \begin{bmatrix} x \\ y \end{bmatrix} = \begin{bmatrix} -3 \\ 5 \end{bmatrix}$$

$$\left(\begin{bmatrix} 3 & -13 \\ -2 & 9 \end{bmatrix} \cdot \begin{bmatrix} 9 & 13 \\ 2 & 3 \end{bmatrix} \right) \cdot \begin{bmatrix} x \\ y \end{bmatrix} = \begin{bmatrix} 3 & -13 \\ -2 & 9 \end{bmatrix} \cdot \begin{bmatrix} -3 \\ 5 \end{bmatrix}$$

$$\begin{bmatrix} 1 & 0 \\ 0 & 1 \end{bmatrix} \cdot \begin{bmatrix} x \\ y \end{bmatrix} = \begin{bmatrix} x \\ y \end{bmatrix} = \begin{bmatrix} -74 \\ 51 \end{bmatrix}$$

The solution set is $\{(-74, 51)\}$

29.
$$\begin{bmatrix} 3 & 8 \\ 5 & 12 \end{bmatrix} \cdot \begin{bmatrix} x \\ y \end{bmatrix} = \begin{bmatrix} -1 \\ 17 \end{bmatrix}$$

$$\left(\begin{bmatrix} -3 & 2 \\ \frac{5}{4} & \frac{-3}{4} \end{bmatrix} \cdot \begin{bmatrix} 3 & 8 \\ 5 & 12 \end{bmatrix} \right) \cdot \begin{bmatrix} x \\ y \end{bmatrix} = \begin{bmatrix} -3 & 2 \\ \frac{5}{4} & \frac{-3}{4} \end{bmatrix} \cdot \begin{bmatrix} -1 \\ 17 \end{bmatrix}$$

$$\begin{bmatrix} 1 & 0 \\ 0 & 1 \end{bmatrix} \cdot \begin{bmatrix} x \\ y \end{bmatrix} = \begin{bmatrix} x \\ y \end{bmatrix} = \begin{bmatrix} 37 \\ -14 \end{bmatrix}$$

The solution set is $\{(37, -14)\}$

31.
$$\begin{bmatrix} 5 & -5 & 4 \\ 0 & 3 & 1 \\ 2 & -1 & 2 \end{bmatrix} \cdot \begin{bmatrix} x \\ y \\ z \end{bmatrix} = \begin{bmatrix} 3 \\ 1 \\ 5 \end{bmatrix}$$

$$\left(\begin{bmatrix} 7 & 6 & -17 \\ 2 & 2 & -5 \\ -6 & -5 & 15 \end{bmatrix} \cdot \begin{bmatrix} 5 & -5 & 4 \\ 0 & 3 & 1 \\ 2 & -1 & 2 \end{bmatrix}\right) \cdot \begin{bmatrix} x \\ y \\ z \end{bmatrix} = \begin{bmatrix} 7 & 6 & -17 \\ 2 & 2 & -5 \\ -6 & -5 & 15 \end{bmatrix} \cdot \begin{bmatrix} 3 \\ 1 \\ 5 \end{bmatrix}$$

$$\begin{bmatrix} 1 & 0 & 0 \\ 0 & 1 & 0 \\ 0 & 0 & 1 \end{bmatrix} \cdot \begin{bmatrix} x \\ y \\ z \end{bmatrix} = \begin{bmatrix} x \\ y \\ z \end{bmatrix} = \begin{bmatrix} -58 \\ -17 \\ 52 \end{bmatrix}$$

The solution set is $\{(-58, -17, 52)\}$

33.
$$\begin{bmatrix} 7 & 6 & -17 \\ 2 & 2 & -5 \\ -6 & -5 & 15 \end{bmatrix} \cdot \begin{bmatrix} x \\ y \\ z \end{bmatrix} = \begin{bmatrix} 0 \\ 3 \\ 7 \end{bmatrix}$$

$$\begin{bmatrix} x \\ y \\ z \end{bmatrix} = \begin{bmatrix} 5 & -5 & 4 \\ 0 & 3 & 1 \\ 2 & -1 & 2 \end{bmatrix} \cdot \begin{bmatrix} 0 \\ 3 \\ 7 \end{bmatrix} = \begin{bmatrix} 13 \\ 16 \\ 11 \end{bmatrix}$$

The solution set is $\{(13, 16, 11)\}$.

35. The matrix has no inverse and the system has no solutions.

37.
$$\begin{bmatrix} -7 & 1 & 1 \\ 3 & 2 & 3 \\ 5 & 1 & 2 \end{bmatrix} \cdot \begin{bmatrix} x \\ y \\ z \end{bmatrix} = \begin{bmatrix} 3 \\ 1 \\ 5 \end{bmatrix}$$

$$\begin{bmatrix} x \\ y \\ z \end{bmatrix} = \frac{1}{5}\begin{bmatrix} -1 & 1 & -1 \\ -9 & 19 & -24 \\ 7 & -12 & 17 \end{bmatrix} \cdot \begin{bmatrix} 3 \\ 1 \\ 5 \end{bmatrix} = \frac{1}{5}\begin{bmatrix} -7 \\ -128 \\ 94 \end{bmatrix} = \begin{bmatrix} -7/5 \\ -128/5 \\ 94/5 \end{bmatrix}$$

The solution set is $\{(-7/5, -128/5, 94/5)\}$.

39.
$$\begin{bmatrix} 2 & -1 & 7 \\ 3 & 9 & 0 \\ 5 & 0 & 2 \end{bmatrix} \cdot \begin{bmatrix} x \\ y \\ z \end{bmatrix} = \begin{bmatrix} 8 \\ 7 \\ 1 \end{bmatrix}$$

$$\begin{bmatrix} x \\ y \\ z \end{bmatrix} = \frac{1}{273} \begin{bmatrix} -18 & -2 & 63 \\ 6 & 31 & -21 \\ 45 & 5 & -21 \end{bmatrix} \cdot \begin{bmatrix} 8 \\ 7 \\ 1 \end{bmatrix} = \begin{bmatrix} -95/273 \\ 244/273 \\ 374/273 \end{bmatrix}$$

The solution set is $\{(-95/273,\ 244/273,\ 374/273)\}$.

41.
$$\begin{bmatrix} 7 & -8 & -5 & 6 \\ 1 & -4 & 0 & 1 \\ 3 & -11 & 3 & 4 \\ 4 & -6 & -1 & 4 \end{bmatrix} \cdot \begin{bmatrix} w \\ x \\ y \\ z \end{bmatrix} = \begin{bmatrix} 1 \\ 3 \\ 1 \\ 7 \end{bmatrix}$$

$$\begin{bmatrix} w \\ x \\ y \\ z \end{bmatrix} = \begin{bmatrix} 35 & -54 & 34 & -73 \\ 1 & -2 & 1 & -2 \\ 10 & -16 & 10 & -21 \\ -31 & 47 & -30 & 65 \end{bmatrix} \cdot \begin{bmatrix} 1 \\ 3 \\ 1 \\ 7 \end{bmatrix} = \begin{bmatrix} -604 \\ -18 \\ -175 \\ 535 \end{bmatrix}$$

The solution set is $\{(-604,\ -18,\ -175,\ 535)\}$.

43.
$$\begin{bmatrix} 0 & 1 & 3 & 1 \\ 5 & 1 & -2 & 5 \\ 1 & -2 & 6 & 3 \\ 2 & 3 & 2 & 3 \end{bmatrix} \cdot \begin{bmatrix} w \\ x \\ y \\ z \end{bmatrix} = \begin{bmatrix} 8 \\ 3 \\ 5 \\ 7 \end{bmatrix}$$

$$\begin{bmatrix} w \\ x \\ y \\ z \end{bmatrix} = \begin{bmatrix} 128 & 35 & -28 & -73 \\ 4 & 1 & -1 & -2 \\ 37 & 10 & -8 & -21 \\ -114 & -31 & 25 & 65 \end{bmatrix} \cdot \begin{bmatrix} 8 \\ 3 \\ 5 \\ 7 \end{bmatrix} = \begin{bmatrix} 478 \\ 16 \\ 139 \\ -425 \end{bmatrix}$$

The solution set is $\{(478,\ 16,\ 139,\ -425)\}$.

45. Let x be the pounds of kind A which are mixed with y pounds of kind B. At the central store the value of the nuts is

$$3.60x + 4.60y = 200(3.80) = 760, \text{ where } x + y = 200.$$

At the northern store the value is

$$3.60x + 4.60y = 300(4.00) = 1200, \text{ where } x + y = 300.$$

At the southern store, the value is

$$3.60x + 4.60y = 250(4.20) = 1050, \text{ where } x + y = 250.$$

The augmented matrix for the system is

$$\left[\begin{array}{cc|c|c|c} 3.6 & 4.6 & 760 & 1200 & 1050 \\ 1 & 1 & 200 & 300 & 250 \end{array}\right]$$

First we find that the inverse of the matrix $\begin{bmatrix} 3.6 & 4.6 \\ 1 & 1 \end{bmatrix}$ is $\begin{bmatrix} -1 & 4.6 \\ 1 & -3.6 \end{bmatrix}$.

$$\begin{bmatrix} -1 & 4.6 \\ 1 & -3.6 \end{bmatrix}\begin{bmatrix} 760 \\ 200 \end{bmatrix} = \begin{bmatrix} 160 \\ 40 \end{bmatrix}, \quad \begin{bmatrix} -1 & 4.6 \\ 1 & -3.6 \end{bmatrix}\begin{bmatrix} 1200 \\ 300 \end{bmatrix} = \begin{bmatrix} 180 \\ 120 \end{bmatrix} \text{ and}$$

$$\begin{bmatrix} -1 & 4.6 \\ 1 & -3.6 \end{bmatrix}\begin{bmatrix} 1050 \\ 250 \end{bmatrix} = \begin{bmatrix} 100 \\ 150 \end{bmatrix}.$$

Hence, at the central store 160 pounds of kind A are mixed with 40 pounds of kind B. At the northern store 180 pounds of kind A are mixed with 120 pounds of kind B. At the southern store 100 pounds of kind A are mixed with 150 pounds of kind B.

47. Let x be the number of tickets for section A and y for section B, where $x + y = 6,000$.

$$\text{On weekdays, } 7x + 10y = 48,000.$$
$$\text{On Saturday, } 7x + 10y = 57,000.$$
$$\text{On Sunday, } 7x + 10y = 51,000.$$

The augmented matrix for the system is

$$\left[\begin{array}{cc|c|c|c} 7 & 10 & 48000 & 57000 & 51000 \\ 1 & 1 & 6000 & 6000 & 6000 \end{array}\right]$$

The inverse of $\begin{bmatrix} 7 & 10 \\ 1 & 1 \end{bmatrix}$ is $\begin{bmatrix} -1/3 & 10/3 \\ 1/3 & -7/3 \end{bmatrix}$.

$$\begin{bmatrix} -1/3 & 10/3 \\ 1/3 & -7/3 \end{bmatrix}\cdot\begin{bmatrix} 48,000 \\ 6000 \end{bmatrix} = \begin{bmatrix} 4000 \\ 2000 \end{bmatrix}, \quad \begin{bmatrix} -1/3 & 10/3 \\ 1/3 & -7/3 \end{bmatrix}\cdot\begin{bmatrix} 57,000 \\ 6000 \end{bmatrix} = \begin{bmatrix} 1000 \\ 5000 \end{bmatrix} \text{ and}$$

$$\begin{bmatrix} -1/3 & 10/3 \\ 1/3 & -7/3 \end{bmatrix}\cdot\begin{bmatrix} 51,000 \\ 6000 \end{bmatrix} = \begin{bmatrix} 3000 \\ 3000 \end{bmatrix}.$$

On weekdays sell 4,000 seats in section A and 2000 in B.
On Saturday sell 1,000 seats in section A and 5000 in B.
On Sunday sell 3000 seats in section A and 3000 in B.

49.

$\begin{cases}\text{Assembly Time} \\ \text{Testing Time}\end{cases}$

	Monday	Tuesday	Wednesday	Thursday	Friday

$\begin{cases} 5x + 3y = 68 \\ 2x + y = 26 \end{cases}$ $\begin{cases} 5x + 3y = 66 \\ 2x + y = 25 \end{cases}$ $\begin{cases} 5x + 3y = 70 \\ 2x + y = 27 \end{cases}$ $\begin{cases} 5x + 3y = 71 \\ 2x + y = 27 \end{cases}$ $\begin{cases} 5x + 3y = 69 \\ 2x + y = 26 \end{cases}$

First we find the inverse of $\begin{bmatrix} 5 & 3 \\ 2 & 1 \end{bmatrix}$

$\left[\begin{array}{cc|cc} 5 & 3 & 1 & 0 \\ 2 & 1 & 0 & 1 \end{array}\right]$ $2R_1 - 5R_2 \rightarrow R_2$ $\left[\begin{array}{cc|cc} 5 & 3 & 1 & 0 \\ 0 & 1 & 2 & -5 \end{array}\right]$ $R_1 - 3R_2 \rightarrow R_1$

$\left[\begin{array}{cc|cc} 5 & 0 & -5 & 15 \\ 0 & 1 & 2 & -5 \end{array}\right]$ $R_1/5 \rightarrow R_1$ $\left[\begin{array}{cc|cc} 1 & 0 & -1 & 3 \\ 0 & 1 & 2 & -5 \end{array}\right]$.

The inverse matrix is $\begin{bmatrix} -1 & 3 \\ 2 & -5 \end{bmatrix}$.

Solving the systems,

On Monday, $\begin{bmatrix} x \\ y \end{bmatrix} = \begin{bmatrix} -1 & 3 \\ 2 & -5 \end{bmatrix} \cdot \begin{bmatrix} 68 \\ 26 \end{bmatrix} = \begin{bmatrix} 10 \\ 6 \end{bmatrix}$.

On Tuesday, $\begin{bmatrix} x \\ y \end{bmatrix} = \begin{bmatrix} -1 & 3 \\ 2 & -5 \end{bmatrix} \cdot \begin{bmatrix} 66 \\ 25 \end{bmatrix} = \begin{bmatrix} 9 \\ 7 \end{bmatrix}$.

On Wednesday, $\begin{bmatrix} x \\ y \end{bmatrix} = \begin{bmatrix} -1 & 3 \\ 2 & -5 \end{bmatrix} \cdot \begin{bmatrix} 70 \\ 27 \end{bmatrix} = \begin{bmatrix} 11 \\ 5 \end{bmatrix}$.

On Thursday, $\begin{bmatrix} x \\ y \end{bmatrix} = \begin{bmatrix} -1 & 3 \\ 2 & -5 \end{bmatrix} \cdot \begin{bmatrix} 71 \\ 27 \end{bmatrix} = \begin{bmatrix} 10 \\ 7 \end{bmatrix}$.

On Friday, $\begin{bmatrix} x \\ y \end{bmatrix} = \begin{bmatrix} -1 & 3 \\ 2 & -5 \end{bmatrix} \cdot \begin{bmatrix} 69 \\ 26 \end{bmatrix} = \begin{bmatrix} 9 \\ 8 \end{bmatrix}$.

The corresponding solution sets are $\{(10, 6)\}$, $\{(9, 7)\}$, $\{(11, 5)\}$, $\{(10, 7)\}$ and $\{(9, 8)\}$.

51.

	Week 1	Week 2	Week 3	Week 4

$\begin{cases} 7x + 4y = 200 \\ 2x + y = 55 \end{cases}$ $\begin{cases} 7x + 4y = 218 \\ 2x + y = 60 \end{cases}$ $\begin{cases} 7x + 4y = 227 \\ 2x + y = 62 \end{cases}$ $\begin{cases} 7x + 4y = 247 \\ 2x + y = 68 \end{cases}$

where the first equation in each system gives the assembly time needed and the second equation gives the painting time.

First we find the inverse of the coefficient matrix:

$$\begin{bmatrix} 7 & 4 & | & 1 & 0 \\ 2 & 1 & | & 0 & 1 \end{bmatrix} \quad 2R_1 - 7R_2 \to R_2 \quad \begin{bmatrix} 7 & 4 & | & 1 & 0 \\ 0 & 1 & | & 2 & -7 \end{bmatrix} \quad R_1 - 4R_2 \to R_1$$

$$\begin{bmatrix} 7 & 0 & | & -7 & 28 \\ 0 & 1 & | & 2 & -7 \end{bmatrix} \quad R_1/7 \to R_1 \quad \begin{bmatrix} 1 & 0 & | & -1 & 4 \\ 0 & 1 & | & 2 & -7 \end{bmatrix}.$$

The inverse matrix is $\begin{bmatrix} -1 & 4 \\ 2 & -7 \end{bmatrix}$.

The solution to the matrix equation $\begin{bmatrix} 7 & 4 \\ 2 & 1 \end{bmatrix} \cdot \begin{bmatrix} x \\ y \end{bmatrix} = \begin{bmatrix} a \\ b \end{bmatrix}$ is

$$\begin{bmatrix} x \\ y \end{bmatrix} = \begin{bmatrix} -1 & 4 \\ 2 & -7 \end{bmatrix} \cdot \begin{bmatrix} a \\ b \end{bmatrix} = \begin{bmatrix} -a + 4b \\ 2a - 7b \end{bmatrix}.$$ Thus for week 1, the solution is

$x = -200 + 4(55) = 20$, $y = 2(200) - 7(55) = 15$.
For week 2, $x = -218 + 4(60) = 22$ and $y = 2(218) - 7(60) = 16$.
For week 3, $x = -227 + 4(62) = 21$ and $y = 2(227) - 7(62) = 20$.
For week 4, $x = -247 + 4(68) = 25$ and $y = 2(247) - 7(68) = 18$.

53.

Monday	Wednesday	Friday
$\begin{cases} 50x + 35y + 20z = 405 \\ 60x + 40y + 35z = 520 \\ 55x + 45y + 30z = 510 \end{cases}$	$\begin{cases} 50x + 35y + 20z = 470 \\ 60x + 40y + 35z = 585 \\ 55x + 45y + 30z = 580 \end{cases}$	$\begin{cases} 50x + 35y + 20z = 505 \\ 60x + 40y + 35z = 655 \\ 55x + 45y + 30z = 615 \end{cases}$

where the first equation in each system gives the number of units of kind A transported, the second equation gives the number of units of kind B, and the last equation gives the number of units of kind C.

First we find the inverse of the coefficient matrix:

$$\begin{bmatrix} 50 & 35 & 20 & | & 1 & 0 & 0 \\ 60 & 40 & 35 & | & 0 & 1 & 0 \\ 55 & 45 & 30 & | & 0 & 0 & 1 \end{bmatrix} \quad \begin{matrix} 6R_1 - 5R_2 \to R_2 \\ 11R_1 - 10R_3 \to R_3 \end{matrix} \quad \begin{bmatrix} 50 & 35 & 20 & | & 1 & 0 & 0 \\ 0 & 10 & -55 & | & 6 & -5 & 0 \\ 0 & -65 & -80 & | & 11 & 0 & -10 \end{bmatrix}$$

$$\begin{matrix} 2R_1 - 7R_2 \to R_1 \\ 13R_2 + 2R_3 \to R_3 \end{matrix} \quad \begin{bmatrix} 100 & 0 & 425 & | & -40 & 35 & 0 \\ 0 & 10 & -55 & | & 6 & -5 & 0 \\ 0 & 0 & -875 & | & 100 & -65 & -20 \end{bmatrix} \quad \begin{matrix} 35R_1 + 17R_3 \to R_1 \\ 175R_2 - 11R_3 \to R_2 \end{matrix}$$

$$\left[\begin{array}{ccc|ccc} 3500 & 0 & 0 & 300 & 120 & -340 \\ 0 & 1750 & 0 & -50 & -160 & 220 \\ 0 & 0 & -875 & 100 & -65 & -20 \end{array}\right] \begin{array}{l} R_1/3500 \to R_1 \\ R_2/1750 \to R_2 \\ -R_3/875 \to R_3 \end{array}$$

$$\left[\begin{array}{ccc|ccc} 1 & 0 & 0 & 300/3500 & 120/3500 & -340/3500 \\ 0 & 1 & 0 & -50/1750 & -160/1750 & 220/1750 \\ 0 & 0 & 1 & -100/875 & 65/875 & 20/875 \end{array}\right].$$

The inverse matrix is $\dfrac{1}{3500} \left[\begin{array}{ccc} 300 & 120 & -340 \\ -100 & -320 & 440 \\ -400 & 260 & 80 \end{array}\right]$

The solution to the vector equation $\left[\begin{array}{ccc} 50 & 35 & 20 \\ 60 & 40 & 35 \\ 55 & 45 & 30 \end{array}\right] \cdot \left[\begin{array}{c} x \\ y \\ z \end{array}\right] = \left[\begin{array}{c} a \\ b \\ c \end{array}\right]$ is

$$\left[\begin{array}{c} x \\ y \\ z \end{array}\right] = \frac{1}{3500} \left[\begin{array}{ccc} 300 & 120 & -340 \\ -100 & -320 & 440 \\ -400 & 260 & 80 \end{array}\right] \cdot \left[\begin{array}{c} a \\ b \\ c \end{array}\right] = \left[\begin{array}{c} 300a + 120b - 340c \\ -100a - 320b + 440c \\ -400a + 260b + 80c \end{array}\right].$$

Thus the solution for Monday is

$$x = (300(405) + 120(520) - 340(510)) \frac{1}{3500} = 3$$

$$y = \frac{1}{3500} (-100(405) - 320(520) - 440(510)) = 5$$

$$z = \frac{1}{3500} (-400(405) + 260(520) + 80(510)) = 4$$

For Wednesday

$$x = \frac{1}{3500} (300(470) + 120(585) - 340(580)) = 4$$

$$y = \frac{1}{3500} (-100(470) - 320(585) - 440(580)) = 6$$

$$z = \frac{1}{3500} (-400(470) + 260(585) + 80(580)) = 3$$

For Friday

$$x = \frac{1}{3500} (300(505) + 120(655) - 340(615)) = 6$$

$$y = \frac{1}{3500} (-100(505) - 320(655) - 440(615)) = 3$$

$$z = \frac{1}{3500} (-400(505) + 260(655) + 80(615)) = 5$$

55. First we find the inverse of $\begin{bmatrix} 2 & 5 \\ 3 & 7 \end{bmatrix}$.

$$\begin{bmatrix} 2 & 5 & | & 1 & 0 \\ 3 & 7 & | & 0 & 1 \end{bmatrix} \quad 3R_1 - 2R_2 \to R_2 \quad \begin{bmatrix} 2 & 5 & | & 1 & 0 \\ 0 & 1 & | & 3 & -2 \end{bmatrix} \quad R_1 - 5R_2 \to R_2$$

$$\begin{bmatrix} 2 & 0 & | & -14 & 10 \\ 0 & 1 & | & 3 & -2 \end{bmatrix} \quad R_1/2 \to R_1 \quad \begin{bmatrix} 1 & 0 & | & -7 & 5 \\ 0 & 1 & | & 3 & -2 \end{bmatrix}$$

The inverse is $\begin{bmatrix} -7 & 5 \\ 3 & -2 \end{bmatrix}$.

Writing the system as a matrix equation

$$\begin{bmatrix} 2 & 5 \\ 3 & 7 \end{bmatrix} \cdot \begin{bmatrix} x \\ y \end{bmatrix} = \begin{bmatrix} 0 \\ 0 \end{bmatrix}$$

$$\left(\begin{bmatrix} -7 & 5 \\ 3 & -2 \end{bmatrix} \cdot \begin{bmatrix} 2 & 5 \\ 3 & 7 \end{bmatrix} \right) \cdot \begin{bmatrix} x \\ y \end{bmatrix} = \begin{bmatrix} 1 & 0 \\ 0 & 1 \end{bmatrix} \cdot \begin{bmatrix} x \\ y \end{bmatrix} = \begin{bmatrix} x \\ y \end{bmatrix} = \begin{bmatrix} 0 \\ 0 \end{bmatrix}.$$

Hence the only solution to the system is $\{(0, 0)\}$.

EXERCISE SET 3.5   DETERMINANTS

1. $\det \begin{bmatrix} 2 & 3 \\ 5 & 4 \end{bmatrix} = 2(4) - 3(5) = 8 - 15 = -7$

3. $\det \begin{bmatrix} 5 & 7 \\ 4 & 6 \end{bmatrix} = 5(6) - 4(7) = 30 - 28 = 2$

5. $\det \begin{bmatrix} 5 & 6 & 3 \\ 4 & 5 & -1 \\ 1 & 0 & 0 \end{bmatrix} = 5(5)(0) + 6(-1)(1) + 3(4)(0) - 3(5)(1) - 0(-1)(5) - 0(4)(6) = -21$

7. $\det \begin{bmatrix} 4 & -1 & 3 \\ 1 & 5 & -1 \\ 1 & 3 & -2 \end{bmatrix} = -40 + 1 + 9 - 15 + 12 - 2 = -35$

9.

$$\det\begin{bmatrix} 1 & -2 & 1 \\ 2 & -1 & -1 \\ 1 & 1 & 4 \end{bmatrix} = -4 + 2 + 2 + 1 + 1 + 16 = 18$$

11.

$$\det\begin{bmatrix} 4 & -2 & 0 \\ 1 & 2 & 2 \\ 1 & 2 & 3 \end{bmatrix} = 4\begin{vmatrix} 2 & 2 \\ 2 & 3 \end{vmatrix} - \begin{vmatrix} -2 & 0 \\ 2 & 3 \end{vmatrix} + \begin{vmatrix} -2 & 0 \\ 2 & 2 \end{vmatrix} = 4(2) - (-6) - 4 = 10$$

13.

$$\det\begin{bmatrix} 2 & -1 & 2 \\ 4 & 1 & 5 \\ 11 & -8 & 10 \end{bmatrix} = 2\begin{vmatrix} 1 & 5 \\ -8 & 10 \end{vmatrix} - 4\begin{vmatrix} -1 & 2 \\ -8 & 10 \end{vmatrix} + 11\begin{vmatrix} -1 & 2 \\ 1 & 5 \end{vmatrix} = 100 - 24 - 77 = -1$$

15.

$$\begin{bmatrix} 7 & -8 & -5 & 6 \\ 3 & -12 & 0 & 3 \\ 3 & -11 & 3 & 4 \\ 4 & -6 & -1 & 4 \end{bmatrix} \quad \begin{array}{l} \frac{1}{3}R_2 \to R_2 \\ R_1 - 5R_4 \to R_1 \\ R_3 + 3R_4 \to R_3 \end{array} \quad \begin{vmatrix} -13 & 22 & 0 & -14 \\ 1 & -4 & 0 & 1 \\ 15 & -29 & 0 & 16 \\ 4 & -6 & -1 & 4 \end{vmatrix} (3)$$

$$= (-1)^7 (3)(-1)\begin{vmatrix} -13 & 22 & -14 \\ 1 & -4 & 1 \\ 15 & -29 & 16 \end{vmatrix}$$

$$= 3\left\{-13\begin{vmatrix} -4 & 1 \\ -29 & 16 \end{vmatrix} - \begin{vmatrix} 22 & -14 \\ -29 & 16 \end{vmatrix} + 15\begin{vmatrix} 22 & -14 \\ -4 & 1 \end{vmatrix}\right\}$$

$$= 3(455 + 54 - 510) = -3$$

17.

$$\begin{vmatrix} 1 & x & -2 \\ 2 & 3 & 2 \\ 4 & 9 & -2 \end{vmatrix} = \begin{vmatrix} 3 & 2 \\ 9 & -2 \end{vmatrix} - x\begin{vmatrix} 2 & 2 \\ 4 & -2 \end{vmatrix} - 2\begin{vmatrix} 2 & 3 \\ 4 & 9 \end{vmatrix}$$

$$= -24 + 12x - 12 = 0 \text{ if } 12x = 36 \text{ or } x = 3.$$

19.

$$\begin{vmatrix} x & 2x+1 & x \\ x^2 & x+1 & 3 \\ 1 & 0 & 0 \end{vmatrix} = -(2x+1)\begin{vmatrix} x^2 & 3 \\ 1 & 0 \end{vmatrix} + (x+1)\begin{vmatrix} x & x \\ 1 & 0 \end{vmatrix}$$

$$= (2x+1)3 - (x+1)x = -3$$

$$6x + 3 - x^2 - x = -3$$
$$x^2 - 5x - 6 = 0$$
$$(x - 6)(x - 1) = 0$$
$$x = 6 \quad \text{or} \quad x = -1$$

21. $\det A_x = \begin{vmatrix} -2 & 2 \\ 3 & 5 \end{vmatrix} = -10 - 6 = -16$

$\det A_y = \begin{vmatrix} 3 & -2 \\ 7 & 3 \end{vmatrix} = 9 + 14 = 23$

$\det A = \begin{vmatrix} 3 & 2 \\ 7 & 5 \end{vmatrix} = 15 - 14 = 1$

$x = \dfrac{\det A_x}{\det A} = -16 \qquad y = \dfrac{\det A_y}{\det A} = 23$

23. $\det A_x = \begin{vmatrix} 7 & -3 \\ 11 & 7 \end{vmatrix} = 49 + 33 = 82$

$\det A_y = \begin{vmatrix} 5 & 7 \\ 2 & 11 \end{vmatrix} = 55 - 14 = 41$

$\det A = \begin{vmatrix} 5 & -3 \\ 2 & 7 \end{vmatrix} = 35 + 6 = 41$

$x = \dfrac{82}{41} = 2 \qquad y = \dfrac{41}{41} = 1$

25. $\det A_x = \begin{vmatrix} 34 & 13 \\ -19 & -3 \end{vmatrix} = -102 + 247 = 145 \qquad \det A_y = \begin{vmatrix} 1 & 34 \\ 2 & -19 \end{vmatrix} = -19 - 68 = -87$

$\det A = \begin{vmatrix} 1 & 13 \\ 2 & -3 \end{vmatrix} = -3 - 26 = -29$

$x = 145/(-29) = -5, \qquad y = -87/(-29) = 3$

27. $\det A_x = \begin{vmatrix} 15 & 3 \\ -16 & -7 \end{vmatrix} = -105 + 48 = -57$ $\qquad$ $\det A_y = \begin{vmatrix} -4 & 15 \\ 3 & -16 \end{vmatrix} = 64 - 45 = 19$

$\det A = \begin{vmatrix} -4 & 3 \\ 3 & -7 \end{vmatrix} = 28 - 9 = 19$ $\qquad x = -57/19 = -3, \qquad y = \dfrac{19}{19} = 1$

29. $\det A_x = \begin{vmatrix} 64 & 7 \\ 26 & 5 \end{vmatrix} = 320 - 182 = 138$ $\qquad \det A_y = \begin{vmatrix} -5 & 64 \\ 3 & 26 \end{vmatrix} = -130 - 192 = -322$

$\det A = \begin{vmatrix} -5 & 7 \\ 3 & 5 \end{vmatrix} = -25 - 21 = -46$ $\qquad x = -138/46 = -3, \qquad y = 322/46 = 7$

31.
$\det A_x = \begin{vmatrix} -1 & 3 & -5 \\ -3 & -2 & 4 \\ -1 & 4 & -3 \end{vmatrix} = -\begin{vmatrix} -2 & 4 \\ 4 & -3 \end{vmatrix} + 3\begin{vmatrix} 3 & -5 \\ 4 & -3 \end{vmatrix} - \begin{vmatrix} 3 & -5 \\ -2 & 4 \end{vmatrix}$

$= -(-10) + 3(11) - 2 = 41$

$\det A_y = \begin{vmatrix} 2 & -1 & -5 \\ 3 & -3 & 4 \\ 6 & -1 & -3 \end{vmatrix} = 2\begin{vmatrix} -3 & 4 \\ -1 & -3 \end{vmatrix} - 3\begin{vmatrix} -1 & -5 \\ -1 & -3 \end{vmatrix} + 6\begin{vmatrix} -1 & -5 \\ -3 & 4 \end{vmatrix}$

$= 2(13) - 3(-2) + 6(-19) = 26 + 6 - 114 = -82$

$\det A_z = \begin{vmatrix} 2 & 3 & -1 \\ 3 & -2 & -3 \\ 6 & 4 & -1 \end{vmatrix} = 2\begin{vmatrix} -2 & -3 \\ 4 & -1 \end{vmatrix} - 3\begin{vmatrix} 3 & -1 \\ 4 & -1 \end{vmatrix} + 6\begin{vmatrix} 3 & -1 \\ -2 & -3 \end{vmatrix}$

$= 2(14) - 3 + 6(-11) = 28 - 3 - 66 = -41$

$\det A = \begin{vmatrix} 2 & 3 & -5 \\ 3 & -2 & 4 \\ 6 & 4 & -3 \end{vmatrix} = 2\begin{vmatrix} -2 & 4 \\ 4 & -3 \end{vmatrix} - 3\begin{vmatrix} 3 & -5 \\ 4 & -3 \end{vmatrix} + 6\begin{vmatrix} 3 & -5 \\ -2 & 4 \end{vmatrix}$

$= 2(-10) - 3(11) + 6(2) = -20 - 33 + 12 = -41$

$x = -41/41 = -1 \qquad y = 82/41 = 2 \qquad z = 41/41 = 1$

33.

$$\det A_x = \begin{vmatrix} 8 & 3 & 2 \\ 1 & 4 & -3 \\ 11 & -2 & 7 \end{vmatrix} = 8 \begin{vmatrix} 4 & -3 \\ -2 & 7 \end{vmatrix} - \begin{vmatrix} 3 & 2 \\ -2 & 7 \end{vmatrix} + 11 \begin{vmatrix} 3 & 2 \\ 4 & -3 \end{vmatrix}$$

$$= 8(22) - 25 + 11(-17) = -36$$

$$\det A_y = \begin{vmatrix} 2 & 8 & 2 \\ -1 & 1 & -3 \\ 3 & 11 & 7 \end{vmatrix} = 2 \begin{vmatrix} 1 & -3 \\ 11 & 7 \end{vmatrix} + \begin{vmatrix} 8 & 2 \\ 11 & 7 \end{vmatrix} + 3 \begin{vmatrix} 8 & 2 \\ 1 & -3 \end{vmatrix}$$

$$= 2(40) + 34 + 3(-26) = 36$$

$$\det A_z = \begin{vmatrix} 2 & 3 & 8 \\ -1 & 4 & 1 \\ 3 & -2 & 11 \end{vmatrix} = 2 \begin{vmatrix} 4 & 1 \\ -2 & 11 \end{vmatrix} + \begin{vmatrix} 3 & 8 \\ -2 & 11 \end{vmatrix} + 3 \begin{vmatrix} 3 & 8 \\ 4 & 1 \end{vmatrix}$$

$$= 2(46) + 49 + 3(-29) = 54$$

$$\det A = \begin{vmatrix} 2 & 3 & 2 \\ -1 & 4 & -3 \\ 3 & -2 & 7 \end{vmatrix} = 2 \begin{vmatrix} 4 & -3 \\ -2 & 7 \end{vmatrix} + \begin{vmatrix} 3 & 2 \\ -2 & 7 \end{vmatrix} + 3 \begin{vmatrix} 3 & 2 \\ 4 & -3 \end{vmatrix}$$

$$= 2(22) + 25 + 3(-17) = 18$$

$$x = -36/18 = -2 \qquad y = 36/18 = 2 \qquad z = 54/18 = 3$$

35.

$$\det A = \begin{vmatrix} 1 & 1 & 3 \\ 3 & 2 & -1 \\ 11 & 8 & 3 \end{vmatrix} = \begin{vmatrix} 2 & -1 \\ 8 & 3 \end{vmatrix} - 3 \begin{vmatrix} 1 & 3 \\ 8 & 3 \end{vmatrix} + 11 \begin{vmatrix} 1 & 3 \\ 2 & -1 \end{vmatrix}$$

$$= 14 - 3(-21) + 11(-7) = 0.$$

Since this determinant is zero, Cramer's rule cannot be used to solve the system. For an alternate method see the solution to Exercise 5, Section 2.2.

37.

$$\det A = \begin{vmatrix} -2 & 3 & 1 \\ 3 & -2 & 2 \\ 4 & 3 & -1 \end{vmatrix} = -2 \begin{vmatrix} -2 & 2 \\ 3 & -1 \end{vmatrix} - 3 \begin{vmatrix} 3 & 1 \\ 3 & -1 \end{vmatrix} + 4 \begin{vmatrix} 3 & 1 \\ -2 & 2 \end{vmatrix}$$

$$= -2(-4) - 3(-6) + 4(8) = 58$$

$$\det A_x = \begin{vmatrix} 4 & 3 & 1 \\ 14 & -2 & 2 \\ 6 & 3 & -1 \end{vmatrix} = 4\begin{vmatrix} -2 & 2 \\ 3 & -1 \end{vmatrix} - 14\begin{vmatrix} 3 & 1 \\ 3 & -1 \end{vmatrix} + 6\begin{vmatrix} 3 & 1 \\ -2 & 2 \end{vmatrix}$$

$$= 4(-4) - 14(-6) + 6(8) = 116$$

$$\det A_y = \begin{vmatrix} -2 & 4 & 1 \\ 3 & 14 & 2 \\ 4 & 6 & -1 \end{vmatrix} = -2\begin{vmatrix} 14 & 2 \\ 6 & -1 \end{vmatrix} - 3\begin{vmatrix} 4 & 1 \\ 6 & -1 \end{vmatrix} + 4\begin{vmatrix} 4 & 1 \\ 14 & 2 \end{vmatrix}$$

$$= -2(-26) - 3(-10) + 4(-6) = 58$$

$$\det A_z = \begin{vmatrix} -2 & 3 & 4 \\ 3 & -2 & 14 \\ 4 & 3 & 6 \end{vmatrix} = -2\begin{vmatrix} -2 & 14 \\ 3 & 6 \end{vmatrix} - 3\begin{vmatrix} 3 & 4 \\ 3 & 6 \end{vmatrix} + 4\begin{vmatrix} 3 & 4 \\ -2 & 14 \end{vmatrix}$$

$$= -2(-54) - 3(6) + 4(50) = 290$$

$$x = 116/58 = 2 \qquad y = 58/58 = 1 \qquad z = 290/58 = 5.$$

39.
$$\det A = \begin{vmatrix} 3 & 2 & -5 \\ 2 & 3 & -8 \\ -1 & 2 & -3 \end{vmatrix} = 3\begin{vmatrix} 3 & -8 \\ 2 & -3 \end{vmatrix} - 2\begin{vmatrix} 2 & -5 \\ 2 & -3 \end{vmatrix} - \begin{vmatrix} 2 & -5 \\ 3 & -8 \end{vmatrix}$$

$$= 3(7) - 2(4) - (-1) = 14$$

$$\det A_x = \begin{vmatrix} -2 & 2 & -5 \\ 5 & 3 & -8 \\ 8 & 2 & -3 \end{vmatrix} = -2\begin{vmatrix} 3 & -8 \\ 2 & -3 \end{vmatrix} - 5\begin{vmatrix} 2 & -5 \\ 2 & -3 \end{vmatrix} + 8\begin{vmatrix} 2 & -5 \\ 3 & -8 \end{vmatrix}$$

$$= -2(7) - 5(4) + 8(-1) = -42$$

$$\det A_y = \begin{vmatrix} 3 & -2 & -5 \\ 2 & 5 & -8 \\ -1 & 8 & -3 \end{vmatrix} = 3\begin{vmatrix} 5 & -8 \\ 8 & -3 \end{vmatrix} - 2\begin{vmatrix} -2 & -5 \\ 8 & -3 \end{vmatrix} - \begin{vmatrix} -2 & -5 \\ 5 & -8 \end{vmatrix}$$

$$= 3(49) - 2(46) - 41 = 14$$

$$\det A_z = \begin{vmatrix} 3 & 2 & -2 \\ 2 & 3 & 5 \\ -1 & 2 & 8 \end{vmatrix} = 3\begin{vmatrix} 3 & 5 \\ 2 & 8 \end{vmatrix} - 2\begin{vmatrix} 2 & -2 \\ 2 & 8 \end{vmatrix} - \begin{vmatrix} 2 & -2 \\ 3 & 5 \end{vmatrix}$$

$$= 3(14) - 2(20) - 16 = -14$$

$$x = -42/14 = -3 \qquad y = 14/14 = 1 \qquad z = -14/14 = -1.$$

41.

$$\det A = \begin{vmatrix} 1 & 1 & 1 \\ 2 & 1 & 3 \\ 1 & -1 & 5 \end{vmatrix} = \begin{vmatrix} 1 & 3 \\ -1 & 5 \end{vmatrix} - 2 \begin{vmatrix} 1 & 1 \\ -1 & 5 \end{vmatrix} + \begin{vmatrix} 1 & 1 \\ 1 & 3 \end{vmatrix}$$

$$= 8 - 2(6) + 2 = -2$$

$$\det A_x = \begin{vmatrix} 6 & 1 & 1 \\ 12 & 1 & 3 \\ 10 & -1 & 5 \end{vmatrix} = 6 \begin{vmatrix} 1 & 3 \\ -1 & 5 \end{vmatrix} - \begin{vmatrix} 12 & 3 \\ 10 & 5 \end{vmatrix} + \begin{vmatrix} 12 & 1 \\ 10 & -1 \end{vmatrix}$$

$$= 6(8) - 30 - 22 = -4$$

$$\det A_y = \begin{vmatrix} 1 & 6 & 1 \\ 2 & 12 & 3 \\ 1 & 10 & 5 \end{vmatrix} = \begin{vmatrix} 12 & 3 \\ 10 & 5 \end{vmatrix} - 2 \begin{vmatrix} 6 & 1 \\ 10 & 5 \end{vmatrix} + \begin{vmatrix} 6 & 1 \\ 12 & 3 \end{vmatrix}$$

$$= 30 - 2(20) + 6 = -4$$

$$\det A_z = \begin{vmatrix} 1 & 1 & 6 \\ 2 & 1 & 12 \\ 1 & -1 & 10 \end{vmatrix} = \begin{vmatrix} 1 & 12 \\ -1 & 10 \end{vmatrix} - 2 \begin{vmatrix} 1 & 6 \\ -1 & 10 \end{vmatrix} + \begin{vmatrix} 1 & 6 \\ 1 & 12 \end{vmatrix}$$

$$= 22 - 2(16) + 6 = -4$$

$$x = -4/(-2) = 2 \qquad y = 2 \qquad z = 2.$$

43.

$$\det A = \begin{vmatrix} 1 & 1 & 0 \\ 1 & 0 & 1 \\ 3 & -1 & 2 \end{vmatrix} = \begin{vmatrix} 0 & 1 \\ -1 & 2 \end{vmatrix} - \begin{vmatrix} 1 & 1 \\ 3 & 2 \end{vmatrix} = 1 - (-1) = 2$$

$$\det A_x = \begin{vmatrix} -4 & 1 & 0 \\ 1 & 0 & 1 \\ 4 & -1 & 2 \end{vmatrix} = -4 \begin{vmatrix} 0 & 1 \\ -1 & 2 \end{vmatrix} - \begin{vmatrix} 1 & 1 \\ 4 & 2 \end{vmatrix} = -4 - (2-4) = -4 + 2 = -2$$

$$\det A_y = \begin{vmatrix} 1 & -4 & 0 \\ 1 & 1 & 1 \\ 3 & 4 & 2 \end{vmatrix} = \begin{vmatrix} 1 & 1 \\ 4 & 2 \end{vmatrix} + 4\begin{vmatrix} 1 & 1 \\ 3 & 2 \end{vmatrix} = -2 + 4(-1) = -6$$

$$\det A_z = \begin{vmatrix} 1 & 1 & -4 \\ 1 & 0 & 1 \\ 3 & -1 & 4 \end{vmatrix} = -\begin{vmatrix} 1 & -4 \\ -1 & 4 \end{vmatrix} - \begin{vmatrix} 1 & 1 \\ 3 & -1 \end{vmatrix} = 0 - (-4) = 4$$

$x = -2/2 = -1 \qquad y = -6/(2) = -3 \qquad z = 4/2 = 2.$

45.

$$\det A = \begin{vmatrix} 3 & 1 & 1 \\ 1 & 1 & 0 \\ 7 & 3 & 3 \end{vmatrix} = \begin{vmatrix} 1 & 1 \\ 7 & 3 \end{vmatrix} + 3\begin{vmatrix} 3 & 1 \\ 1 & 1 \end{vmatrix} = 3 - 7 + 3(3-1) = 2$$

$$\det A_x = \begin{vmatrix} 0 & 1 & 1 \\ 1 & 1 & 0 \\ 2 & 3 & 3 \end{vmatrix} = -\begin{vmatrix} 1 & 1 \\ 3 & 3 \end{vmatrix} + 2\begin{vmatrix} 1 & 1 \\ 1 & 0 \end{vmatrix} = 0 - 2 = -2$$

$$\det A_y = \begin{vmatrix} 3 & 0 & 1 \\ 1 & 1 & 0 \\ 7 & 2 & 3 \end{vmatrix} = 1\begin{vmatrix} 1 & 1 \\ 7 & 2 \end{vmatrix} + 3\begin{vmatrix} 3 & 0 \\ 1 & 1 \end{vmatrix} = 2 - 7 + 9 = 4$$

$$\det A_z = \begin{vmatrix} 3 & 1 & 0 \\ 1 & 1 & 1 \\ 7 & 3 & 2 \end{vmatrix} = -\begin{vmatrix} 3 & 1 \\ 7 & 3 \end{vmatrix} + 2\begin{vmatrix} 3 & 1 \\ 1 & 1 \end{vmatrix} = -2 + 4 = 2$$

$x = -2/2 = -1 \qquad y = 4/2 = 2 \qquad z = 2/2 = 1.$

For Problems 47-55, the systems of equations are set up in the solutions to Exercises 29-37 of Section 2.2.

47. The system is $\begin{cases} x + y + z = 400 \\ 420x + 525y + 600z = 180,000 \\ x - 3y - 3z = 0 \end{cases}$

$$\det A = \begin{vmatrix} 1 & 1 & 1 \\ 420 & 525 & 600 \\ 1 & -3 & -3 \end{vmatrix} = \begin{vmatrix} 525 & 600 \\ -3 & -3 \end{vmatrix} - 420\begin{vmatrix} 1 & 1 \\ -3 & -3 \end{vmatrix} + \begin{vmatrix} 1 & 1 \\ 525 & 600 \end{vmatrix}$$

$$= 225 + 0 + 75 = 300$$

$$\det A_x = \begin{vmatrix} 400 & 1 & 1 \\ 180{,}000 & 525 & 600 \\ 0 & -3 & -3 \end{vmatrix} = 400 \begin{vmatrix} 525 & 600 \\ -3 & -3 \end{vmatrix} - 180{,}000 \begin{vmatrix} 1 & 1 \\ -3 & -3 \end{vmatrix} = 90{,}000$$

$$\det A_y = \begin{vmatrix} 1 & 400 & 1 \\ 420 & 180{,}000 & 600 \\ 1 & 0 & -3 \end{vmatrix} = 400 \begin{vmatrix} 400 & 1 \\ 180{,}000 & 600 \end{vmatrix} - 3 \begin{vmatrix} 1 & 400 \\ 420 & 180{,}000 \end{vmatrix}$$

$$= 60{,}000 - 36{,}000 = 24{,}000$$

$$\det A_z = \begin{vmatrix} 1 & 1 & 400 \\ 420 & 525 & 180{,}000 \\ 1 & -3 & 0 \end{vmatrix} = \begin{vmatrix} 1 & 400 \\ 525 & 180{,}000 \end{vmatrix} + 3 \begin{vmatrix} 1 & 400 \\ 420 & 180{,}000 \end{vmatrix}$$

$$= -30{,}000 + 36{,}000 = 6000$$

$x = 90{,}000/300 = 300$
$y = 24{,}000/300 = 80$
$z = 6000/300 = 20$

49. The system is $\begin{cases} x + y + z = 50{,}000 \\ x - y - z = 0 \\ 62x + 90y + 110z = 4{,}000{,}000 \end{cases}$

$$\det A = \begin{vmatrix} 1 & 1 & 1 \\ 1 & -1 & -1 \\ 62 & 90 & 110 \end{vmatrix} = 62 \begin{vmatrix} 1 & 1 \\ -1 & -1 \end{vmatrix} - 90 \begin{vmatrix} 1 & 1 \\ 1 & -1 \end{vmatrix} + 110 \begin{vmatrix} 1 & 1 \\ 1 & -1 \end{vmatrix}$$

$$= 180 - 220 = -40$$

$$\det A_x = \begin{vmatrix} 50{,}000 & 1 & 1 \\ 0 & -1 & -1 \\ 4{,}000{,}000 & 90 & 110 \end{vmatrix} = 50{,}000 \begin{vmatrix} -1 & -1 \\ 90 & 110 \end{vmatrix} + 4{,}000{,}000 \begin{vmatrix} 1 & 1 \\ -1 & -1 \end{vmatrix}$$

$$= 50{,}000(-20) = -1{,}000{,}000$$

$$\det A_y = \begin{vmatrix} 1 & 50{,}000 & 1 \\ 1 & 0 & -1 \\ 62 & 4{,}000{,}000 & 110 \end{vmatrix} = - \begin{vmatrix} 50{,}000 & 1 \\ 4{,}000{,}000 & 110 \end{vmatrix} + \begin{vmatrix} 1 & 50{,}000 \\ 62 & 4{,}000{,}000 \end{vmatrix}$$

$$= -600{,}000$$

$$\det A_z = \begin{vmatrix} 1 & 1 & 50{,}000 \\ 1 & -1 & 0 \\ 62 & 90 & 4{,}000{,}000 \end{vmatrix} = 50{,}000 \begin{vmatrix} 1 & -1 \\ 62 & 90 \end{vmatrix} + 4{,}000{,}000 \begin{vmatrix} 1 & 1 \\ 1 & -1 \end{vmatrix}$$

$$= -400{,}000$$

$x = 1{,}000{,}000/40 = 25{,}000$
$y = 600{,}000/40 = 15{,}000$
$z = 400{,}000/40 = 10{,}000$

51. The system is $\begin{cases} 2x + 3y + z = 57.75 \\ x + 2y + 3z = 67.50 \\ 2x + 3y + 3z = 85.25 \end{cases}$

$$\det A = \begin{vmatrix} 2 & 3 & 1 \\ 1 & 2 & 3 \\ 2 & 3 & 3 \end{vmatrix} = 2 \begin{vmatrix} 2 & 3 \\ 3 & 3 \end{vmatrix} - \begin{vmatrix} 3 & 1 \\ 3 & 3 \end{vmatrix} + 2 \begin{vmatrix} 3 & 1 \\ 2 & 3 \end{vmatrix}$$

$$= 2(-3) - 6 + 2(7) = 2$$

$$\det A_x = \begin{vmatrix} 57.75 & 3 & 1 \\ 67.50 & 2 & 3 \\ 85.25 & 3 & 3 \end{vmatrix} = 57.75 \begin{vmatrix} 2 & 3 \\ 3 & 3 \end{vmatrix} - 67.50 \begin{vmatrix} 3 & 1 \\ 3 & 3 \end{vmatrix} + 85.25 \begin{vmatrix} 3 & 1 \\ 2 & 3 \end{vmatrix}$$

$$= 57.75(-3) - 67.50(6) + 85.25(7) = 18.5$$

$$\det A_y = \begin{vmatrix} 2 & 57.75 & 1 \\ 1 & 67.50 & 3 \\ 2 & 85.25 & 3 \end{vmatrix} = -57.75 \begin{vmatrix} 1 & 3 \\ 2 & 3 \end{vmatrix} + 67.50 \begin{vmatrix} 2 & 1 \\ 2 & 3 \end{vmatrix} - 85.25 \begin{vmatrix} 2 & 1 \\ 1 & 3 \end{vmatrix}$$

$$= -57.75(-3) + 67.50(4) - 85.25(5) = 17$$

$$\det A_z = \begin{vmatrix} 2 & 3 & 57.75 \\ 1 & 2 & 67.50 \\ 2 & 3 & 85.25 \end{vmatrix} = 57.75 \begin{vmatrix} 1 & 2 \\ 2 & 3 \end{vmatrix} - 67.50 \begin{vmatrix} 2 & 3 \\ 2 & 3 \end{vmatrix} + 85.25 \begin{vmatrix} 2 & 3 \\ 1 & 2 \end{vmatrix}$$

$$= 57.75(-1) + 0 + 85.25(1) = 27.5$$

$x = 18.5/2 = 9.25$
$y = 17/2 = 8.50$
$z = 27.5/2 = 13.75$

53. The system is $\begin{cases} 3x + 2y + 5z = 51 \\ 5x + 6y + 4z = 67 \\ 4x + 7y + 6z = 77 \end{cases}$

$$\det A = \begin{vmatrix} 3 & 2 & 5 \\ 5 & 6 & 4 \\ 4 & 7 & 6 \end{vmatrix} = 3\begin{vmatrix} 6 & 4 \\ 7 & 6 \end{vmatrix} - 5\begin{vmatrix} 2 & 5 \\ 7 & 6 \end{vmatrix} + 4\begin{vmatrix} 2 & 5 \\ 6 & 4 \end{vmatrix}$$

$$= 3(8) - 5(-23) + 4(-22) = 51$$

$$\det A_x = \begin{vmatrix} 51 & 2 & 5 \\ 67 & 6 & 4 \\ 77 & 7 & 6 \end{vmatrix} = 51\begin{vmatrix} 6 & 4 \\ 7 & 6 \end{vmatrix} - 67\begin{vmatrix} 2 & 5 \\ 7 & 6 \end{vmatrix} + 77\begin{vmatrix} 2 & 5 \\ 6 & 4 \end{vmatrix}$$

$$= 51(8) - 67(-23) + 77(-22) = 255$$

$$\det A_y = \begin{vmatrix} 3 & 51 & 5 \\ 5 & 67 & 4 \\ 4 & 77 & 6 \end{vmatrix} = -51\begin{vmatrix} 5 & 4 \\ 4 & 6 \end{vmatrix} + 67\begin{vmatrix} 3 & 5 \\ 4 & 6 \end{vmatrix} - 77\begin{vmatrix} 3 & 5 \\ 5 & 4 \end{vmatrix}$$

$$= -51(14) + 67(-2) - 77(-13) = 153$$

$$\det A_z = \begin{vmatrix} 3 & 2 & 51 \\ 5 & 6 & 67 \\ 4 & 7 & 77 \end{vmatrix} = 51\begin{vmatrix} 5 & 6 \\ 4 & 7 \end{vmatrix} - 67\begin{vmatrix} 3 & 2 \\ 4 & 7 \end{vmatrix} + 77\begin{vmatrix} 3 & 2 \\ 5 & 6 \end{vmatrix}$$

$$= 51(11) - 67(13) + 77(8) = 306$$

$x = 255/51 = 5$
$y = 153/51 = 3$
$z = 306/51 = 6$

55. The system is $\begin{cases} x - y \quad\;\;\; = -4 \\ -x \quad\;\; + z = -24 \\ x - 1.5y + z = -10 \end{cases}$

$$\det A = \begin{vmatrix} 1 & -1 & 0 \\ -1 & 0 & 1 \\ 1 & -1.5 & 1 \end{vmatrix} = \begin{vmatrix} 0 & 1 \\ -1.5 & 1 \end{vmatrix} + \begin{vmatrix} -1 & 1 \\ 1 & 1 \end{vmatrix} = 1.5 - 2 = -.5$$

(continued)

(problem #55, continued)

$$\det A_x = \begin{vmatrix} -4 & -1 & 0 \\ -24 & 0 & 1 \\ -10 & -1.5 & 1 \end{vmatrix} = -4 \begin{vmatrix} 0 & 1 \\ -1.5 & 1 \end{vmatrix} + \begin{vmatrix} -24 & 1 \\ -10 & 1 \end{vmatrix} = -6 - 14 = -20$$

$$\det A_y = \begin{vmatrix} 1 & -4 & 0 \\ -1 & -24 & 1 \\ 1 & -10 & 1 \end{vmatrix} = \begin{vmatrix} -24 & 1 \\ -10 & 1 \end{vmatrix} + 4 \begin{vmatrix} -1 & 1 \\ 1 & 1 \end{vmatrix} = -14 - 8 = -22$$

$$\det A_z = \begin{vmatrix} 1 & -1 & -4 \\ -1 & 0 & -24 \\ 1 & -1.5 & -10 \end{vmatrix} = \begin{vmatrix} -1 & -4 \\ -1.5 & -10 \end{vmatrix} + 24 \begin{vmatrix} 1 & -1 \\ 1 & -1.5 \end{vmatrix} = 4 - 12 = -8$$

$x = 20/.5 = 40$
$y = 22/.5 = 44$
$z = 8/.5 = 16$

57. $\det \begin{bmatrix} 1 & x & y \\ 1 & 2 & 3 \\ 1 & 5 & 6 \end{bmatrix} = 0$ is the equation of the line in determinant form.

Simplifying, $1 \begin{vmatrix} 2 & 3 \\ 5 & 6 \end{vmatrix} - \begin{vmatrix} x & y \\ 5 & 6 \end{vmatrix} + \begin{vmatrix} x & y \\ 2 & 3 \end{vmatrix} = 0$

$-3 - (6x - 5y) + 3x - 2y = 0$
$-3x + 3y = 3 \qquad \text{or} \qquad -x + y = 1$

59. The equation of the line in determinant form is $\det \begin{bmatrix} 1 & x & y \\ 1 & 1 & 7 \\ 1 & 3 & -6 \end{bmatrix} = 0$

Simplifying, $\begin{vmatrix} 1 & 7 \\ 3 & -6 \end{vmatrix} - \begin{vmatrix} x & y \\ 3 & -6 \end{vmatrix} + \begin{vmatrix} x & y \\ 1 & 7 \end{vmatrix} = 0$

$-27 - (-6x - 3y) + (7x - y) = 0$
$-27 + 6x + 3y + 7x - y = 0$
$13x + 2y = 27$

61. The equation of the line in determinant form is $\det \begin{bmatrix} 1 & x & y \\ 1 & 4 & -2 \\ 1 & -5 & -3 \end{bmatrix} = 0$

Simplifying, $\begin{vmatrix} 4 & -2 \\ -5 & -3 \end{vmatrix} - \begin{vmatrix} x & y \\ -5 & -3 \end{vmatrix} + \begin{vmatrix} x & y \\ 4 & -2 \end{vmatrix} = 0$

$-22 - (-3x - 5y) + (-2x) - 4y = 0$

$x - 9y = 22$

63. If $A = \begin{bmatrix} a_{11} & a_{12} \\ a_{21} & a_{22} \end{bmatrix}$ and det $A = 3$,

$$\det B = \det 5A = \begin{bmatrix} 5a_{11} & 5a_{12} \\ 5a_{21} & 5a_{22} \end{bmatrix} = 5 \cdot 5 \det A = 25(3) = 75.$$

65. $\det B = \det(-3A) = (-3)^4 \det A = 81(-2) = -162$

67. $\det B = \det A^T = \det A = -6$

69. $\begin{vmatrix} 1 & 3 & 1 \\ 1 & 5 & -2 \\ 1 & -3 & -2 \end{vmatrix} = \begin{vmatrix} 5 & -2 \\ -3 & -2 \end{vmatrix} - \begin{vmatrix} 3 & 1 \\ -3 & -2 \end{vmatrix} + \begin{vmatrix} 3 & 1 \\ 5 & -2 \end{vmatrix} = -16 + 3 - 11 = -24$

The area of the triangle is $\dfrac{|-24|}{2} = 12$ square units.

71. $\begin{vmatrix} 1 & 3 & 7 \\ 1 & 3 & -1 \\ 1 & -4 & -6 \end{vmatrix} = \begin{vmatrix} 3 & -1 \\ -4 & -6 \end{vmatrix} - \begin{vmatrix} 3 & 7 \\ -4 & -6 \end{vmatrix} + \begin{vmatrix} 3 & 7 \\ 3 & -1 \end{vmatrix} = -22 - 10 - 24 = -56$

The area of the triangle is $\dfrac{|-56|}{2} = 28$ square units.

EXERCISE SET 3.6 INVERSE OF A SQUARE MATRIX USING ADJOINTS

1. $\det \begin{bmatrix} 1 & 0 \\ 1 & 0 \end{bmatrix} = 0 - 0 = 0.$ Therefore $\begin{bmatrix} 1 & 0 \\ 1 & 0 \end{bmatrix}$ has no inverse.

3. $\det \begin{bmatrix} 3 & 2 \\ 9 & 6 \end{bmatrix} = 18 - 18 = 0.$ Hence the matrix has no inverse.

5.

$$\det \begin{bmatrix} 5 & 7 \\ 2 & 3 \end{bmatrix} = 15 - 14 = 1.$$

$$c_{11} = \det [3] = 3 \quad c_{12} = -\det [2] = -2 \quad c_{21} = -\det [7] = -7 \quad c_{22} = \det [5] = 5$$

$$\text{adj} \begin{bmatrix} 5 & 7 \\ 2 & 3 \end{bmatrix} = \begin{bmatrix} 3 & -2 \\ -7 & 5 \end{bmatrix}$$

$$\begin{bmatrix} 5 & 7 \\ 2 & 3 \end{bmatrix}^{-1} = \frac{1}{1} \begin{bmatrix} 3 & -7 \\ -2 & 5 \end{bmatrix} = \begin{bmatrix} 3 & -7 \\ -2 & 5 \end{bmatrix}.$$

7.

$$\det \begin{bmatrix} 9 & 13 \\ 2 & 3 \end{bmatrix} = 27 - 26 = 1 \quad \text{adj} \begin{bmatrix} 9 & 13 \\ 2 & 3 \end{bmatrix} = \begin{bmatrix} 3 & -2 \\ -13 & 9 \end{bmatrix}$$

$$\begin{bmatrix} 9 & 13 \\ 2 & 3 \end{bmatrix}^{-1} = \frac{1}{1} \begin{bmatrix} 3 & -13 \\ -2 & 9 \end{bmatrix} = \begin{bmatrix} 3 & -13 \\ -2 & 9 \end{bmatrix}.$$

9.

$$\det \begin{bmatrix} 3 & 8 \\ 5 & 12 \end{bmatrix} = -4 \quad \text{adj} \begin{bmatrix} 3 & 8 \\ 5 & 12 \end{bmatrix} = \begin{bmatrix} 12 & -5 \\ -8 & 3 \end{bmatrix}$$

$$\begin{bmatrix} 3 & 8 \\ 5 & 12 \end{bmatrix}^{-1} = \begin{bmatrix} 12/(-4) & -8/(-4) \\ -5/(-4) & 3/(-4) \end{bmatrix} = \begin{bmatrix} -3 & 2 \\ 5/4 & -3/4 \end{bmatrix}.$$

11.

$$\det \begin{bmatrix} 5 & -5 & 4 \\ 0 & 3 & 1 \\ 2 & -1 & 2 \end{bmatrix} = 5 \begin{vmatrix} 3 & 1 \\ -1 & 2 \end{vmatrix} + 2 \begin{vmatrix} -5 & 4 \\ 3 & 1 \end{vmatrix} = 5(7) + 2(-17) = 1$$

$$c_{11} = \begin{vmatrix} 3 & 1 \\ -1 & 2 \end{vmatrix} = 7 \quad c_{12} = - \begin{vmatrix} 0 & 1 \\ 2 & 2 \end{vmatrix} = 2 \quad c_{13} = \begin{vmatrix} 0 & 3 \\ 2 & -1 \end{vmatrix} = -6$$

$$c_{21} = - \begin{vmatrix} -5 & 4 \\ -1 & 2 \end{vmatrix} = -(-10 + 4) = 6 \quad c_{22} = \begin{vmatrix} 5 & 4 \\ 2 & 2 \end{vmatrix} = 2 \quad c_{23} = - \begin{vmatrix} 5 & -5 \\ 2 & -1 \end{vmatrix} = -5$$

$$c_{31} = \begin{vmatrix} -5 & 4 \\ 3 & 1 \end{vmatrix} = -17 \quad c_{32} = - \begin{vmatrix} 5 & 4 \\ 0 & 1 \end{vmatrix} = -5 \quad c_{33} = \begin{vmatrix} 5 & -5 \\ 0 & 3 \end{vmatrix} = 15$$

$$\text{adj} \begin{bmatrix} 5 & -5 & 4 \\ 0 & 3 & 1 \\ 2 & -1 & 2 \end{bmatrix} = \begin{bmatrix} 7 & 2 & -6 \\ 6 & 2 & -5 \\ -17 & -5 & 15 \end{bmatrix}$$

$$\begin{bmatrix} 5 & -5 & 4 \\ 0 & 3 & 1 \\ 2 & -1 & 2 \end{bmatrix}^{-1} = \begin{bmatrix} 7 & 6 & -17 \\ 2 & 2 & -5 \\ -6 & -5 & 15 \end{bmatrix}$$

13.

$$\text{Let } A = \begin{bmatrix} 7 & 6 & -17 \\ 2 & 2 & -5 \\ -6 & -5 & 15 \end{bmatrix}.$$

$$c_{11} = \begin{vmatrix} 2 & -5 \\ -5 & 15 \end{vmatrix} = 5 \qquad c_{12} = - \begin{vmatrix} 2 & -5 \\ -6 & 15 \end{vmatrix} = 0 \qquad c_{13} = \begin{vmatrix} 2 & 2 \\ -6 & -5 \end{vmatrix} = 2$$

$$\det A = 7(5) + 6(0) - 17(2) = 1$$

$$c_{21} = - \begin{vmatrix} 6 & -17 \\ -5 & 15 \end{vmatrix} = -5 \qquad c_{22} = \begin{vmatrix} 7 & -17 \\ -6 & 15 \end{vmatrix} = 3 \qquad c_{23} = - \begin{vmatrix} 7 & 6 \\ -6 & -5 \end{vmatrix} = -1$$

$$c_{31} = \begin{vmatrix} 6 & -17 \\ 2 & -5 \end{vmatrix} = 4 \qquad c_{32} = - \begin{vmatrix} 7 & -17 \\ 2 & -5 \end{vmatrix} = 1 \qquad c_{33} = \begin{vmatrix} 7 & 6 \\ 2 & 2 \end{vmatrix} = 2$$

$$A^{-1} = \begin{bmatrix} 5 & -5 & 4 \\ 0 & 3 & 1 \\ 2 & -1 & 2 \end{bmatrix}$$

15.

$$\text{Let } A = \begin{bmatrix} 1 & 0 & -2 \\ 2 & 5 & -6 \\ 4 & 5 & -10 \end{bmatrix}$$

$$c_{11} = \begin{vmatrix} 5 & -6 \\ 5 & -10 \end{vmatrix} = -20 \qquad c_{12} = - \begin{vmatrix} 2 & -6 \\ 4 & -10 \end{vmatrix} = -4 \qquad c_{13} = \begin{vmatrix} 2 & 5 \\ 4 & 5 \end{vmatrix} = -10$$

$$\det A = 1(-20) + (-2)(-10) = 0$$

Since the determinant is zero, A^{-1} does not exist.

17. Let $A = \begin{bmatrix} -7 & 1 & 1 \\ 3 & 2 & 3 \\ 5 & 1 & 2 \end{bmatrix}$

$c_{11} = \begin{vmatrix} 2 & 3 \\ 1 & 2 \end{vmatrix} = 1 \qquad c_{12} = -\begin{vmatrix} 3 & 3 \\ 5 & 2 \end{vmatrix} = 9 \qquad c_{13} = \begin{vmatrix} 3 & 2 \\ 5 & 1 \end{vmatrix} = -7$

$\det A = -7(1) + 1(9) - 1(7) = -5$

$c_{21} = -\begin{vmatrix} 1 & 1 \\ 1 & 2 \end{vmatrix} = -1 \qquad c_{22} = \begin{vmatrix} -7 & 1 \\ 5 & 2 \end{vmatrix} = -19 \qquad c_{23} = -\begin{vmatrix} -7 & 1 \\ 5 & 1 \end{vmatrix} = 12$

$c_{31} = \begin{vmatrix} 1 & 1 \\ 2 & 3 \end{vmatrix} = 1 \qquad c_{32} = -\begin{vmatrix} -7 & 1 \\ 3 & 3 \end{vmatrix} = -24 \qquad c_{33} = \begin{vmatrix} -7 & 1 \\ 3 & 2 \end{vmatrix} = -17$

$A^{-1} = \dfrac{-1}{5} \begin{bmatrix} 1 & -1 & 1 \\ 9 & -19 & -24 \\ -7 & 12 & -17 \end{bmatrix} = .2 \begin{bmatrix} -1 & 1 & -1 \\ -9 & 19 & -24 \\ 7 & -12 & 17 \end{bmatrix}$

19. Let $A = \begin{bmatrix} 2 & -1 & 7 \\ 3 & 9 & 0 \\ 5 & 0 & 2 \end{bmatrix}$

$c_{11} = \begin{vmatrix} 9 & 0 \\ 0 & 2 \end{vmatrix} = 18 \qquad c_{12} = -\begin{vmatrix} 3 & 0 \\ 5 & 2 \end{vmatrix} = -6 \qquad c_{13} = \begin{vmatrix} 3 & 9 \\ 5 & 0 \end{vmatrix} = -45$

$\det A = 18(2) + 1(6) - 7(45) = -273$

$c_{21} = -\begin{vmatrix} -1 & 7 \\ 0 & 2 \end{vmatrix} = 2 \qquad c_{22} = \begin{vmatrix} 2 & 7 \\ 5 & 2 \end{vmatrix} = -31 \qquad c_{23} = -\begin{vmatrix} 2 & -1 \\ 5 & 0 \end{vmatrix} = -5$

$c_{31} = \begin{vmatrix} -1 & 7 \\ 9 & 0 \end{vmatrix} = -63 \qquad c_{32} = -\begin{vmatrix} 2 & 7 \\ 3 & 0 \end{vmatrix} = 21 \qquad c_{33} = \begin{vmatrix} 2 & -1 \\ 3 & 9 \end{vmatrix} = 2$

$\text{adj } A = \begin{bmatrix} 18 & -6 & -45 \\ 2 & -31 & -5 \\ -63 & 21 & 21 \end{bmatrix} \qquad A^{-1} = \dfrac{1}{273} \begin{bmatrix} -18 & -2 & 63 \\ 6 & 31 & -21 \\ 45 & 5 & -21 \end{bmatrix}$

21. Let $A = \begin{bmatrix} 7 & -8 & -5 & 6 \\ 1 & -4 & 0 & 1 \\ 3 & -11 & 3 & 4 \\ 4 & -6 & -1 & 4 \end{bmatrix}$

$$c_{11} = \begin{vmatrix} -4 & 0 & 1 \\ -11 & 3 & 4 \\ -6 & -1 & 4 \end{vmatrix} = -35 \qquad c_{12} = -\begin{vmatrix} 1 & 0 & 1 \\ 3 & 3 & 4 \\ 4 & -1 & 4 \end{vmatrix} = -1$$

$$c_{13} = \begin{vmatrix} 1 & -4 & 1 \\ 3 & 3 & 4 \\ 4 & -6 & 4 \end{vmatrix} = -10 \qquad c_{14} = -\begin{vmatrix} 1 & -4 & 0 \\ 3 & -11 & 3 \\ 4 & -6 & -1 \end{vmatrix} = 31$$

$\det A = 7(-35) - 1(-8) - 5(-10) + 6(31) = -1$

$$c_{21} = -\begin{vmatrix} -8 & -5 & 6 \\ -11 & 3 & 4 \\ -6 & -1 & 4 \end{vmatrix} = 54 \qquad c_{22} = \begin{vmatrix} 7 & -5 & 6 \\ 3 & 3 & 4 \\ 4 & -1 & 4 \end{vmatrix} = 2$$

$$c_{23} = -\begin{vmatrix} 7 & -8 & 6 \\ 3 & -11 & 4 \\ 4 & -1 & 4 \end{vmatrix} = 16 \qquad c_{24} = \begin{vmatrix} 7 & -8 & -5 \\ 3 & -11 & 3 \\ 4 & -6 & -1 \end{vmatrix} = -47$$

$$c_{31} = \begin{vmatrix} -8 & -5 & 6 \\ -4 & 0 & 1 \\ -6 & -1 & 4 \end{vmatrix} = -34 \qquad c_{32} = -\begin{vmatrix} 7 & -5 & 6 \\ 1 & 0 & 1 \\ 4 & -1 & 4 \end{vmatrix} = -1$$

$$c_{33} = \begin{vmatrix} 7 & -8 & 6 \\ 1 & -4 & 1 \\ 4 & -6 & 4 \end{vmatrix} = -10 \qquad c_{34} = -\begin{vmatrix} 7 & -8 & -5 \\ 1 & -4 & 0 \\ 4 & -6 & -1 \end{vmatrix} = 30$$

$$c_{41} = -\begin{vmatrix} -8 & -5 & 6 \\ -4 & 0 & 1 \\ -11 & 3 & 4 \end{vmatrix} = 73 \qquad c_{42} = \begin{vmatrix} 7 & -5 & 6 \\ 1 & 0 & 1 \\ 3 & 3 & 4 \end{vmatrix} = 2$$

$$c_{43} = -\begin{vmatrix} 7 & -8 & 6 \\ 1 & -4 & 1 \\ 3 & -11 & 4 \end{vmatrix} = 21 \qquad c_{44} = \begin{vmatrix} 7 & -8 & -5 \\ 1 & -4 & 0 \\ 3 & -11 & 3 \end{vmatrix} = -65$$

$$\text{adj } A = \begin{bmatrix} -35 & -1 & -10 & 31 \\ 54 & 2 & 16 & -47 \\ -34 & -1 & -10 & 30 \\ 73 & 2 & 21 & -65 \end{bmatrix} \qquad A^{-1} = \begin{bmatrix} 35 & -54 & 34 & -73 \\ 1 & -2 & 1 & -2 \\ 10 & -16 & 10 & -21 \\ -31 & 47 & -30 & 65 \end{bmatrix}$$

25. Let A be a matrix in which the ith row and the jth row are identical. Let B be the matrix formed by interchanging the ith and j^{th} rows of A. Since corresponding rows of A and B are all identical, A = B, and det A = det B. However, det B = −det A because two rows of A were interchanged to obtain B. Hence det A = −det A and det A = 0.

27. Let A be an $n \times n$ matrix, A(adj A) is the $n \times n$ matrix with the ij^{th} element

$*$ $a_{i1}C_{j1} + a_{i2}C_{j2} + \cdots + a_{in}C_{jn}.$

(adj A)A is the $n \times n$ matrix with ij^{th} element

$**$ $C_{i1}a_{j1} + C_{i2}a_{j2} + \cdots + C_{in}a_{jn}.$

If $i \neq j$, both these two sums are zero by Exercise 26, and if $i = j$, the sums are identical. Hence A(adj A) = (adj A)A.

(det A)I_n is the $n \times n$ matrix with ij^{th} element 0 if $i \neq j$. If $i = j$, the ij^{th} element is det A, which by definition is given by $a_{i1}C_{i1} + a_{i2}C_{i2} + \cdots + a_{in}C_{in}.$ The latter sum is identical to sums $(*)$ and $(**)$ when $i = j$. Hence A(adj A) = (adj A)A = (det A)I_n as the corresponding elements of these three $n \times n$ matrices are all identical.

EXERCISE SET 3.7 APPLICATIONS IN ECONOMICS AND CRYPTOGRAPHY: INPUT-OUTPUT ANALYSIS AND CODED MESSAGES

1. $I_2 - B = \begin{bmatrix} 1 & 0 \\ 0 & 1 \end{bmatrix} - \begin{bmatrix} .25 & .50 \\ .40 & .40 \end{bmatrix} = \begin{bmatrix} .75 & -.50 \\ -.40 & .60 \end{bmatrix}$

Finding $(I_2 - B)^{-1}$,

$\left[\begin{array}{cc|cc} .75 & -.5 & 1 & 0 \\ -.40 & .6 & 0 & 1 \end{array} \right] \quad 6R_1 + 5R_2 \to R_1 \quad \left[\begin{array}{cc|cc} 2.5 & 0 & 6 & 5 \\ -.4 & .6 & 0 & 1 \end{array} \right]$

$.8R_1 + 5R_2 \to R_2 \quad \left[\begin{array}{cc|cc} 2.5 & 0 & 6 & 5 \\ 0 & 3 & 4.8 & 9 \end{array} \right] \quad \begin{array}{c} R_1/2.5 \to R_1 \\ R_3/3 \to R_3 \end{array} \quad \left[\begin{array}{cc} 2.4 & 2 \\ 1.6 & 3 \end{array} \right]$

$(I_2 - B)^{-1} = \begin{bmatrix} 2.4 & 2 \\ 1.6 & 3 \end{bmatrix}$

(a) $X = \begin{bmatrix} 2.4 & 2 \\ 1.6 & 3 \end{bmatrix} \cdot \begin{bmatrix} 240 \\ 360 \end{bmatrix} = \begin{bmatrix} 1296 \\ 1464 \end{bmatrix}$

Industry I should produce 1296 units and Industry II should produce 1464 units.

(b) $X = \begin{bmatrix} 2.4 & 2 \\ 1.6 & 3 \end{bmatrix} \cdot \begin{bmatrix} 320 \\ 450 \end{bmatrix} = \begin{bmatrix} 1668 \\ 1862 \end{bmatrix}$

Industry I should produce 1668 units and Industry II should produce 1862 units.

(c) $\quad X = \begin{bmatrix} 2.4 & 2 \\ 1.6 & 3 \end{bmatrix} \cdot \begin{bmatrix} 380 \\ 290 \end{bmatrix} = \begin{bmatrix} 1492 \\ 1478 \end{bmatrix}$

Industry I should produce 1492 units and Industry II should produce 1478 units.

3. $\quad I - B = \begin{bmatrix} .40 & -.40 \\ -.35 & .60 \end{bmatrix}$. Finding the inverse of $I - B$, adj $(I - B) = \begin{bmatrix} .65 & .35 \\ .40 & .40 \end{bmatrix}^T$.

det $(I - B) = -.6(.4) - .4(.35) = .1$

$(I - B)^{-1} = \dfrac{1}{.1} \begin{bmatrix} .60 & .40 \\ .35 & .40 \end{bmatrix} = \begin{bmatrix} 6 & 4 \\ 3.5 & 4 \end{bmatrix}$

(a) $\quad \begin{bmatrix} 6 & 4 \\ 3.5 & 4 \end{bmatrix} \cdot \begin{bmatrix} 580 \\ 490 \end{bmatrix} = \begin{bmatrix} 5440 \\ 3990 \end{bmatrix}$

Industry I should produce 5440 units, Industry II should produce 3990 units.

(b) $\quad \begin{bmatrix} 6 & 4 \\ 3.5 & 4 \end{bmatrix} \cdot \begin{bmatrix} 620 \\ 600 \end{bmatrix} = \begin{bmatrix} 6120 \\ 4570 \end{bmatrix}$

Industry I should produce 6120 units, Industry II should produce 4570 units.

(c) $\quad \begin{bmatrix} 6 & 4 \\ 3.5 & 4 \end{bmatrix} \cdot \begin{bmatrix} 710 \\ 850 \end{bmatrix} = \begin{bmatrix} 7660 \\ 5885 \end{bmatrix}$

Industry I should produce 7660 units, Industry II should produce 5885 units.

5. $\quad B = \begin{bmatrix} .20 & .46 \\ .50 & .40 \end{bmatrix} \qquad I - B = \begin{bmatrix} .80 & -.46 \\ -.50 & .60 \end{bmatrix} \qquad \det(I - B) = .25$

adj$(I - B) = \begin{bmatrix} .60 & .50 \\ .46 & .80 \end{bmatrix}^T = \begin{bmatrix} .60 & .46 \\ .50 & .80 \end{bmatrix} \qquad (I - B)^{-1} = \begin{bmatrix} 2.4 & 1.84 \\ 2 & 3.2 \end{bmatrix}$

(a) $\quad \begin{bmatrix} 2.4 & 1.84 \\ 2 & 3.2 \end{bmatrix} \cdot \begin{bmatrix} 780 \\ 820 \end{bmatrix} = \begin{bmatrix} 3380.8 \\ 4184 \end{bmatrix}$

Industry I should produce 3380.8 units, Industry II should produce 4184 units.

(b) $\quad \begin{bmatrix} 2.4 & 1.84 \\ 2 & 3.2 \end{bmatrix} \cdot \begin{bmatrix} 690 \\ 850 \end{bmatrix} = \begin{bmatrix} 3220 \\ 4100 \end{bmatrix}$

Industry I should produce 3220 units, Industry II should produce 4100 units.

(c) $\begin{bmatrix} 2.4 & 1.84 \\ 2 & 3.2 \end{bmatrix} \cdot \begin{bmatrix} 950 \\ 670 \end{bmatrix} = \begin{bmatrix} 3512.8 \\ 4044 \end{bmatrix}$

Industry I should produce 3512.8 units, Industry II should produce 4044 units.

7. $B = \begin{bmatrix} .30 & .25 & .30 \\ .20 & .40 & .20 \\ .40 & .30 & .20 \end{bmatrix}$ $I - B = \begin{bmatrix} .70 & -.25 & -.30 \\ -.20 & .60 & -.20 \\ -.40 & -.30 & .80 \end{bmatrix}$ $\det (I - B) = .144$

$\text{adj } (I - B) = \begin{bmatrix} .42 & .24 & .30 \\ .29 & .44 & .31 \\ .23 & .20 & .37 \end{bmatrix}$ $(I - B)^{-1} = \frac{1}{.144} \begin{bmatrix} .42 & .29 & .23 \\ .24 & .44 & .20 \\ .30 & .31 & .37 \end{bmatrix}$

(a) $\frac{1}{.144} \begin{bmatrix} .42 & .29 & .23 \\ .24 & .44 & .20 \\ .30 & .31 & .37 \end{bmatrix} \cdot \begin{bmatrix} 550 \\ 800 \\ 470 \end{bmatrix} = \begin{bmatrix} 571.1 \\ 578 \\ 586.9 \end{bmatrix} \frac{1}{.144} = \begin{bmatrix} 3966 \\ 4014 \\ 4076 \end{bmatrix}$

Industries I, II and III must produce approximately 3966, 4743 and 4076 units respectively.

(b) $\frac{1}{.144} \begin{bmatrix} .42 & .29 & .23 \\ .24 & .44 & .20 \\ .30 & .31 & .37 \end{bmatrix} \cdot \begin{bmatrix} 650 \\ 745 \\ 475 \end{bmatrix} = \begin{bmatrix} 4155 \\ 4019 \\ 4178 \end{bmatrix}$

Industries I, II and III must produce approximately 4155, 4019 and 4178 units respectively.

(c) $\frac{1}{.144} \begin{bmatrix} .42 & .29 & .23 \\ .24 & .44 & .20 \\ .30 & .31 & .37 \end{bmatrix} \cdot \begin{bmatrix} 470 \\ 750 \\ 840 \end{bmatrix} = \begin{bmatrix} 4223 \\ 4242 \\ 4752 \end{bmatrix}$

Industries I, II and III must produce approximately 4223, 4242 and 4752 units respectively.

9. $B = \begin{bmatrix} .1 & .4 & .3 \\ .5 & .25 & .2 \\ .2 & .3 & .4 \end{bmatrix}$ $I - B = \begin{bmatrix} .9 & -.4 & -.3 \\ -.5 & .75 & -.2 \\ -.2 & -.3 & .6 \end{bmatrix}$ $\det (I - B) = .125 = \frac{1}{8}$

$$\text{adj } (I - B) = \begin{bmatrix} .39 & .33 & .305 \\ .34 & .48 & .33 \\ .3 & .35 & .475 \end{bmatrix} \qquad (I - B)^{-1} = 8 \begin{bmatrix} .39 & .33 & .305 \\ .34 & .48 & .33 \\ .3 & .35 & .475 \end{bmatrix}$$

(a) $\quad 8 \begin{bmatrix} .39 & .33 & .305 \\ .34 & .48 & .33 \\ .3 & .35 & .475 \end{bmatrix} \cdot \begin{bmatrix} 380 \\ 540 \\ 730 \end{bmatrix} = \begin{bmatrix} 4392 \\ 5034 \\ 5198 \end{bmatrix}$

The units required of Industries I, II and III are 4392, 5034 and 5198 respectively.

(b) $\quad 8 \begin{bmatrix} .39 & .33 & .305 \\ .34 & .48 & .33 \\ .3 & .35 & .475 \end{bmatrix} \cdot \begin{bmatrix} 670 \\ 890 \\ 500 \end{bmatrix} = \begin{bmatrix} 5660 \\ 6560 \\ 6000 \end{bmatrix}$

5660, 6560, 6000 units are required of the respective industries.

(c) $\quad 8 \begin{bmatrix} .39 & .33 & .305 \\ .34 & .48 & .33 \\ .3 & .35 & .475 \end{bmatrix} \cdot \begin{bmatrix} 600 \\ 785 \\ 435 \end{bmatrix} = \begin{bmatrix} 5006 \\ 5795 \\ 5291 \end{bmatrix}$

5006, 5795 and 5291 units are required of the respective industries.

11. The technological coefficient matrix is $B = \begin{bmatrix} 455/1820 & 1050/2100 \\ 728/1820 & 840/2100 \end{bmatrix} = \begin{bmatrix} .25 & .5 \\ .4 & .4 \end{bmatrix}$

$$I_3 - B = \begin{bmatrix} .75 & -.5 \\ -.4 & .6 \end{bmatrix} \quad \text{adj}(I_3 - B) = \begin{bmatrix} .6 & .4 \\ .5 & .75 \end{bmatrix} \qquad \det(I_3 - B) = .25$$

$$(I_3 - B)^{-1} = 4 \begin{bmatrix} .6 & .5 \\ .4 & .75 \end{bmatrix} = \begin{bmatrix} 2.4 & 2 \\ 1.6 & 3 \end{bmatrix}$$

(a) $\begin{bmatrix} 2.4 & 2 \\ 1.6 & 3 \end{bmatrix} \cdot \begin{bmatrix} 450 \\ 640 \end{bmatrix} = \begin{bmatrix} 2360 \\ 2640 \end{bmatrix}$

Industries I and II must produce 2,360 and 2,640 millions of dollars per year, respectively.

(b) $\begin{bmatrix} 2.4 & 2 \\ 1.6 & 3 \end{bmatrix} \cdot \begin{bmatrix} 650 \\ 780 \end{bmatrix} = \begin{bmatrix} 3120 \\ 3380 \end{bmatrix}$

The industries must produce 3,120 and 3,380 millions of dollars per year, respectively.

(c) $\begin{bmatrix} 2.4 & 2 \\ 1.6 & 3 \end{bmatrix} \cdot \begin{bmatrix} 760 \\ 860 \end{bmatrix} = \begin{bmatrix} 3544 \\ 3796 \end{bmatrix}$

The industries must produce 3544 and 3796 millions of dollars per year, respectively.

13. The technological matrix is $B = \begin{bmatrix} 1512/2520 & 720/1800 \\ 882/2520 & 720/1800 \end{bmatrix} = \begin{bmatrix} .6 & .4 \\ .35 & .4 \end{bmatrix}$

$I - B = \begin{bmatrix} .4 & -.4 \\ -.35 & .6 \end{bmatrix}$ $\det (I - B) = .1$

$\text{adj } (I - B) = \begin{bmatrix} .6 & .35 \\ .4 & .4 \end{bmatrix}$ $(I - B)^{-1} = \begin{bmatrix} 6 & 4 \\ 3.5 & 4 \end{bmatrix}$

(a) $\begin{bmatrix} 6 & 4 \\ 3.5 & 4 \end{bmatrix} \cdot \begin{bmatrix} 250 \\ 340 \end{bmatrix} = \begin{bmatrix} 2860 \\ 2235 \end{bmatrix}$

Industries I and II must produce 2860 and 2235 millions of dollars per year respectively.

(b) $\begin{bmatrix} 6 & 4 \\ 3.5 & 4 \end{bmatrix} \cdot \begin{bmatrix} 380 \\ 580 \end{bmatrix} = \begin{bmatrix} 4600 \\ 3650 \end{bmatrix}$

The industries must produce 4600 and 3650 millions of dollars per year respectively.

(c) $\begin{bmatrix} 6 & 4 \\ 3.5 & 4 \end{bmatrix} \cdot \begin{bmatrix} 460 \\ 420 \end{bmatrix} = \begin{bmatrix} 4440 \\ 3290 \end{bmatrix}$

The industries must produce 4440 and 3290 millions of dollars per year respectively.

15. The technological matrix is

$B = \begin{bmatrix} .2 & .46 \\ .5 & .4 \end{bmatrix}$ $I - B = \begin{bmatrix} .8 & -.46 \\ -.5 & .6 \end{bmatrix}$ $\det (I - B) = .25$

$\text{adj } (I - B) = \begin{bmatrix} .6 & .5 \\ .46 & .8 \end{bmatrix}$ $(I - B)^{-1} = \begin{bmatrix} 2.4 & 1.84 \\ 2 & 3.2 \end{bmatrix}$

(a) $\begin{bmatrix} 2.4 & 1.84 \\ 2 & 3.2 \end{bmatrix} \cdot \begin{bmatrix} 850 \\ 740 \end{bmatrix} = \begin{bmatrix} 3401.6 \\ 4068 \end{bmatrix}$

Industries I and II must produce 3401.6 and 4068 millions of dollars per year respectively.

(b) $\begin{bmatrix} 2.4 & 1.84 \\ 2 & 3.2 \end{bmatrix} \cdot \begin{bmatrix} 920 \\ 710 \end{bmatrix} = \begin{bmatrix} 3514.4 \\ 4112 \end{bmatrix}$

The industries must produce 3514.4 and 4112 millions of dollars per year respectively.

(c) $\begin{bmatrix} 2.4 & 1.84 \\ 2 & 3.2 \end{bmatrix} \cdot \begin{bmatrix} 670 \\ 840 \end{bmatrix} = \begin{bmatrix} 3153.6 \\ 4028 \end{bmatrix}$

The industries must produce 3153.6 and 4028 millions of dollars per year respectively.

17. The technological matrix is B = $\begin{bmatrix} 1560/5200 & 1025/4100 & 1380/4600 \\ 1040/5200 & 1640/4100 & 920/4600 \\ 2080/5200 & 1230/4100 & 920/4600 \end{bmatrix}$

$= \begin{bmatrix} .3 & .25 & .3 \\ .2 & .4 & .2 \\ .4 & .3 & .2 \end{bmatrix}$ I − B = $\begin{bmatrix} .7 & -.25 & -.3 \\ -.2 & .6 & -.2 \\ -.4 & -.3 & .8 \end{bmatrix}$ det (I − B) = .144

adj (I − B) = $\begin{bmatrix} .42 & .24 & .30 \\ .29 & .44 & .31 \\ .23 & .20 & .37 \end{bmatrix}$ $(I - B)^{-1} = \frac{1}{.144} \begin{bmatrix} .42 & .29 & .23 \\ .24 & .44 & .20 \\ .30 & .31 & .37 \end{bmatrix}$

(a) $\frac{1}{.144} \begin{bmatrix} .42 & .29 & .23 \\ .24 & .44 & .20 \\ .30 & .31 & .37 \end{bmatrix} \cdot \begin{bmatrix} 1358.5 \\ 425 \\ 481 \end{bmatrix} = \begin{bmatrix} 5586.46 \\ 4230.83 \\ 4981.04 \end{bmatrix}$

Industries I, II and III must produce approximately 5286, 4231 and 4981 million dollars per year respectively.

(b) $\frac{1}{.144} \begin{bmatrix} .42 & .29 & .23 \\ .24 & .44 & .20 \\ .30 & .31 & .37 \end{bmatrix} \cdot \begin{bmatrix} 950 \\ 850 \\ 720 \end{bmatrix} = \begin{bmatrix} 5633 \\ 5181 \\ 5659 \end{bmatrix}$

The industries must produce approximately 5633, 5181 and 5659 million dollars per year, respectively.

19. The technological matrix is B = $\begin{bmatrix} 420/4200 & 1760/4400 & 1140/3800 \\ 2100/4200 & 1100/4400 & 760/3800 \\ 840/4200 & 1320/4400 & 1520/3800 \end{bmatrix}$

$$= \begin{bmatrix} .1 & .4 & .3 \\ .5 & .25 & .2 \\ .2 & .3 & .4 \end{bmatrix} \qquad I - B = \begin{bmatrix} .9 & -.4 & -.3 \\ -.5 & .75 & -.2 \\ -.2 & -.3 & .6 \end{bmatrix} \qquad \det(I - B) = \tfrac{1}{8}$$

$$\text{adj}(I - B) = \begin{bmatrix} .39 & .34 & .30 \\ .33 & .48 & .35 \\ .305 & .33 & .475 \end{bmatrix} \qquad (I-B)^{-1} = \begin{bmatrix} 3.12 & 2.64 & 2.44 \\ 2.72 & 3.84 & 2.64 \\ 2.40 & 2.80 & 3.80 \end{bmatrix}$$

(a)
$$\begin{bmatrix} 3.12 & 2.64 & 2.44 \\ 2.72 & 3.84 & 2.64 \\ 2.40 & 2.80 & 3.80 \end{bmatrix} \cdot \begin{bmatrix} 1012 \\ 352 \\ 216 \end{bmatrix} = \begin{bmatrix} 4614 \\ 4675 \\ 4235 \end{bmatrix}$$

In millions of dollars per year, Industries I, II and III must produce approximately 4614, 4675 and 4235, respectively.

(b)
$$\begin{bmatrix} 3.12 & 2.64 & 2.44 \\ 2.72 & 3.84 & 2.64 \\ 2.40 & 2.80 & 3.80 \end{bmatrix} \cdot \begin{bmatrix} 750 \\ 1250 \\ 480 \end{bmatrix} = \begin{bmatrix} 6811.2 \\ 8107.2 \\ 7124 \end{bmatrix}$$

The industries must produce approximately 6811, 8107 and 7124 millions of dollars per year, respectively.

21. $A = \begin{bmatrix} 1 & 0 & 2 \\ 2 & 1 & 4 \\ 3 & 5 & 5 \end{bmatrix} \qquad \det A = \begin{vmatrix} 1 & 4 \\ 5 & 5 \end{vmatrix} + 2 \begin{vmatrix} 2 & 1 \\ 3 & 5 \end{vmatrix} = -15 + 2(7) = -1$

$$\text{adj}(A) = \begin{bmatrix} -15 & 10 & -2 \\ 2 & -1 & 0 \\ 7 & -5 & 1 \end{bmatrix}^{T} \qquad A^{-1} = \begin{bmatrix} 15 & -10 & 2 \\ -2 & 1 & 0 \\ -7 & 5 & -1 \end{bmatrix}$$

$$\begin{bmatrix} 15 & -10 & 2 \\ -2 & 1 & 0 \\ -7 & 5 & -1 \end{bmatrix} \cdot \begin{bmatrix} 46 \\ 110 \\ 207 \end{bmatrix} = \begin{bmatrix} 4 \\ 18 \\ 21 \end{bmatrix} \longleftrightarrow \begin{bmatrix} D \\ R \\ U \end{bmatrix}$$

$$\begin{bmatrix} 15 & -10 & 2 \\ -2 & 1 & 0 \\ -7 & 5 & -1 \end{bmatrix} \cdot \begin{bmatrix} 61 \\ 141 \\ 251 \end{bmatrix} = \begin{bmatrix} 7 \\ 19 \\ 27 \end{bmatrix} \longleftrightarrow \begin{bmatrix} G \\ S \\ \end{bmatrix}$$

$$\begin{bmatrix} 15 & -10 & 2 \\ -2 & 1 & 0 \\ -7 & 5 & -1 \end{bmatrix} \cdot \begin{bmatrix} 35 \\ 79 \\ 138 \end{bmatrix} = \begin{bmatrix} 11 \\ 9 \\ 12 \end{bmatrix} \longleftrightarrow \begin{bmatrix} K \\ I \\ L \end{bmatrix}$$

$$\begin{bmatrix} 15 & -10 & 2 \\ -2 & 1 & 0 \\ -7 & 5 & -1 \end{bmatrix} \cdot \begin{bmatrix} 66 \\ 159 \\ 306 \end{bmatrix} = \begin{bmatrix} 12 \\ 27 \\ 27 \end{bmatrix} \longleftrightarrow \begin{bmatrix} L \\ \\ \end{bmatrix}$$

The message is "DRUGS KILL "

23. Let $A = \begin{bmatrix} 1 & 0 & 1 \\ 0 & 1 & 1 \\ 2 & 0 & 3 \end{bmatrix}$. Find A^{-1}.

$$\left[\begin{array}{ccc|ccc} 1 & 0 & 1 & 1 & 0 & 0 \\ 0 & 1 & 1 & 0 & 1 & 0 \\ 2 & 0 & 3 & 0 & 0 & 1 \end{array}\right] \quad 2R_1 - R_3 \to R_3 \quad \left[\begin{array}{ccc|ccc} 1 & 0 & 1 & 1 & 0 & 0 \\ 0 & 1 & 1 & 0 & 1 & 0 \\ 0 & 0 & -1 & 2 & 0 & -1 \end{array}\right]$$

$$\begin{array}{c} R_1 + R_3 \to R_1 \\ R_2 + R_3 \to R_2 \\ -R_3 \to R_3 \end{array} \left[\begin{array}{ccc|ccc} 1 & 0 & 0 & 3 & 0 & -1 \\ 0 & 1 & 0 & 2 & 1 & -1 \\ 0 & 0 & 1 & -2 & 0 & 1 \end{array}\right] \quad A^{-1} = \begin{bmatrix} 3 & 0 & -1 \\ 2 & 1 & -1 \\ -2 & 0 & 1 \end{bmatrix}$$

$$A^{-1} \cdot \begin{bmatrix} x \\ y \\ z \end{bmatrix} = \begin{bmatrix} 3 & 0 & -1 \\ 2 & 1 & -1 \\ -2 & 0 & 1 \end{bmatrix} \cdot \begin{bmatrix} x \\ y \\ z \end{bmatrix} = \begin{bmatrix} 3x - z \\ 2x + y - z \\ -2x + z \end{bmatrix}$$

$$A^{-1} \cdot \begin{bmatrix} 40 \\ 41 \\ 101 \end{bmatrix} = \begin{bmatrix} 120 - 101 \\ 80 + 41 - 101 \\ -80 + 101 \end{bmatrix} = \begin{bmatrix} 19 \\ 20 \\ 21 \end{bmatrix} \longleftrightarrow \begin{bmatrix} S \\ T \\ U \end{bmatrix}$$

$$A^{-1} \cdot \begin{bmatrix} 31 \\ 52 \\ 89 \end{bmatrix} = \begin{bmatrix} 93 - 89 \\ 62 + 52 - 89 \\ -62 + 89 \end{bmatrix} = \begin{bmatrix} 4 \\ 25 \\ 27 \end{bmatrix} \longleftrightarrow \begin{bmatrix} D \\ Y \\ \end{bmatrix}$$

$$A^{-1} \cdot \begin{bmatrix} 10 \\ 27 \\ 25 \end{bmatrix} = \begin{bmatrix} 5 \\ 22 \\ 5 \end{bmatrix} \longleftrightarrow \begin{bmatrix} E \\ V \\ E \end{bmatrix} \qquad A^{-1} \cdot \begin{bmatrix} 45 \\ 52 \\ 117 \end{bmatrix} = \begin{bmatrix} 18 \\ 25 \\ 27 \end{bmatrix} \longleftrightarrow \begin{bmatrix} R \\ Y \\ \end{bmatrix}$$

$$A^{-1} \cdot \begin{bmatrix} 29 \\ 26 \\ 83 \end{bmatrix} = \begin{bmatrix} 4 \\ 1 \\ 25 \end{bmatrix} \longleftrightarrow \begin{bmatrix} D \\ A \\ Y \end{bmatrix}$$ The message is "STUDY EVERY DAY"

25. The message to be sent is

5, 4, 21, 3, 1, 20, 9, 15, 14, 27, 9, 19, 27, 1, 27, 16, 18, 9, 22, 9, 12, 5, 7, 5.

$$A \cdot \begin{bmatrix} x \\ y \\ z \end{bmatrix} = \begin{bmatrix} 1 & 0 & 0 \\ 3 & 1 & 2 \\ 2 & 0 & 1 \end{bmatrix} \cdot \begin{bmatrix} x \\ y \\ z \end{bmatrix} = \begin{bmatrix} x \\ 3x + y - 2z \\ 2x + z \end{bmatrix}$$

$$A \cdot \begin{bmatrix} 5 \\ 4 \\ 21 \end{bmatrix} = \begin{bmatrix} 5 \\ 61 \\ 31 \end{bmatrix} \qquad A \cdot \begin{bmatrix} 3 \\ 1 \\ 20 \end{bmatrix} = \begin{bmatrix} 3 \\ 50 \\ 26 \end{bmatrix} \qquad A \cdot \begin{bmatrix} 9 \\ 15 \\ 14 \end{bmatrix} = \begin{bmatrix} 9 \\ 70 \\ 32 \end{bmatrix}$$

$$A \cdot \begin{bmatrix} 27 \\ 9 \\ 19 \end{bmatrix} = \begin{bmatrix} 27 \\ 128 \\ 73 \end{bmatrix} \qquad A \cdot \begin{bmatrix} 27 \\ 1 \\ 27 \end{bmatrix} = \begin{bmatrix} 27 \\ 136 \\ 81 \end{bmatrix} \qquad A \cdot \begin{bmatrix} 16 \\ 18 \\ 9 \end{bmatrix} = \begin{bmatrix} 16 \\ 84 \\ 41 \end{bmatrix}$$

$$A \cdot \begin{bmatrix} 22 \\ 9 \\ 12 \end{bmatrix} = \begin{bmatrix} 22 \\ 99 \\ 56 \end{bmatrix} \qquad A \cdot \begin{bmatrix} 5 \\ 7 \\ 5 \end{bmatrix} = \begin{bmatrix} 5 \\ 32 \\ 15 \end{bmatrix}$$

The encoded message is 5, 61, 31, 3, 50, 26, 9, 70, 32, 27, 128, 73, 27, 136, 81, 16, 84, 41, 22, 99, 56, 5, 32, 15.

EXERCISE SET 3.8 MARKOV CHAINS

1. $\frac{1}{3} + \frac{1}{6} + \frac{1}{6} \neq 1$. Not a state vector.

3. $-.3$ is negative. Not a state vector.

5. A state vector. All components are positive and their sum is 1.

7. Not a transition matrix since the sums of the row entries are not both 1.

9. Not a transition matrix since one of the entries is negative.

11. A transition matrix since all entries are positive and the 3 sums of the row entries are each 1.

13. Let $T = \begin{bmatrix} .2 & .8 \\ 1 & 0 \end{bmatrix}$ $T^2 = \begin{bmatrix} .2 & .8 \\ 1 & 0 \end{bmatrix} \cdot \begin{bmatrix} .2 & .8 \\ 1 & 0 \end{bmatrix} = \begin{bmatrix} .84 & .16 \\ .2 & .8 \end{bmatrix}$

Since all entries of T^2 are positive, T is regular.

15. Let $T = \begin{bmatrix} 0 & 1 \\ 1 & 0 \end{bmatrix}.$ $T^2 = \begin{bmatrix} 0 & 1 \\ 1 & 0 \end{bmatrix} \cdot \begin{bmatrix} 0 & 1 \\ 1 & 0 \end{bmatrix} = \begin{bmatrix} 1 & 0 \\ 0 & 1 \end{bmatrix} = I_2$

$T^3 = T \cdot \begin{bmatrix} 1 & 0 \\ 0 & 1 \end{bmatrix} = T$ and for all $n > 2$, $T^n = T$. Hence T is not regular.

17. $T^2 = \begin{bmatrix} .2 & .4 & .4 \\ .1 & .7 & .2 \\ .6 & 0 & .4 \end{bmatrix} \cdot \begin{bmatrix} .2 & .4 & .4 \\ .1 & .7 & .2 \\ .6 & 0 & .4 \end{bmatrix} = \begin{bmatrix} .32 & .36 & .32 \\ .21 & .53 & .26 \\ .36 & .24 & .40 \end{bmatrix}.$

Since all entries of T^2 are positive, T is regular.

19. The matrix is regular if there is a vector $\begin{bmatrix} x & y \end{bmatrix}$ such that

$\begin{bmatrix} x & y \end{bmatrix} \cdot \begin{bmatrix} .3 & .7 \\ .3 & .7 \end{bmatrix} = \begin{bmatrix} x & y \end{bmatrix}.$ $\begin{bmatrix} x & y \end{bmatrix}$ is the steady-state vector.

This leads to the system

$\begin{cases} .3x + .3y = x \\ .7x + .7y = y \\ \quad x + y = 1 \end{cases}$ $\begin{cases} .7x - .3y = 0 \\ \quad x + y = 1 \end{cases}$ $\begin{cases} .7x - .3y = 0 \\ .7x + .7y = .7 \end{cases}$ $\begin{cases} 1.0y = .7 \\ x + y = 1 \end{cases}$ $\begin{cases} y = .7 \\ x = .3 \end{cases}$

The steady-state vector is $\begin{bmatrix} .3 & .7 \end{bmatrix}.$

21. $\begin{bmatrix} x & y \end{bmatrix} \cdot \begin{bmatrix} .78 & .22 \\ .18 & .82 \end{bmatrix} = \begin{bmatrix} x & y \end{bmatrix}$ if

$\begin{cases} .78x + .18y = x \\ .22x + .82y = y \\ \quad x + y = 1 \end{cases}$ $\begin{cases} .22x - .18y = 0 \\ \quad x + y = 1 \end{cases}$ $\begin{cases} 22x - 18y = 0 \\ 22x + 22y = 22 \end{cases}$ $\begin{cases} 40y = 22 \\ x + y = 1 \end{cases}$ $\begin{cases} y = .55 \\ x = .45 \end{cases}$

The steady-state vector is $\begin{bmatrix} .45 & .55 \end{bmatrix}.$

23. If $\begin{bmatrix} x & y & z \end{bmatrix}$ is a steady-state vector, $x + y + z = 1$

and $\begin{bmatrix} x & y & z \end{bmatrix} \begin{bmatrix} .15 & .1 & .75 \\ .25 & .3 & .45 \\ .3125 & .55 & .1375 \end{bmatrix} = \begin{bmatrix} x & y & z \end{bmatrix}$ or $\begin{array}{l} .15x + .25y + .3125z = x \\ .1x + .3y + .55z = y \\ .75x + .45y + .1375z = z. \end{array}$

This leads to the system $\begin{cases} .85x - .25y - .3125z = 0 \\ -.1x + .7y - .55z = 0 \\ -.75x - .45y + .8625z = 0 \\ x + y + z = 1. \end{cases}$

Solving,

$$\left[\begin{array}{ccc|c} 8500 & -2500 & -3125 & 0 \\ -10 & 70 & -55 & 0 \\ -7500 & -4500 & 8625 & 0 \\ 1 & 1 & 1 & 1 \end{array}\right] \quad \begin{array}{l} R_1/5 \to R_4 \\ R_2/5 \to R_2 \\ R_3/5 \to R_3 \\ R_4 \to R_1 \end{array}$$

$$\left[\begin{array}{ccc|c} 1 & 1 & 1 & 1 \\ -2 & 14 & -11 & 0 \\ -1500 & -900 & 1725 & 0 \\ 1700 & -500 & -625 & 0 \end{array}\right] \quad \begin{array}{l} 2R_1 + R_2 \to R_2 \\ 1500R_1 + R_3 \to R_3 \\ 1700R_1 - R_4 \to R_4 \end{array}$$

$$\left[\begin{array}{ccc|c} 1 & 1 & 1 & 1 \\ 0 & 16 & -9 & 2 \\ 0 & 600 & 3225 & 1500 \\ 0 & 2200 & 2325 & 1700 \end{array}\right] \quad \begin{array}{l} 11R_3 - 3R_4 \to R_4 \\ 16R_1 - R_2 \to R_1 \\ 150R_2 - 4R_3 \to R_4 \end{array}$$

$$\left[\begin{array}{ccc|c} 16 & 0 & 25 & 14 \\ 0 & 16 & -9 & 2 \\ 0 & 0 & -14,250 & -5700 \\ 0 & 0 & 28,500 & 11,400 \end{array}\right] \quad \begin{array}{l} 2R_3 + R_4 \to R_4 \\ -R_3/14,250 \to R_4 \end{array}$$

$$\left[\begin{array}{ccc|c} 16 & 0 & 25 & 14 \\ 0 & 16 & -9 & 2 \\ 0 & 0 & 1 & .4 \\ 0 & 0 & 0 & 0 \end{array}\right] \quad \begin{array}{l} R_1 - 25R_3 \to R_1 \\ R_2 + 9R_3 \to R_2 \end{array} \quad \left[\begin{array}{ccc|c} 16 & 0 & 0 & 4 \\ 0 & 16 & 0 & 5.6 \\ 0 & 0 & 1 & .4 \\ 0 & 0 & 0 & 0 \end{array}\right]$$

$$\begin{array}{l} R_1/16 \to R_1 \\ R_2/16 \to R_2 \end{array} \quad \left[\begin{array}{ccc|c} 1 & 0 & 0 & .25 \\ 0 & 1 & 0 & .35 \\ 0 & 0 & 1 & .4 \\ 0 & 0 & 0 & 0 \end{array}\right]. \text{ The steady-state vector is } \begin{bmatrix} .25 & .35 & .4 \end{bmatrix}.$$

25. $v_1 = \begin{bmatrix} .15 & .85 \end{bmatrix} \cdot \begin{bmatrix} .91 & .09 \\ .03 & .97 \end{bmatrix} = \begin{bmatrix} .162 & .838 \end{bmatrix}$

$v_2 = \begin{bmatrix} .162 & .838 \end{bmatrix} \cdot \begin{bmatrix} .91 & .09 \\ .03 & .97 \end{bmatrix} = \begin{bmatrix} .17256 & .82744 \end{bmatrix}$

Guess the steady-state vector is $\begin{bmatrix} .2 & .8 \end{bmatrix}$.

Checking, $\begin{bmatrix} .2 & .8 \end{bmatrix} \cdot \begin{bmatrix} .91 & .09 \\ .03 & .97 \end{bmatrix} = \begin{bmatrix} .206 & .794 \end{bmatrix}$. $\begin{bmatrix} .2 & .8 \end{bmatrix}$ does not check.

$v_3 = \begin{bmatrix} .17256 & .82744 \end{bmatrix} \cdot \begin{bmatrix} .91 & .09 \\ .03 & .97 \end{bmatrix} = \begin{bmatrix} .1818528 & .8181472 \end{bmatrix}$.

Guess $\begin{bmatrix} .25 & .75 \end{bmatrix}$ is the steady-state vector.

$\begin{bmatrix} .25 & .75 \end{bmatrix} \cdot \begin{bmatrix} .91 & .09 \\ .03 & .97 \end{bmatrix} = \begin{bmatrix} .25 & .75 \end{bmatrix}$. It checks!

27. $v_1 = \begin{bmatrix} .25 & .75 \end{bmatrix} \cdot \begin{bmatrix} .585 & .415 \\ .085 & .915 \end{bmatrix} = \begin{bmatrix} .21 & .79 \end{bmatrix}$

$v_2 = \begin{bmatrix} .21 & .79 \end{bmatrix} \cdot \begin{bmatrix} .585 & .415 \\ .085 & .915 \end{bmatrix} = \begin{bmatrix} .19 & .81 \end{bmatrix}$

$v_3 = \begin{bmatrix} .19 & .81 \end{bmatrix} \cdot \begin{bmatrix} .585 & .415 \\ .085 & .915 \end{bmatrix} = \begin{bmatrix} .18 & .82 \end{bmatrix}$

$v_4 = \begin{bmatrix} .18 & .82 \end{bmatrix} \cdot \begin{bmatrix} .585 & .415 \\ .085 & .915 \end{bmatrix} = \begin{bmatrix} .175 & .825 \end{bmatrix}$

Guess the steady-state vector is $\begin{bmatrix} .17 & .83 \end{bmatrix}$.

Checking, $\begin{bmatrix} .17 & .83 \end{bmatrix} \cdot \begin{bmatrix} .585 & .415 \\ .085 & .915 \end{bmatrix} = \begin{bmatrix} .17 & .83 \end{bmatrix}$. It checks!

29. $v_1 = \begin{bmatrix} .35 & .52 & .13 \end{bmatrix} \cdot \begin{bmatrix} .3 & .5 & .2 \\ .2 & .7 & .1 \\ .1 & .4 & .5 \end{bmatrix} = \begin{bmatrix} .222 & .591 & .187 \end{bmatrix}$

Guess the steady-state vector is $\begin{bmatrix} .2 & .6 & .2 \end{bmatrix}$.

$$\begin{bmatrix} .2 & .6 & .2 \end{bmatrix} \cdot \begin{bmatrix} .3 & .5 & .2 \\ .2 & .7 & .1 \\ .1 & .4 & .5 \end{bmatrix} = \begin{bmatrix} .2 & .6 & .2 \end{bmatrix}. \text{ It checks.}$$

31. **(a)** Given $v_0 = \begin{bmatrix} 1 & 0 & 0 \end{bmatrix}$ we find v_4.

$$v_1 = \begin{bmatrix} 1 & 0 & 0 \end{bmatrix} \cdot \begin{bmatrix} .25 & .30 & .45 \\ .30 & .40 & .30 \\ .55 & .20 & .25 \end{bmatrix} = \begin{bmatrix} .25 & .30 & .45 \end{bmatrix}$$

$$v_2 = \begin{bmatrix} .25 & .30 & .45 \end{bmatrix} \cdot \begin{bmatrix} .25 & .30 & .45 \\ .30 & .40 & .30 \\ .55 & .20 & .25 \end{bmatrix} = \begin{bmatrix} .4 & .285 & .315 \end{bmatrix}$$

$$v_3 = \begin{bmatrix} .4 & .285 & .315 \end{bmatrix} \cdot \begin{bmatrix} .25 & .30 & .45 \\ .30 & .40 & .30 \\ .55 & .20 & .25 \end{bmatrix} = \begin{bmatrix} .35875 & .297 & .34425 \end{bmatrix}$$

$$v_4 = \begin{bmatrix} .35875 & .297 & .34425 \end{bmatrix} \cdot \begin{bmatrix} .25 & .30 & .45 \\ .30 & .40 & .30 \\ .55 & .20 & .25 \end{bmatrix} = \begin{bmatrix} .368125 & .295275 & .3366 \end{bmatrix}$$

The stock has approximately a 36.8% chance of increasing, a 29.5% chance of staying the same and a 33.7% chance of decreasing on January 9.

(b) Given $v_0 = \begin{bmatrix} 0 & 1 & 0 \end{bmatrix}$ we find v_5 where $T = \begin{bmatrix} .25 & .30 & .45 \\ .30 & .40 & .30 \\ .55 & .20 & .25 \end{bmatrix}$

$$v_1 = \begin{bmatrix} 0 & 1 & 0 \end{bmatrix} \cdot T = \begin{bmatrix} .3 & .4 & .3 \end{bmatrix}$$

$$v_2 = \begin{bmatrix} .3 & .4 & .3 \end{bmatrix} \cdot T = \begin{bmatrix} .36 & .31 & .33 \end{bmatrix}$$

$$v_3 = \begin{bmatrix} .36 & .31 & .33 \end{bmatrix} \cdot T = \begin{bmatrix} .3645 & .298 & .3375 \end{bmatrix}$$

$$v_4 = \begin{bmatrix} .3645 & .298 & .3375 \end{bmatrix} \cdot T = \begin{bmatrix} .36615 & .29605 & .3378 \end{bmatrix}$$

$$v_5 = \begin{bmatrix} .36615 & .29605 & .3378 \end{bmatrix} \cdot T = \begin{bmatrix} .3661425 & .295825 & .3380325 \end{bmatrix}$$

The stock has approximately a 36.6% chance of increasing, 33.8% chance of decreasing and 29.6% chance of staying the same on March 28.

(c) Given $v_0 = \begin{bmatrix} 0 & 0 & 1 \end{bmatrix}$, find v_6.

$$v_1 = \begin{bmatrix} 0 & 0 & 1 \end{bmatrix} \cdot T = \begin{bmatrix} .55 & .20 & .25 \end{bmatrix}$$

$$v_2 = \begin{bmatrix} .55 & .20 & .25 \end{bmatrix} \cdot T = \begin{bmatrix} .335 & .295 & .37 \end{bmatrix}$$

$$v_3 = \begin{bmatrix} .335 & .295 & .37 \end{bmatrix} \cdot T = \begin{bmatrix} .37575 & .2925 & .33175 \end{bmatrix}$$

$$v_4 = \begin{bmatrix} .37575 & .2925 & .33175 \end{bmatrix} \cdot T = \begin{bmatrix} .36415 & .296075 & .339775 \end{bmatrix}$$

$$v_5 = \begin{bmatrix} .36415 & .296075 & .339775 \end{bmatrix} \cdot T = \begin{bmatrix} .36673625 & .29563 & .33763375 \end{bmatrix}$$

$$v_6 = \begin{bmatrix} .36673625 & .29563 & .33763375 \end{bmatrix} \cdot T = \begin{bmatrix} .366071625 & .295799625 & .33812875 \end{bmatrix}$$

The stock has approximately a 36.6% chance of increasing, a 29.6% chance of staying the same, and a 33.8% chance of decreasing on September 29.

33. We are given $v_0 = \begin{bmatrix} 1 & 0 & 0 \end{bmatrix}$ and $T = \begin{bmatrix} .65 & .20 & .15 \\ .25 & .60 & .15 \\ .35 & .25 & .40 \end{bmatrix}$

(a) $v_1 = v_0 \cdot T = \begin{bmatrix} .65 & .20 & .15 \end{bmatrix}$

$v_2 = v_1 \cdot T = \begin{bmatrix} .525 & .2875 & .1875 \end{bmatrix}$.

The chance party B will take control in 1993 is 28.75%.

(b) $v_3 = v_2 \cdot T = \begin{bmatrix} .47875 & .324375 & .196875 \end{bmatrix}$.

The chance A will take control in 1996 is 47.875%.

(c) The chance party C will take control in 1999 is $v_3 \cdot \begin{bmatrix} .15 & .15 & .40 \end{bmatrix} = .19921875$ or approximately 19.9%.

35. (a) Given $v_0 = \begin{bmatrix} 0 & 0 & 1 \end{bmatrix}$ we find v_3 for $T = \begin{bmatrix} .6 & .3 & .1 \\ .3 & .55 & .15 \\ .2 & .2 & .6 \end{bmatrix}$

$v_1 = \begin{bmatrix} .2 & .2 & .6 \end{bmatrix}$

$v_2 = \begin{bmatrix} .2 & .2 & .6 \end{bmatrix} \cdot T = \begin{bmatrix} .3 & .29 & .41 \end{bmatrix}$

$v_3 = \begin{bmatrix} .3 & .29 & .41 \end{bmatrix} \cdot T = \begin{bmatrix} .349 & .3315 & .3195 \end{bmatrix}$.

The chance of a sunny Thursday is 31.95%.

(b) $v_4 = \begin{bmatrix} .349 & .3315 & .3195 \end{bmatrix} \cdot T = \begin{bmatrix} .37275 & .350925 & .276325 \end{bmatrix}$.

The chance of a cloudy Friday is 35.0925%.

(c) The chance of a rainy Saturday is

$$v_4 \cdot \begin{bmatrix} .6 & .3 & .1 \end{bmatrix} = .35656 \text{ or } 35.656\%.$$

37. $vT = \begin{bmatrix} x & y \end{bmatrix} \cdot \begin{bmatrix} a & b \\ c & d \end{bmatrix} = \begin{bmatrix} ax + cy & bx + dy \end{bmatrix}$.

Since x, y, a, b, c and d are all non-negative, $ax + cy$ and $bx + dy$ are non-negative. $(ax + cy) + (bx + dy) = (a + b)x + (c + d)y$ where $a + b = 1$ and $c + d = 1$ as T is a transition matrix. Furthermore $x + y = 1$ as v is a state vector. Hence $(ax + cy) + (bx + dy) = x + y = 1$. vT is a state vector as both entries are non-negative and the sum of the two entries is one.

39. If the ith row of a transition matrix T is $\begin{bmatrix} c_{i1} & c_{i2} & \cdots & c_{in} \end{bmatrix}$ and $c_{ii} = 1$, all other entries in this row must be zero. That is, $c_{ij} = 0$ if $i \neq j$. The ij^{th} element of T^2 is $c_{i1}c_{1j} + c_{i2}c_{2j} + \cdots + c_{in}c_{nj} = c_{ii}c_{ij} = c_{ij}$. Thus the ith row of T^2 is the same as the ith row of T and its entries are all zero except for the entry on the main diagonal. The same holds for the ith row of T^n. Hence T is not regular, as the only entry in the ith row which is not zero is the entry on the main diagonal.

EXERCISE SET 3.9   CHAPTER REVIEW

1. $[a_{ij}] = \begin{bmatrix} 2 - 3 & 2 - 6 \\ 4 - 3 & 4 - 6 \end{bmatrix} = \begin{bmatrix} -1 & -4 \\ 1 & -2 \end{bmatrix}$

3. $[c_{ij}] = \begin{bmatrix} 0 & -1^2 + 6 & -1^2 + 9 & -1^2 + 12 \\ -2^2 + 3 & 0 & -2^2 + 9 & -2^2 + 12 \\ -3^2 + 3 & -3^2 + 6 & 0 & -3^2 + 12 \\ -4^2 + 3 & -4^2 + 6 & -4^2 + 9 & 0 \end{bmatrix} = \begin{bmatrix} 0 & 5 & 8 & 11 \\ -1 & 0 & 5 & 8 \\ -6 & -3 & 0 & 3 \\ -13 & -10 & -7 & 0 \end{bmatrix}$

5. Equating corresponding elements of the matrices,

(a) $x = 3$, $y = 4$, $z = 7$

(b) $x = 6$, $y = 2$, $z = 3$

(c) $\begin{aligned} x + 2 &= 2x - 1 \\ y - 3 &= 1 + 2y \\ z + 4 &= 5z \end{aligned}$ or $\begin{aligned} x &= 3 \\ y &= -4 \\ z &= 1 \end{aligned}$

7. (a)   Let $A = \begin{bmatrix} 1 & 2 \\ 3 & 4 \end{bmatrix}$ and $B = \begin{bmatrix} 5 & 6 \\ 7 & 8 \end{bmatrix}$.

$AB = \begin{bmatrix} 1 & 2 \\ 3 & 4 \end{bmatrix} \cdot \begin{bmatrix} 5 & 6 \\ 7 & 8 \end{bmatrix} = \begin{bmatrix} 21 & 22 \\ 43 & 50 \end{bmatrix}$.

$BA = \begin{bmatrix} 5 & 6 \\ 7 & 8 \end{bmatrix} \cdot \begin{bmatrix} 1 & 2 \\ 3 & 4 \end{bmatrix} = \begin{bmatrix} 18 & 34 \\ 31 & 46 \end{bmatrix}$. $AB \neq BA$

(b) Let $A = \begin{bmatrix} 2 & 0 \\ 0 & 2 \end{bmatrix}$ and $B = \begin{bmatrix} 1/2 & 0 \\ 0 & 1/2 \end{bmatrix}$.

$$AB = \begin{bmatrix} 2 & 0 \\ 0 & 2 \end{bmatrix} \cdot \begin{bmatrix} 1/2 & 0 \\ 0 & 1/2 \end{bmatrix} = \begin{bmatrix} 1 & 0 \\ 0 & 1 \end{bmatrix}$$

$$= \begin{bmatrix} 1/2 & 0 \\ 0 & 1/2 \end{bmatrix} \cdot \begin{bmatrix} 2 & 0 \\ 0 & 2 \end{bmatrix} = BA$$

9. (a)

$$\left[\begin{array}{ccc|c} 3 & 4 & -5 & 4 \\ -4 & -2 & 5 & 6 \end{array} \right] \quad 4R_1 + 3R_2 \to R_2$$

$$\left[\begin{array}{ccc|c} 3 & 4 & -5 & 4 \\ 0 & 10 & -5 & 34 \end{array} \right] \quad 5R_1 - 2R_2 \to R_1$$

$$\left[\begin{array}{ccc|c} 15 & 0 & -15 & -48 \\ 0 & 10 & -5 & 34 \end{array} \right] \quad \begin{array}{l} R_1/15 \to R_1 \\ R_2/10 \to R_2 \end{array}$$

$$\left[\begin{array}{ccc|c} 1 & 0 & -1 & -3.2 \\ 0 & 1 & -.5 & 3.4 \end{array} \right]$$

The corresponding equations are $\begin{array}{l} x - z = -3.2 \\ y - .5z = 3.4 \end{array}$.

Let $z = a$, $y = 3.4 + .5a$ and $x = a - 3.2$.
The solution set is $\{(a - 3.2,\ 3.4 + .5a,\ a) \mid a \text{ is a real number}\}$.

(b) $\left[\begin{array}{ccc|c} 5 & 3 & -4 & -5 \\ -6 & 4 & -3 & -25 \\ 1 & 4 & -6 & -20 \end{array} \right] \quad \begin{array}{l} R_2 + 6R_3 \to R_2 \\ R_1 - 5R_3 \to R_3 \end{array} \quad \left[\begin{array}{ccc|c} 5 & 3 & -4 & -5 \\ 0 & 28 & -39 & -145 \\ 0 & -17 & 26 & 95 \end{array} \right]$

$\begin{array}{l} 17R_1 + 3R_3 \to R_1 \\ 17R_2 + 28R_3 \to R_3 \end{array} \left[\begin{array}{ccc|c} 85 & 0 & 10 & 200 \\ 0 & 28 & -39 & -145 \\ 0 & 0 & 65 & 195 \end{array} \right] \quad \begin{array}{l} R_3/65 \to R_3 \\ R_1/5 \to R_1 \end{array} \left[\begin{array}{ccc|c} 17 & 0 & 2 & 40 \\ 0 & 28 & -39 & -145 \\ 0 & 0 & 1 & 3 \end{array} \right]$

$\begin{array}{l} R_1 - 2R_2 \to R_1 \\ R_2 + 39R_3 \to R_3 \end{array} \left[\begin{array}{ccc|c} 17 & 0 & 0 & 34 \\ 0 & 28 & 0 & -28 \\ 0 & 0 & 1 & 3 \end{array} \right] \quad \begin{array}{l} R_1/17 \to R_1 \\ R_2/28 \to R_2 \end{array} \left[\begin{array}{ccc|c} 1 & 0 & 0 & 2 \\ 0 & 1 & 0 & -1 \\ 0 & 0 & 1 & 3 \end{array} \right]$

The solution set is $\{(2, -1, 3)\}$

11.

$$\left[\begin{array}{ccc|c|c|c} 2 & 6 & 3 & 3 & 0 & -5 \\ 2 & 7 & 5 & -4 & -3 & 1 \\ 1 & 3 & 2 & 7 & 5 & -3 \end{array}\right]$$

$$\begin{array}{c} R_1 - R_2 \to R_2 \\ R_2 - 2R_3 \to R_3 \end{array}$$

$$\left[\begin{array}{ccc|c|c|c} 2 & 6 & 3 & 3 & 0 & -5 \\ 0 & -1 & -2 & 7 & 3 & -6 \\ 0 & 1 & 1 & -18 & -13 & 7 \end{array}\right]$$

$$\begin{array}{c} R_2 + R_3 \to R_3 \\ R_1 - 6R_3 \to R_1 \end{array}$$

$$\left[\begin{array}{ccc|c|c|c} 2 & 0 & -3 & 111 & 78 & -47 \\ 0 & -1 & -2 & 7 & 3 & -6 \\ 0 & 1 & -1 & -11 & -10 & 1 \end{array}\right]$$

$$\begin{array}{c} R_1 - 3R_3 \to R_1 \\ R_2 - 2R_3 \to R_2 \end{array}$$

$$\left[\begin{array}{ccc|c|c|c} 2 & 0 & 0 & 144 & 108 & -50 \\ 0 & -1 & 0 & 29 & 23 & -8 \\ 0 & 0 & -1 & -11 & -10 & 1 \end{array}\right]$$

$$\begin{array}{c} R_1/2 \to R_1 \\ -R_2 \to R_2 \\ -R_3 \to R_3 \end{array}$$

$$\left[\begin{array}{ccc|c|c|c} 1 & 0 & 0 & 72 & 54 & -25 \\ 0 & 1 & 0 & -29 & -23 & 8 \\ 0 & 0 & 1 & 11 & -10 & -1 \end{array}\right]$$

The solution sets are

(a) $\{(72, -29, 11)\}$

(b) $\{(54, -23, 10)\}$

(c) $\{(-25, 8, -1)\}$

13. (a)

$$\left[\begin{array}{cc} 9 & 11 \\ 4 & 5 \end{array}\right] \cdot \left[\begin{array}{c} x \\ y \end{array}\right] = \left[\begin{array}{c} 3 \\ -2 \end{array}\right]$$

$$\left[\begin{array}{cc} 9 & 11 \\ 4 & 5 \end{array}\right]^{-1} = \left[\begin{array}{cc} 5 & -11 \\ -4 & 9 \end{array}\right]$$

$$\left[\begin{array}{c} x \\ y \end{array}\right] = \left[\begin{array}{cc} 5 & -11 \\ -4 & 9 \end{array}\right] \cdot \left[\begin{array}{c} 3 \\ -2 \end{array}\right] = \left[\begin{array}{c} 37 \\ -30 \end{array}\right]$$

The solution set is $\{(37, -30)\}$

(b)

$$\left[\begin{array}{ccc} 2 & 4 & 3 \\ 6 & 9 & 8 \\ 1 & 2 & 2 \end{array}\right] \cdot \left[\begin{array}{c} x \\ y \\ z \end{array}\right] = \left[\begin{array}{c} -3 \\ 6 \\ 18 \end{array}\right],$$

The inverse of $\left[\begin{array}{ccc} 2 & 4 & 3 \\ 6 & 9 & 8 \\ 1 & 2 & 2 \end{array}\right]$ is $-\dfrac{1}{3}\left[\begin{array}{ccc} 2 & -2 & 5 \\ -4 & 1 & 2 \\ 3 & 0 & -6 \end{array}\right].$

$$\begin{bmatrix} x \\ y \\ z \end{bmatrix} = -\frac{1}{3} \begin{bmatrix} 2 & -2 & 5 \\ -4 & 1 & 2 \\ 3 & 0 & -6 \end{bmatrix} \cdot \begin{bmatrix} -3 \\ 6 \\ 18 \end{bmatrix} = \begin{bmatrix} -24 \\ -18 \\ 39 \end{bmatrix}.$$

The solution set is $\{(-24, -18, 39)\}$.

(c)
$$\begin{bmatrix} 2 & 4 & 1 \\ 3 & 7 & 3 \\ 5 & 10 & 3 \end{bmatrix} \cdot \begin{bmatrix} x \\ y \\ z \end{bmatrix} = \begin{bmatrix} -3 \\ 2 \\ 4 \end{bmatrix}.$$

Since the inverse of $\begin{bmatrix} 2 & 4 & 1 \\ 3 & 7 & 3 \\ 5 & 10 & 3 \end{bmatrix}$ is $\begin{bmatrix} -9 & -2 & 5 \\ 6 & 1 & -3 \\ -5 & 0 & 2 \end{bmatrix}$,

$$\begin{bmatrix} x \\ y \\ z \end{bmatrix} = \begin{bmatrix} -9 & -2 & 5 \\ 6 & 1 & -3 \\ -5 & 0 & 2 \end{bmatrix} \cdot \begin{bmatrix} -3 \\ 2 \\ 4 \end{bmatrix} = \begin{bmatrix} 43 \\ -28 \\ 23 \end{bmatrix}$$

The solution set is $\{(43, -28, 23)\}$.

15. (a) $\det A = \begin{vmatrix} 2 & 5 \\ 3 & -2 \end{vmatrix} = -19$ $\qquad \det A_x = \begin{vmatrix} -11 & 5 \\ 12 & -2 \end{vmatrix} = -38$

$\det A_y = \begin{vmatrix} 2 & -11 \\ 3 & 12 \end{vmatrix} = 57 \qquad x = -38/(-19) = 2, \quad y = 57/(-19) = -3$

The solution set is $\{(2, -3)\}$

(b) $\det A = \begin{vmatrix} 2 & 3 & -4 \\ 3 & -2 & 5 \\ 5 & 1 & -3 \end{vmatrix} = 2 \begin{vmatrix} -2 & 5 \\ 1 & -3 \end{vmatrix} - 3 \begin{vmatrix} 3 & -4 \\ 1 & -3 \end{vmatrix} + 5 \begin{vmatrix} 3 & -4 \\ -2 & 5 \end{vmatrix}$

$\qquad = 2(1) - 3(-5) + 5(7) = 52$

$\det A_x = \begin{vmatrix} -12 & 3 & -4 \\ 13 & -2 & 5 \\ -15 & 1 & -3 \end{vmatrix} = -12 \begin{vmatrix} -2 & 5 \\ 1 & -3 \end{vmatrix} - 13 \begin{vmatrix} 3 & -4 \\ 1 & -3 \end{vmatrix} - 15 \begin{vmatrix} 3 & -4 \\ -2 & 5 \end{vmatrix}$

$\qquad = -12(1) - 13(-5) - 15(7) = -52$

$$\det A_y = \begin{vmatrix} 2 & -12 & -4 \\ 3 & 13 & 5 \\ 5 & -15 & -3 \end{vmatrix} = 2 \begin{vmatrix} 13 & 5 \\ -15 & -3 \end{vmatrix} + 3 \begin{vmatrix} 12 & -4 \\ -15 & -3 \end{vmatrix} + 5 \begin{vmatrix} -12 & -4 \\ 13 & 5 \end{vmatrix}$$

$$= 2(36) + 3(24) + 5(-8) = 104$$

$$\det A_z = \begin{vmatrix} 2 & 3 & -12 \\ 3 & -2 & 13 \\ 5 & 1 & -15 \end{vmatrix} = 2 \begin{vmatrix} -2 & 13 \\ 1 & -15 \end{vmatrix} - 3 \begin{vmatrix} 3 & -12 \\ 1 & -15 \end{vmatrix} + 5 \begin{vmatrix} 3 & -12 \\ -2 & 13 \end{vmatrix}$$

$$= 2(17) - 3(-33) + 5(15) = 208$$

$$x = -\frac{52}{52} = -1 \qquad y = \frac{104}{52} = 2 \qquad z = \frac{208}{52} = 4$$

The solution set is $\{(-1, 2, 4)\}$

17. The technological matrix is $B = \begin{bmatrix} .3 & .25 & .3 \\ .2 & .4 & .2 \\ .4 & .3 & .2 \end{bmatrix}$ $\qquad I - B = \begin{bmatrix} .7 & -.25 & -.3 \\ -.2 & .6 & -.2 \\ -.4 & -.3 & .8 \end{bmatrix}$

$$\det (I - B) = .144, \quad \text{adj} (I - B) = \begin{bmatrix} .42 & .24 & .30 \\ .29 & .44 & .31 \\ .23 & .20 & .37 \end{bmatrix}, \quad (I - B)^{-1} = \frac{1}{.144} \begin{bmatrix} .42 & .29 & .23 \\ .24 & .44 & .20 \\ .30 & .31 & .37 \end{bmatrix}$$

(a) $\quad \dfrac{1}{.144} \begin{bmatrix} .42 & .29 & .23 \\ .24 & .44 & .20 \\ .30 & .31 & .37 \end{bmatrix} \cdot \begin{bmatrix} 660 \\ 960 \\ 564 \end{bmatrix} = \begin{bmatrix} 4759 \\ 4817 \\ 4891 \end{bmatrix}$

Industries I, II and III must produce approximately 4759, 4817 and 4891 million dollars per year respectively.

(b) $\quad \dfrac{1}{.144} \begin{bmatrix} .42 & .29 & .23 \\ .24 & .44 & .20 \\ .30 & .31 & .37 \end{bmatrix} \cdot \begin{bmatrix} 910 \\ 1043 \\ 665 \end{bmatrix} = \begin{bmatrix} 5817 \\ 5627 \\ 5850 \end{bmatrix}$

The industries must produce approximately 5817, 5627 and 5850 million dollars per year, respectively.

(c) $\quad \dfrac{1}{.144} \begin{bmatrix} .42 & .29 & .23 \\ .24 & .44 & .20 \\ .30 & .31 & .37 \end{bmatrix} \cdot \begin{bmatrix} 940 \\ 1500 \\ 1680 \end{bmatrix} = \begin{bmatrix} 8446 \\ 8483 \\ 9504 \end{bmatrix}$

The industries must produce approximately 8446, 8483 and 9504 million dollars per year, respectively.

19. $B = \begin{bmatrix} 3 & 1 \\ 5 & 2 \end{bmatrix}$, $\quad \det B = 1$, $\quad \text{adj } B = \begin{bmatrix} 2 & -5 \\ -1 & 3 \end{bmatrix}$, $\quad B^{-1} = \begin{bmatrix} 2 & -1 \\ -5 & 3 \end{bmatrix}$.

$\begin{bmatrix} 2 & -1 \\ -5 & 3 \end{bmatrix} \cdot \begin{bmatrix} 36 \\ 65 \end{bmatrix} = \begin{bmatrix} 7 \\ 15 \end{bmatrix}$ $\quad \begin{bmatrix} 2 & -1 \\ -5 & 3 \end{bmatrix} \cdot \begin{bmatrix} 49 \\ 83 \end{bmatrix} = \begin{bmatrix} 15 \\ 4 \end{bmatrix}$

$\begin{bmatrix} 2 & -1 \\ -5 & 3 \end{bmatrix} \cdot \begin{bmatrix} 93 \\ 159 \end{bmatrix} = \begin{bmatrix} 27 \\ 12 \end{bmatrix}$ $\quad \begin{bmatrix} 2 & -1 \\ -5 & 3 \end{bmatrix} \cdot \begin{bmatrix} 66 \\ 111 \end{bmatrix} = \begin{bmatrix} 21 \\ 3 \end{bmatrix}$

$\begin{bmatrix} 2 & -1 \\ -5 & 3 \end{bmatrix} \cdot \begin{bmatrix} 60 \\ 109 \end{bmatrix} = \begin{bmatrix} 11 \\ 27 \end{bmatrix}$ $\quad$ The message is "GOOD LUCK"

21. (a) Not a transition matrix. The sums of the entries for the two rows are not one.

(b) A transition matrix.

(c) Not a transition matrix since it is not square.

(d) A transition matrix.

23. $\begin{bmatrix} x & y \end{bmatrix} \cdot \begin{bmatrix} .74 & .26 \\ .24 & .76 \end{bmatrix} = \begin{bmatrix} x & y \end{bmatrix}$ is the steady state vector. This leads to the system

$\begin{cases} .74x + .24y = x \\ x + y = 1 \end{cases}$ $\quad \begin{cases} .26x - .24y = 0 \\ x + y = 1 \end{cases}$ $\quad \begin{matrix} E_1 - .26E_2 \to E_1 \\ E_2 \leftrightarrow E_1 \end{matrix}$

$\begin{cases} x + y = 1 \\ -.5y = -.26 \end{cases}$ $\quad -E_2/.5 \to E_2$ $\quad \begin{cases} x + y = 1 \\ y = .52 \end{cases}$ $\quad E_1 - E_2 \to E_1$

$\begin{cases} x = .48 \\ y = .52 \end{cases}$. The steady-state vector is $\begin{bmatrix} .48 & .52 \end{bmatrix}$.

25. (a) Given $v_0 = \begin{bmatrix} 0 & 1 & 0 \end{bmatrix}$, find v_2, where $T = \begin{bmatrix} .55 & .35 & .10 \\ .50 & .45 & .05 \\ .15 & .25 & .60 \end{bmatrix}$

$v_1 = \begin{bmatrix} 0 & 1 & 0 \end{bmatrix} \cdot T = \begin{bmatrix} .50 & .45 & .05 \end{bmatrix}$

$v_2 = \begin{bmatrix} .50 & .45 & .05 \end{bmatrix} \cdot T = \begin{bmatrix} .5075 & .39 & .1025 \end{bmatrix}$

The chance of a sunny Wednesday is 10.25%.

(b) $v_3 = \begin{bmatrix} .5075 & .39 & .1025 \end{bmatrix} \cdot T = \begin{bmatrix} .4895 & .37875 & .13175 \end{bmatrix}$

The chance of a cloudy Thursday is 37.875%

(c) The chance of a rainy Friday is $v_3 \cdot \begin{bmatrix} .55 & .50 & .15 \end{bmatrix} = .4783625$ or 47.83625%.

27. (a) $8x + 5y$ is the amount of assembly time and $3x + 2y$ is the amount of testing time required daily.

The systems of equations are

$\begin{cases} 8x + 5y = 70 \\ 3x + 2y = 20 \end{cases}$ $\begin{cases} 8x + 5y = 75 \\ 3x + 2y = 30 \end{cases}$ $\begin{cases} 8x + 5y = 64 \\ 3x + 2y = 28 \end{cases}$ $\begin{cases} 8x + 5y = 75 \\ 3x + 2y = 33 \end{cases}$

$\begin{cases} 8x + 5y = 62 \\ 3x + 2y = 25 \end{cases}$

(b)

$\begin{bmatrix} 8 & 5 & | & 70 & 75 & 64 & 75 & 62 \\ 3 & 2 & | & 20 & 30 & 28 & 33 & 25 \end{bmatrix}$ $\quad 3R_1 - 8R_2 \rightarrow R_2$

$\begin{bmatrix} 8 & 5 & | & 70 & 75 & 64 & 75 & 62 \\ 0 & -1 & | & 50 & -15 & -32 & -39 & -14 \end{bmatrix}$ $\quad \begin{matrix} R_1 + 5R_2 \rightarrow R_1 \\ -R_2 \rightarrow R_2 \end{matrix}$

$\begin{bmatrix} 8 & 0 & | & 320 & 0 & -96 & -120 & -8 \\ 0 & 1 & | & -50 & 15 & 32 & 39 & 14 \end{bmatrix}$ $\quad R_1/8 \rightarrow R_1$

$\begin{bmatrix} 1 & 0 & | & 40 & 0 & -12 & -15 & -1 \\ 0 & 1 & | & -50 & 15 & 32 & 39 & 14 \end{bmatrix}$

The solutions of the system are $\{(40, -50)\}$, $\{(0, 15)\}$, $\{(-12, 32)\}$, $\{(-15, 39)\}$ and $\{(-1, 14)\}$ respectively

However, only on Tuesday is the solution feasible for this problem.

CHAPTER 4

LINEAR PROGRAMMING

EXERCISE SET 4.1 THE GEOMETRIC APPROACH

1. point $z=2x+5y+3$

(0,0)   3
(10,0)   23   The maximum value of z is 59
(8,8)   59   The minimum value is 3
(0,11)   58

3. point $z=5x+3y-7$

(0,9)   20
(0,0) -7 The maximum value of z is 68.
(6,12)   59   The minimum value is -7.
(9,10)   68
(8,0)   33

5. point $z=.25x+.45y-.75$

(5,18)   8.6
(15,10)   7.5   The maximum value of z is 22.65.
(35,20)   17   The minimum value is 7.5
(18,42)   22.65

7.

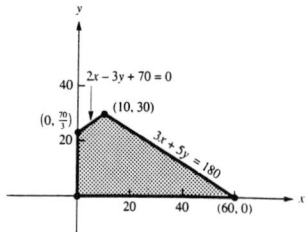

point $z=4x+7y$

(10,30)   250
(0,0)   0
(60,0)   240
$(0,\frac{70}{3})$ $490/3=163.\overline{3}$

The maximum value of z is 250.
The minimum value is 0.

9.

point $z=5x-13y$

(5,17.4) -201.2
(40.6, 12)   47
(12, 23) -239
(5, 12) -131

The maximum value of z is 47.
The minimum is -239.

11.

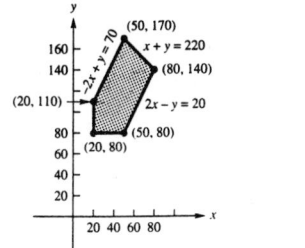

point	$z=2x+2y-13$
(50, 170)	427
(80, 140)	427
(50, 80)	247
(20, 80)	187
(20, 110)	247

The maximum value of z is 427.
The minimum value of z is 187.

13.

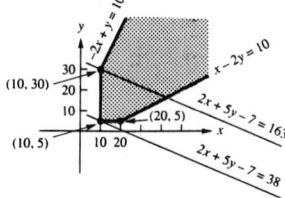

point	$z=2x+5y-7$
(10, 5)	38
(20, 5)	58
(10, 30)	163

Consider the lines $2x + 5y - 7 = 163$ and $2x + 5y - 7 = 38$. The slope of each is $-.4$. If $c < 38$, the graph of $2x + 5y - 7 = c$ will not intersect the region. If $c > 163$, the graph of $2x + 5y - 7 = c$ will intersect the region. Hence z has no maximum value but has a minimum value of 38.

15.

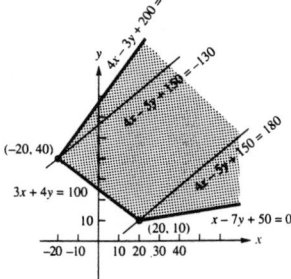

point	$z=4x-5y+150$
(20, 10)	180
(−20, 40)	−130

The graphs of $4x - 5y + 150 = 180$ and $4x - 5y + 150 = -130$ are shown.

If $c < -130$, the graph of $4x - 5y + 150 = c$ will intersect the region.
If $c > 180$, the graph of $4x - 5y + 150 = c$ will intersect the region.
There is no maximum or minimum value for z.

17. If $y=0$, $z = x$ which increases without bound as x becomes large. Thus z has no maximum. If $x=0$, $z = -y$ which decreases without bound as y becomes large. Thus z has no minimum.

19. $x \geq 150$
$y \geq 100$
$x+y \leq 800$
$x \leq 3y$

profit $= z = 370x+290y$
For the graph of feasible solutions see problem 19 of Exercise Set 2.4.

point	z
(150, 650)	244,000
(150, 100)	84,500
(300, 100)	140,000
(600, 200)	280,000

The profit is maximized when 600 acres are planted in wheat and 200 acres are planted in barley.

21. $x \geq 2500$
$y \geq 3600$
$3x+2y \leq 38000$
$\frac{1}{6}x+\frac{1}{10}y \leq 2000$

profit $= z = 350x+275y$
The graph of the system is shown as the solution to problem 21 of Exercise Set 2.4.

point	z
(2500, 15,250)	5,068,750
(6000, 10,000)	4,850,000
(2500, 3600)	1,865,000
(9840, 3600)	4,434,000

The maximum profit occurs when 2500 units of the deluxe model and 15,250 units of the standard model are produced.

23. $x+y \leq 300$
$x \geq 50$
$y \leq 200$
$x \leq 2y$

The cost $z = 80x+70y$
The graph of the system is shown as the solution to Problem 23 of Exercise Set 2.4.

point	z
(50, 25)	5,750
(200, 100)	23,000
(100, 200)	22,000
(50, 200)	18,000

The cost is minimized when 50 shares of stock A and 25 shares of stock B are purchased.

25. $3x+5y \leq 120$
$x \geq 0$
$y \geq 0$
$4x+7y \leq 166$

The profit $z = 10x+25y$
The graph of the system is shown as the solution to Problem #25 of Exercise Set 2.4.

point	z
(10, 18)	550
(40, 0)	400
$(0, \frac{166}{7})$	592.86
(0, 0)	0

The church's maximum profit occurs for no dolls and 23 blankets.

27. $4x+3y \leq 360$
$x \geq 3$
$y \geq 2$
$2x+5y \leq 32$

The profit $z = 5x+4y$
The graph of the system is shown as the solution to Problem 27 of Exercise Set 2.4.

point	z
(3, 5.2)	35.8
(6, 4)	46
(3, 2)	23
(7.5, 2)	45.5

The maximum profit occurs when 6 baskets and 4 vases are produced.

29. $x \geq 10$
$0 \leq y \leq 100$
$x+y \leq 120$
$.6x+.3y \leq 45$

The revenue $z = 95x+50y$
The graph of the system is shown as the solution to Problem 29 of Exercise Set 2.4.

point	z
(10, 0)	950
(75, 0)	7125
(30, 90)	7350
(20, 100)	6900
(10, 100)	5950

The maximum revenue occurs for 30 first class rooms and 90 regular rooms.

EXERCISE SET 4.2   THE THREE DIMENSIONAL COORDINATE SYSTEM

1. (a)

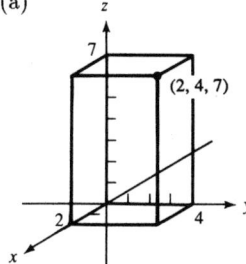

(b)

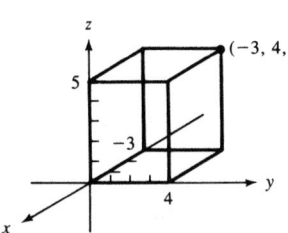

(c)

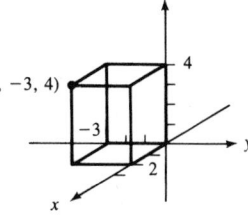

3. (a)

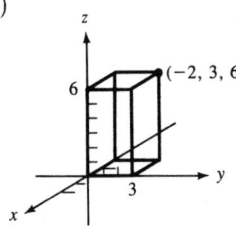

(b)

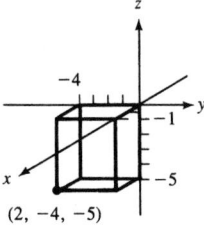

(c)

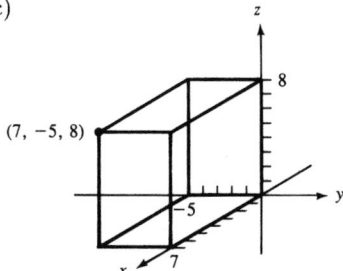

5.

x	y	z
0	0	20
12	0	0
0	30	0

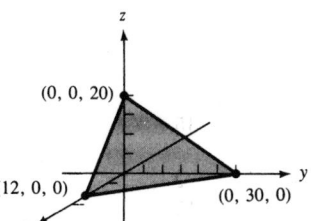

7.

x	y	z
0	0	10
-30	0	0
0	6	0

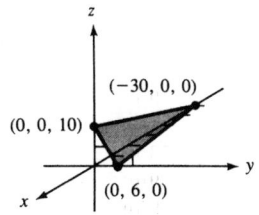

9. **(a)**

$$\begin{bmatrix} 9 & -7 & 3 & | & 9 \\ -11 & 6 & 4 & | & 12 \\ 8 & 4 & -5 & | & 8 \end{bmatrix}$$

$4R_1 - 3R_2 \rightarrow R_2$
$5R_1 + 3R_3 \rightarrow R_1$

$$\begin{bmatrix} 69 & -23 & 0 & | & 69 \\ 69 & -46 & 0 & | & 0 \\ 8 & 4 & -5 & | & 8 \end{bmatrix}$$

$R_1 - R_2 \rightarrow R_2$
$R_1/23 \rightarrow R_3$

$$\begin{bmatrix} 3 & -1 & 0 & | & 3 \\ 0 & 23 & 0 & | & 69 \\ 8 & 4 & -5 & | & 8 \end{bmatrix}$$

$R_2/23 \rightarrow R_3$

$$\begin{bmatrix} 3 & -1 & 0 & | & 3 \\ 0 & 1 & 0 & | & 3 \\ 8 & 4 & -5 & | & 8 \end{bmatrix}$$

$R_1 + R_2 \rightarrow R_1$
$R_3 - 4R_2 \rightarrow R_3$

$$\begin{bmatrix} 3 & 0 & 0 & | & 6 \\ 0 & 1 & 0 & | & 3 \\ 8 & 0 & -5 & | & -4 \end{bmatrix}$$

$R_1/3 \rightarrow R_3$

$$\begin{bmatrix} 1 & 0 & 0 & | & 2 \\ 0 & 1 & 0 & | & 3 \\ 8 & 0 & -5 & | & -4 \end{bmatrix}$$

$-8R_1 + R_3 \rightarrow R_3$

$$\begin{bmatrix} 1 & 0 & 0 & | & 2 \\ 0 & 1 & 0 & | & 3 \\ 0 & 0 & -5 & | & -20 \end{bmatrix}$$

$-R_3/5 \rightarrow R_3$

$$\begin{bmatrix} 1 & 0 & 0 & | & 2 \\ 0 & 1 & 0 & | & 3 \\ 0 & 0 & 1 & | & 4 \end{bmatrix}$$

The point of intersection is (2, 3, 4).

(b)

11. (a)

$$\begin{bmatrix} 20 & -38 & 25 & | & 100 \\ -1 & 1 & 1 & | & 4 \\ 24 & 30 & -33 & | & 120 \end{bmatrix}$$

$R_1 - 25R_2 \rightarrow R_1$
$R_3/3 \rightarrow R_3$

$$\begin{bmatrix} 45 & -63 & 0 & | & 0 \\ -1 & 1 & 1 & | & 4 \\ 8 & 10 & -11 & | & 40 \end{bmatrix}$$

$R_1/3 \rightarrow R_1$

$$\begin{bmatrix} 15 & -21 & 0 & | & 0 \\ -1 & 1 & 1 & | & 4 \\ 8 & 10 & -11 & | & 40 \end{bmatrix}$$

$11R_2 + R_3 \rightarrow R_3$
$R_1/3 \rightarrow R_1$

$$\begin{bmatrix} 5 & -7 & 0 & | & 0 \\ -1 & 1 & 1 & | & 4 \\ -3 & 21 & 0 & | & 84 \end{bmatrix}$$

$R_3/3 \rightarrow R_3$

$$\begin{bmatrix} 5 & -7 & 0 & | & 0 \\ -1 & 1 & 1 & | & 4 \\ -1 & 7 & 0 & | & 28 \end{bmatrix}$$

$R_1 + R_3 \rightarrow R_3$
$R_1 + 7R_2 \rightarrow R_2$

$$\begin{bmatrix} 5 & -7 & 0 & | & 0 \\ -2 & 0 & 7 & | & 28 \\ 4 & 0 & 0 & | & 28 \end{bmatrix}$$

$4R_1 - 5R_3 \rightarrow R_1$
$2R_2 + R_3 \rightarrow R_2$
$R_3/4 \rightarrow R_3$

$$\begin{bmatrix} 0 & -28 & 0 & | & -140 \\ 0 & 0 & 14 & | & 84 \\ 1 & 0 & 0 & | & 7 \end{bmatrix}$$

$-R_1/28 \rightarrow R_1$
$R_2/14 \rightarrow R_2$

$$\begin{bmatrix} 0 & 1 & 0 & | & 5 \\ 0 & 0 & 1 & | & 6 \\ 1 & 0 & 0 & | & 7 \end{bmatrix}$$

The point of intersection is $(7, 5, 6)$.

(b)

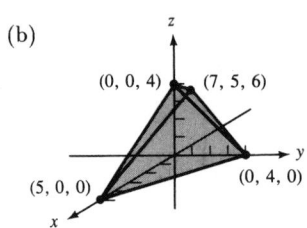

(0, 0, 4)   (7, 5, 6)

(5, 0, 0)   (0, 4, 0)

13. (a)

$$\begin{bmatrix} 7 & -10 & 14 & 42 \\ -13 & 12 & 20 & 60 \\ 5 & 6 & -13 & 30 \end{bmatrix}$$

$$13R_1 + 7R_2 \rightarrow R_2$$
$$5R_1 - 7R_3 \rightarrow R_3$$

$$\begin{bmatrix} 7 & -10 & 14 & 42 \\ 0 & -46 & 322 & 966 \\ 0 & -92 & 161 & 0 \end{bmatrix}$$

$$-2R_2 + R_3 \rightarrow R_3$$

$$\begin{bmatrix} 7 & -10 & 14 & 42 \\ 0 & -46 & 322 & 966 \\ 0 & 0 & -483 & -1932 \end{bmatrix}$$

$$-R_3/483 \rightarrow R_3$$
$$R_2/46 \rightarrow R_2$$

$$\begin{bmatrix} 7 & -10 & 14 & 42 \\ 0 & -1 & 7 & 21 \\ 0 & 0 & 1 & 4 \end{bmatrix}$$

$$R_1 - 10R_2 \rightarrow R_1$$
$$-R_2 + 7R_3 \rightarrow R_2$$

$$\begin{bmatrix} 7 & 0 & -56 & -168 \\ 0 & 1 & 0 & 7 \\ 0 & 0 & 1 & 4 \end{bmatrix}$$

$$R_1/7 \rightarrow R_1$$

$$\begin{bmatrix} 1 & 0 & -8 & -24 \\ 0 & 1 & 0 & 7 \\ 0 & 0 & 1 & 4 \end{bmatrix}$$

$$R_1 + 8R_3 \rightarrow R_1$$

$$\begin{bmatrix} 1 & 0 & 0 & 8 \\ 0 & 1 & 0 & 7 \\ 0 & 0 & 1 & 4 \end{bmatrix}$$

The point of intersection is (8, 7, 4).

(b)

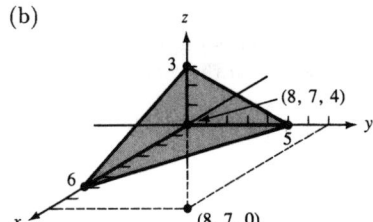

15. The number of subsets each with 2 members of a set with 6 elements is

$$\frac{6!}{2!(6-2)!}=\frac{6!}{2!4!}=\frac{6\cdot5}{2}=15$$

17. $\dfrac{8!}{2!(8-2)!}=\dfrac{8!}{2!6!}=\dfrac{8\cdot7}{2}=28$

19. The number of subsets each with 3 members of a set with 5 elements is $\dfrac{5!}{3!(5-3)!}=\dfrac{120}{12}=10$

21. $\dfrac{9!}{3!(9-3)!}=\dfrac{9!}{3!6!}=\dfrac{9\cdot8\cdot7}{6}=84$

EXERCISE SET 4.3 INTRODUCTION TO THE SIMPLEX METHOD

1.

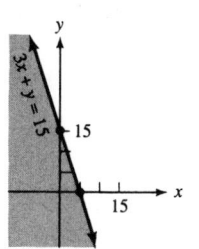

Let $3x+y+s = 15$.
Suppose (a, b, c) satisfies this equation.
If $c > 0$, (a, b) is in the shaded region,
if $c=0$, (a, b) is on the boundary, and
if $c < 0$, (a, b) is outside the region shaded.

3.

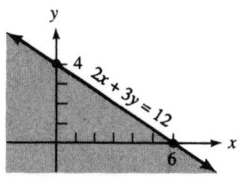

Let $2x+3y+s = 12$.
Suppose (a, b, c) satisfies this equation.
If $c > 0$, (a, b) is in the shaded region,
if $c=0$, (a, b) is on the boundary, and
if $c < 0$, (a, b) is outside the shaded region.

5.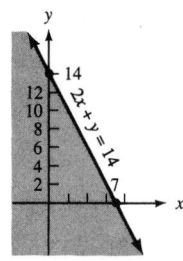

Let $2x+y+s = 14$.
Suppose $(a,\ b,\ c)$ satisfies this equation.
If $c > 0$, $(a,\ b)$ is within the shaded region,
if $c=0$, $(a,\ b)$ is on the boundary, and
if $c < 0$, $(a,\ b)$ is outside the shaded region.

7. For the equation $4x+5y+4z = 40$ we have the points

x	y	z
0	0	10
10	0	0
0	8	0

. The graph of the equation includes the indicated plane.

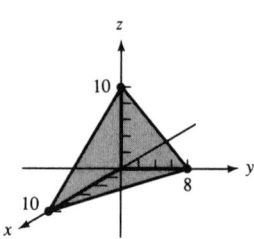

Let $4x+5y+4z+s = 40$.
If $(a,\ b,\ c,\ d)$ is a solution to this equation, if $d > 0$, $(a,\ b,\ c)$ is on the same side of the plane as is the origin, if $d < 0$, $(a,\ b,\ c)$ is on the opposite side of the plane, and if $d=0$, $(a,\ b,\ c)$ is on the plane.

9. For the equation $4x+y+4z = 16$ we have the points

x	y	z
0	0	4
4	0	0
0	16	0

. The graph of the equation includes the shaded region.

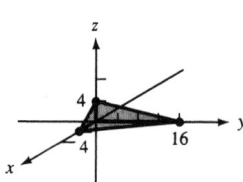

Let $4x+y+4z+s=16$. If $(a,\ b,\ c,\ d)$ is a solution to this equation, if $d > 0$, $(a,\ b,\ c)$ is on the same side of the plane as is the origin, if $d < 0$, $(a,\ b,\ c)$ is on the opposite side of the plane, and if $d=0$, $(a,\ b,\ c)$ is on the plane.

11. The graph of $2x+2y+3z = 12$ includes the points

x	y	z
0	0	4
6	0	0
0	6	0

as shown.

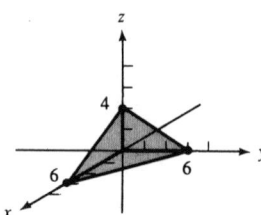

Let $2x+2y+3z+s=12$.

If (a, b, c, d) is a solution to this equation, if $d > 0$, (a, b, c) is on the same side of the plane as is the origin, if $d < 0$, (a, b, c) is on the opposite side of the plane, and if $d=0$, (a, b, c) lies on the plane.

13. (a)

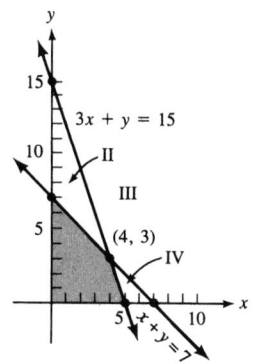

(b) $x+y+r=7$
$3x+y+s=15$

(c) Suppose a and b are non-negative and (a, b, c, d) is a solution to the system of part (b). (a, b) is within the shaded region if c and d are positive.

If c is positive and d is negative (a, b) is in region IV.
If c and d are both negative (a, b) is in region III.
If c is negative and d positive (a, b) is in region II.

If one or two of a, b, c and d are zero, while the others are positive (a, b) is on the boundary of the region and if exactly two of a, b, c and d equal zero (a, b) is one of its four vertices.

15. (a)

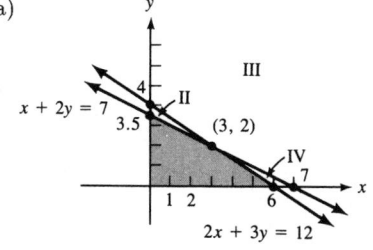

(b) $x+2y+r=7$
$2x+3y+s=12$

(c) Suppose (a, b, c, d) is a solution to the above system and a and b are non-negative. If r and s are positive (a, b) is in the shaded region.

If $c < 0$ and $d > 0$, (a, b) is in region II.
If c and d are negative (a, b) is in region III.
If $c > 0$ and $d < 0$, (a, b) is in region IV.

If one or two of a, b, c and d are zero while the others are positive (a, b) is on the boundary of the shaded region and if exactly two of them are zero (a, b) is one of its four vertices.

17. (a)

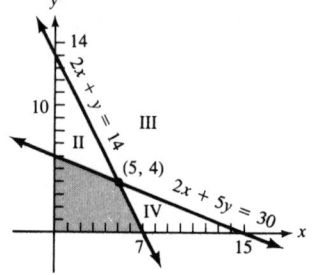

(b) $2x+5y+r=30$
$2x+y+s=14$

(c) Suppose (a, b, c, d) is a solution to the above system and a and b are non-negative.

If c and d are positive, (a, b) is in the shaded region.
If $c < 0$ and $d > 0$, (a, b) is in region II.
If $c < 0$ and $d < 0$, (a, b) is in region III.
If $c > 0$ and $d < 0$, (a, b) is in region IV.

If one or two of a, b, c and d are zero and the others are positive, (a, b) is on the boundary of the shaded region, and if two of them are zero (a, b) is one of the four vertices.

19. (a)

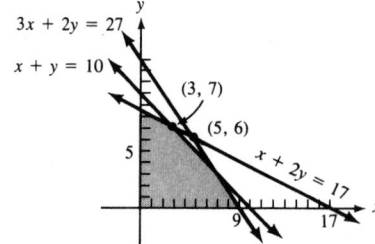

(b) $x+y+r=10$
$x+2y+s=17$
$3x+2y+t=27$

(c) Suppose (a, b, c, d, e) is a solution to the above system and a and b are non-negative.

If c, d and e are positive, (a, b) is in the shaded region. If $a=0$ and $b=0$, $a=0$ and $d=0$, $d=0$ and $c=0$, $c=0$ and $e=0$, or $b=0$ and $e=0$, (a, b) is one of the vertices of the shaded region. Another way to say this is that if exactly two of a, b, c, d and e are zero and the others positive, (a, b) is one of the 5 vertices of the region.

21. (a)

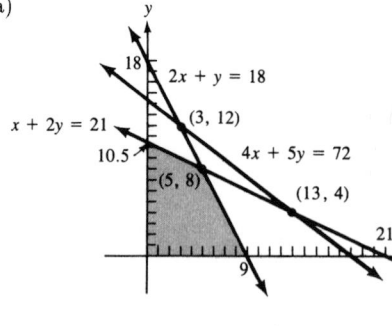

(b) $2x+y+r=18$
$x+2y+s=21$
$4x+5y+t=72$

(c) Suppose (a, b, c, d, e) is a solution to the above system and a and b are non-negative, (a, b) is in the shaded region if c, d and e are positive.

If exactly two of a, b, c, d and e are zero and the others are positive, (a, b) is one of the four vertices.

In Exercises 23, 25 and 27, five planes are included in the system. To find where the planes intersect, $\frac{5!}{3!2!}=10$ systems of equations must be solved.

System	Points of intersection take the form
$\{1, 2, 3\}$	$(0, y, z)$
$\{1, 2, 4\}$	$(x, 0, z)$
$\{1, 2, 5\}$	$(x, y, 0)$
$\{1, 3, 4\}$	$(0, 0, z)$
$\{1, 3, 5\}$	$(0, y, 0)$
$\{1, 4, 5\}$	$(x, 0, 0)$
$\{2, 3, 4\}$	$(0, 0, z)$
$\{2, 3, 5\}$	$(0, y, 0)$
$\{2, 4, 5\}$	$(x, 0, 0)$
$\{3, 4, 5\}$	$(0, 0, 0)$

23. (a)

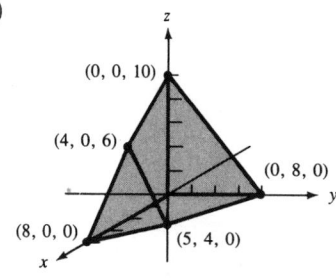

(b) $4x+5y+4z+s = 40$
$12x+9y+8z+t = 96$

(c) Suppose (a, b, c, d, e) is a solution to the system of part (b). If a, b, c, d and e are all positive, (a, b, c) is inside the polyhedron.

Vertices of the polyhedron occur when exactly three of a, b, c, d and e are zero and the other two are positive.

25. (a)

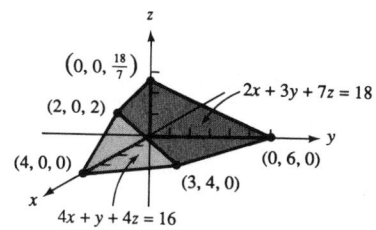

(b) $4x+y+4z+s = 16$
$2x+3y+7z+t = 18$

Suppose (a, b, c, d, e) is a solution to the system of part (b). If a, b, c, d and e are all positive, (a, b, c) lies inside the polyhedron. If exactly three of a, b, c, d and e are zero and the other two positive, (a, b, c) is one of the 6 vertices of the polyhedron.

27. (a)

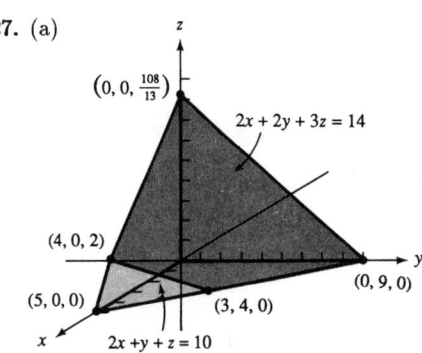

$\left(0, 0, \frac{108}{13}\right)$

$2x + 2y + 3z = 14$

$(4, 0, 2)$

$(5, 0, 0)$ $(3, 4, 0)$

$(0, 9, 0)$

x $2x + y + z = 10$

(b) $2x+y+z+s = 10$
$2x+2y+3z+t = 14$

(c) Suppose (a, b, c, d, e) is a solution to the system of part (b). If a, b, c, d and e are positive, (a, b, c) is inside the polyhedron. If exactly three of a, b, c, d and e are zero and the other two are positive, (a, b, c) is a vertex of the polyhedron.

EXERCISE SET 4.4 MOVING FROM CORNER TO CORNER

1. Letting x_1, x_3 and $x_5 = 0$, $x_4 = 5$ and $x_2 = -3$. The solution is $(0, -3, 0, 5, 0)$.

3. Letting x_2, x_4 and $x_6 = 0$, $x_3 = -2$, $x_1 = 13$, and $x_5 = 17$.
The solution is $(13, 0, -2, 0, 17, 0)$.

5. $(-2, 3, 0, 5, 0, 13)$.

7. Let $\begin{array}{l} x+y+r=7 \\ 3x+y+s=15 \end{array}$. The augmented matrix is

$$\left[\begin{array}{cccc|c} 1 & 1 & 1 & 0 & 7 \\ 3 & 1 & 0 & 1 & 15 \end{array}\right].$$

One solution is $(0, 0, 7, 15)$ and one corner is $(0, 0)$.

Now y becomes a basic variable:

$$\begin{array}{c} \\ \leftarrow r \\ s \end{array} \begin{array}{cccc} x & y & r & s \\ \end{array} \left[\begin{array}{cccc|c} 1 & \boxed{1} & 1 & 0 & 7 \\ 3 & 1 & 0 & 1 & 15 \\ & \uparrow & & & \end{array}\right]$$

$-R_1 + R_2 \rightarrow R_2$

$$\left[\begin{array}{cccc|c} 1 & 1 & 1 & 0 & 7 \\ 2 & 0 & -1 & 1 & 8 \end{array}\right]$$

A second solution is $(0, 7, 0, 8)$ and one corner is $(0, 7)$.

Next x becomes a basic variable:

$$\begin{array}{c} r \\ \leftarrow s \end{array} \begin{array}{cccc} x & y & r & s \\ \end{array} \left[\begin{array}{cccc|c} 1 & 1 & 1 & 0 & 7 \\ \boxed{2} & 0 & -1 & 1 & 8 \\ \uparrow & & & & \end{array}\right]$$

$\frac{1}{2}R_2 \rightarrow R_2$

$$\left[\begin{array}{cccc|c} 1 & 1 & 1 & 0 & 7 \\ 1 & 0 & -\frac{1}{2} & \frac{1}{2} & 4 \end{array}\right]$$

$-R_2 + R_1 \rightarrow R_1$

$$\left[\begin{array}{cccc|c} 0 & 1 & \frac{3}{2} & -\frac{1}{2} & 3 \\ 1 & 0 & -\frac{1}{2} & \frac{1}{2} & 4 \end{array}\right]$$

A third solution is $(4, 3, 0, 0)$ and $(4, 3)$ is a corner.

9. Let $\begin{array}{l} x+2y+r=7 \\ 2x+3y+s=12 \end{array}$. The augmented matrix is

$$\left[\begin{array}{cccc|c} 1 & 2 & 1 & 0 & 7 \\ 2 & 3 & 0 & 1 & 12 \end{array}\right].$$

One solution is $(0, 0, 7, 12)$ and $(0, 0)$ is a corner.

Now y becomes a basic variable:

$$\begin{array}{c} \\ \leftarrow r \\ s \end{array} \begin{array}{cccc} x & y & r & s \\ \end{array}$$
$$\leftarrow r \left[\begin{array}{cccc|c} 1 & \boxed{2} & 1 & 0 & 7 \\ 2 & 3 & 0 & 1 & 12 \end{array}\right] \qquad \tfrac{1}{2}R_2 \to R_2$$
$$\uparrow$$

$$\left[\begin{array}{cccc|c} \tfrac{1}{2} & 1 & \tfrac{1}{2} & 0 & \tfrac{7}{2} \\ 2 & 3 & 0 & 1 & 12 \end{array}\right] \qquad -3R_1 + R_2 \to R_2$$

$$\left[\begin{array}{cccc|c} \tfrac{1}{2} & 1 & \tfrac{1}{2} & 0 & \tfrac{7}{2} \\ \tfrac{1}{2} & 0 & -\tfrac{3}{2} & 1 & \tfrac{3}{2} \end{array}\right]$$

Another solution is $(0, \tfrac{7}{2}, 0, \tfrac{3}{2})$ and $(0, \tfrac{7}{2})$ is a corner.

Next x becomes a basic variable:

$$\begin{array}{cccc} & x & y & r & s \end{array}$$
$$\begin{array}{c} r \\ \leftarrow s \end{array}\left[\begin{array}{cccc|c} \tfrac{1}{2} & 1 & \tfrac{1}{2} & 0 & \tfrac{7}{2} \\ \boxed{1/2} & 0 & -\tfrac{3}{2} & 1 & \tfrac{3}{2} \end{array}\right] \qquad 2R_2 \to R_2$$
$$\uparrow$$

$$\left[\begin{array}{cccc|c} \tfrac{1}{2} & 1 & \tfrac{1}{2} & 0 & \tfrac{7}{2} \\ 1 & 0 & -3 & 2 & 3 \end{array}\right] \qquad -\tfrac{1}{2}R_2 + R_1 \to R_1$$

$$\left[\begin{array}{cccc|c} 0 & 1 & 2 & -1 & 2 \\ 1 & 0 & -3 & 2 & 3 \end{array}\right]$$

Another solution is $(3, 2, 0, 0)$ and corner is $(3, 2)$.

11. Let $\begin{array}{l} 2x+5y+r=30 \\ 2x+y+s=14 \end{array}$. The augmented matrix is

$$\begin{array}{cccc} x & y & r & s \end{array}$$
$$\left[\begin{array}{cccc|c} 2 & 5 & 1 & 0 & 30 \\ 2 & 1 & 0 & 1 & 14 \end{array}\right]$$

and $(0, 0, 30, 14)$ is a solution and $(0, 0)$ a corner.

$$\begin{array}{cccc} & x & y & r & s \end{array}$$
$$\begin{array}{c} r \\ \leftarrow s \end{array}\left[\begin{array}{cccc|c} 2 & 5 & 1 & 0 & 30 \\ \boxed{2} & 1 & 0 & 1 & 14 \end{array}\right] \qquad \tfrac{1}{2}R_2 \to R_2$$
$$\uparrow$$

$$\begin{array}{cccc} x & y & r & s \end{array}$$
$$\left[\begin{array}{cccc|c} 2 & 5 & 1 & 0 & 30 \\ 1 & \tfrac{1}{2} & 0 & \tfrac{1}{2} & 7 \end{array}\right] \qquad -2R_2 + R_1 \to R_1$$

$$- \; 167 \; -$$

$$\begin{array}{cccc} x & y & r & s \\ \end{array}$$

$$\left[\begin{array}{cccc|c} 0 & 4 & 1 & -1 & 16 \\ 1 & \frac{1}{2} & 0 & \frac{1}{2} & 7 \end{array}\right]$$

Another solution is $(7, 0, 16, 0)$ and $(7, 0)$ is a corner.

$$\begin{array}{cccc} x & y & r & s \\ \end{array}$$

$$\begin{array}{c} \leftarrow r \\ s \\ \\ \end{array} \left[\begin{array}{cccc|c} 0 & 4 & 1 & -1 & 16 \\ 1 & \frac{1}{2} & 0 & \frac{1}{2} & 7 \end{array}\right]$$ $\frac{1}{4}R_1 \rightarrow R_1$

$\uparrow$

$$\begin{array}{cccc} x & y & r & s \\ \end{array}$$

$$\left[\begin{array}{cccc|c} 0 & 1 & \frac{1}{4} & -\frac{1}{4} & 4 \\ 1 & \frac{1}{2} & 0 & \frac{1}{2} & 7 \end{array}\right]$$ $-\frac{1}{2}R_1 + R_2 \rightarrow R_2$

$$\begin{array}{cccc} x & y & r & s \\ \end{array}$$

$$\left[\begin{array}{cccc|c} 0 & 1 & \frac{1}{4} & -\frac{1}{4} & 4 \\ 1 & 0 & -\frac{1}{8} & \frac{5}{8} & 5 \end{array}\right]$$

Another solution is $(5, 4, 0, 0)$ and $(5, 4)$ is a corner.

$x+y+r = 10$
13. Let $x+2y+s = 17$. The augmented matrix is
$3x+2y+t = 27$

$$\begin{array}{ccccc} x & y & r & s & t \\ \end{array}$$

$$\left[\begin{array}{ccccc|c} 1 & 1 & 1 & 0 & 0 & 10 \\ 1 & 2 & 0 & 1 & 0 & 17 \\ 3 & 2 & 0 & 0 & 1 & 27 \end{array}\right].$$

One solution is $(0, 0, 10, 17, 27)$ and $(0, 0)$ is a corner.

$$\begin{array}{ccccc} x & y & r & s & t \\ \end{array}$$

$$\begin{array}{c} r \\ s \\ \leftarrow t \\ \\ \end{array} \left[\begin{array}{ccccc|c} 1 & 1 & 1 & 0 & 0 & 10 \\ 1 & 2 & 0 & 1 & 0 & 17 \\ \boxed{3} & 2 & 0 & 0 & 1 & 27 \end{array}\right]$$ $\frac{1}{3}R_3 \rightarrow R_3$

$\uparrow$

$$\left[\begin{array}{ccccc|c} 1 & 1 & 1 & 0 & 0 & 10 \\ 1 & 2 & 0 & 1 & 0 & 17 \\ 1 & \frac{2}{3} & 0 & 0 & \frac{1}{3} & 9 \end{array}\right]$$ $\begin{array}{c} -R_3 + R_1 \rightarrow R_1 \\ -R_3 + R_2 \rightarrow R_2 \end{array}$

$$\left[\begin{array}{ccccc|c} 0 & \frac{1}{3} & 1 & 0 & -\frac{1}{3} & 1 \\ 0 & \frac{4}{3} & 0 & 1 & -\frac{1}{3} & 8 \\ 1 & \frac{2}{3} & 0 & 0 & \frac{1}{3} & 9 \end{array}\right]$$

Another solution is $(9, 0, 1, 8, 0)$ and corner is $(9, 0)$.

Next,

$$
\begin{array}{c}
\leftarrow x \\
r \\
t
\end{array}
\left[
\begin{array}{ccccc|c}
0 & \boxed{1/3} & 1 & 0 & -\frac{1}{3} & 1 \\
0 & \frac{4}{3} & 0 & 1 & -\frac{1}{3} & 8 \\
1 & \frac{2}{3} & 0 & 0 & \frac{1}{3} & 9
\end{array}
\right]
\qquad 3R_1 \to R_1
$$

$$\uparrow$$

$$
\left[
\begin{array}{ccccc|c}
0 & 1 & 3 & 0 & -1 & 3 \\
0 & \frac{4}{3} & 0 & 1 & -\frac{1}{3} & 8 \\
1 & \frac{2}{3} & 0 & 0 & \frac{1}{3} & 9 \\
0 & 1 & 3 & 0 & -1 & 3 \\
0 & 0 & -4 & 1 & 1 & 4 \\
1 & 0 & -2 & 0 & 1 & 7
\end{array}
\right]
$$

$-\frac{4}{3}R_1 + R_2 \to R_2$
$-\frac{2}{3}R_1 + R_3 \to R_3$

A solution is $(7, 3, 0, 4, 0)$
and corner is $(7, 3)$.

Finally,

$$
\begin{array}{c}
x \\
\leftarrow y \\
s
\end{array}
\left[
\begin{array}{ccccc|c}
0 & 1 & 3 & 0 & -1 & 3 \\
0 & 0 & -4 & 1 & \boxed{1} & 4 \\
1 & 0 & -2 & 0 & 1 & 7
\end{array}
\right]
\qquad
\begin{array}{l}
R_1 + R_2 \to R_1 \\
-R_2 + R_3 \to R_3
\end{array}
$$

$$\uparrow$$

$$
\left[
\begin{array}{ccccc|c}
0 & 1 & -1 & 1 & 0 & 7 \\
0 & 0 & -4 & 1 & 1 & 4 \\
1 & 0 & 2 & -1 & 0 & 3
\end{array}
\right]
$$

A solution is $(3, 7, 0, 0, 4)$
and $(3, 7)$ is a corner.

15. Let
$$
\begin{aligned}
2x + y + r &= 18 \\
x + 2y + s &= 21 \\
4x + 5y + t &= 72
\end{aligned}
$$
. The augmented matrix is

$$
\begin{array}{ccccc}
x & y & r & s & t
\end{array}
$$
$$
\left[
\begin{array}{ccccc|c}
2 & 1 & 1 & 0 & 0 & 18 \\
1 & 2 & 0 & 1 & 0 & 21 \\
4 & 5 & 0 & 0 & 1 & 72
\end{array}
\right]
$$

and $(0, 0)$ is a corner.

$$
\begin{array}{c}
\leftarrow r \\
s \\
t
\end{array}
\left[
\begin{array}{ccccc|c}
\boxed{2} & 1 & 1 & 0 & 0 & 18 \\
1 & 2 & 0 & 1 & 0 & 21 \\
4 & 5 & 0 & 0 & 1 & 72
\end{array}
\right]
\qquad \frac{1}{2}R_1 \to R_1
$$

$$\uparrow$$

$$
\left[
\begin{array}{ccccc|c}
1 & \frac{1}{2} & \frac{1}{2} & 0 & 0 & 9 \\
1 & 2 & 0 & 1 & 0 & 21 \\
4 & 5 & 0 & 0 & 1 & 72
\end{array}
\right]
$$

$-R_1 + R_2 \to R_2$
$-4R_1 + R_3 \to R_3$

$$\begin{array}{c} x \\ \leftarrow s \\ t \end{array} \left[\begin{array}{ccccc|c} 1 & \frac{1}{2} & \frac{1}{2} & 0 & 0 & 9 \\ 0 & \boxed{3/2} & -\frac{1}{2} & 1 & 0 & 12 \\ 0 & 3 & -2 & 0 & 1 & 36 \\ & \uparrow & & & & \end{array}\right]$$

Another corner is (9, 0).
$\frac{2}{3}R_2 \rightarrow R_2$

$$\left[\begin{array}{ccccc|c} 1 & \frac{1}{2} & \frac{1}{2} & 0 & 0 & 9 \\ 0 & 1 & -\frac{1}{3} & \frac{2}{3} & 0 & 8 \\ 0 & 3 & -2 & 0 & 1 & 36 \end{array}\right]$$

$-\frac{1}{2}R_2 + R_1 \rightarrow R_1$
$-3R_2 + R_3 \rightarrow R_3$

$$\begin{array}{c} \leftarrow x \\ y \\ t \end{array} \left[\begin{array}{ccccc|c} 1 & 0 & \boxed{2/3} & -\frac{1}{3} & 0 & 5 \\ 0 & 1 & -\frac{1}{3} & \frac{2}{3} & 0 & 8 \\ 0 & 0 & -1 & -2 & 1 & 12 \\ & & \uparrow & & & \end{array}\right]$$

Another corner is (5, 8).
$\frac{3}{2}R_1 \rightarrow R_1$

$$\left[\begin{array}{ccccc|c} \frac{3}{2} & 0 & 1 & -\frac{1}{2} & 0 & \frac{15}{2} \\ 0 & 1 & -\frac{1}{3} & \frac{2}{3} & 0 & 8 \\ 0 & 0 & -1 & -2 & 1 & 12 \end{array}\right]$$

$\frac{1}{3}R_1 + R_2 \rightarrow R_2$
$R_1 + R_3 \rightarrow R_3$

$$\left[\begin{array}{ccccc|c} \frac{3}{2} & 0 & 1 & -\frac{1}{2} & 0 & \frac{15}{2} \\ \frac{1}{2} & 1 & 0 & \frac{1}{2} & 0 & \frac{21}{2} \\ \frac{3}{2} & 0 & 0 & -\frac{5}{2} & 1 & \frac{39}{2} \end{array}\right]$$

A fourth corner is (0, 10.5).

17. Let $\begin{array}{l} 4x+5y+4z+s = 40 \\ 12x+9y+8z+t = 96 \end{array}$. The augmented matrix is

$$\begin{array}{ccccc} x & y & z & s & t \\ \end{array}$$
$$\left[\begin{array}{ccccc|c} 4 & 5 & 4 & 1 & 0 & 40 \\ 12 & 9 & 8 & 0 & 1 & 96 \end{array}\right],$$

and one corner is (0, 0, 0).

$$\begin{array}{c} s \\ \leftarrow t \end{array} \left[\begin{array}{ccccc|c} 4 & 5 & 4 & 1 & 0 & 40 \\ \boxed{12} & 9 & 8 & 0 & 1 & 96 \\ \uparrow & & & & & \end{array}\right]$$

$\frac{1}{12}R_2 \rightarrow R_2$

$$\left[\begin{array}{ccccc|c} 4 & 5 & 4 & 1 & 0 & 40 \\ 1 & \frac{3}{4} & \frac{2}{3} & 0 & \frac{1}{12} & 8 \end{array}\right]$$

$-4R_2 + R_1 \rightarrow R_1$

$$\left[\begin{array}{ccccc|c} 0 & 2 & \frac{4}{3} & 1 & -\frac{1}{3} & 8 \\ 1 & \frac{3}{4} & \frac{2}{3} & 0 & \frac{1}{12} & 8 \end{array}\right]$$

(8, 0, 0) is another corner.

$$\begin{bmatrix} 0 & \boxed{2} & \frac{4}{3} & 1 & -\frac{1}{3} & \Big| & 8 \\ 1 & \frac{3}{4} & \frac{2}{3} & 0 & \frac{1}{12} & \Big| & 8 \end{bmatrix}$$
$\uparrow$

$\frac{1}{2}R_1 \rightarrow R_1$

$$\begin{bmatrix} 0 & 1 & \frac{2}{3} & \frac{1}{2} & -\frac{1}{6} & \Big| & 4 \\ 1 & \frac{3}{4} & \frac{2}{3} & 0 & \frac{1}{12} & \Big| & 8 \end{bmatrix}$$

$-\frac{3}{4}R_1 + R_2 \rightarrow R_2$

$\leftarrow \begin{bmatrix} 0 & 1 & \boxed{2/3} & \frac{1}{2} & -\frac{1}{6} & \Big| & 4 \\ 1 & 0 & \frac{1}{6} & -\frac{3}{8} & \frac{5}{24} & \Big| & 5 \end{bmatrix}$
$\uparrow$

$(5, 4, 0)$ is another corner.
$\frac{3}{2}R_1 \rightarrow R_1$

$$\begin{bmatrix} 0 & \frac{3}{2} & 1 & \frac{3}{4} & -\frac{1}{4} & \Big| & 6 \\ 1 & 0 & \frac{1}{6} & -\frac{3}{8} & \frac{5}{24} & \Big| & 5 \end{bmatrix}$$

$-\frac{1}{6}R_1 + R_2 \rightarrow R_2$

$$\begin{bmatrix} 0 & \frac{3}{2} & 1 & \frac{3}{4} & -\frac{1}{4} & \Big| & 6 \\ 1 & -\frac{1}{4} & 0 & -\frac{1}{2} & \frac{1}{6} & \Big| & 4 \end{bmatrix}$$

$(4, 0, 6)$ is a fourth solution.

19. Let $\begin{aligned} 4x+y+4z+s &= 16 \\ 2x+3y+7z+t &= 18 \end{aligned}$. The augmented matrix is

$$\begin{array}{ccccc} x & y & z & s & t \\ \begin{bmatrix} 4 & 1 & 4 & 1 & 0 & \Big| & 16 \\ 2 & 3 & 7 & 0 & 1 & \Big| & 18 \end{bmatrix} \end{array}$$

and $(0, 0, 0)$ is a corner.

$\leftarrow s \begin{bmatrix} \boxed{4} & 1 & 4 & 1 & 0 & \Big| & 16 \\ 2 & 3 & 7 & 0 & 1 & \Big| & 18 \end{bmatrix}$
$\underset{x}{\uparrow}$

$\frac{1}{4}R_1 \rightarrow R_1$

$$\begin{bmatrix} 1 & \frac{1}{4} & 1 & \frac{1}{4} & 0 & \Big| & 4 \\ 2 & 3 & 7 & 0 & 1 & \Big| & 18 \end{bmatrix}$$

$-2R_1 + R_2 \rightarrow R_2$

$\leftarrow t \begin{bmatrix} 1 & \frac{1}{4} & 1 & \frac{1}{4} & 0 & \Big| & 4 \\ 0 & \frac{5}{2} & \boxed{5} & -\frac{1}{2} & 1 & \Big| & 10 \end{bmatrix}$
$\underset{z}{\uparrow}$

$(4, 0, 0)$ is a corner.
$(1/5)R_2 \rightarrow R_2$

$$\begin{bmatrix} 1 & \frac{1}{4} & 1 & \frac{1}{4} & 0 & \Big| & 4 \\ 0 & \frac{1}{2} & 1 & -\frac{1}{10} & \frac{1}{5} & \Big| & 2 \end{bmatrix}$$

$-R_2 + R_1 \rightarrow R_1$

$$
\begin{array}{c}
\\
\leftarrow z
\end{array}
\left[
\begin{array}{ccccc|c}
1 & -\frac{1}{4} & 0 & \frac{3}{20} & -\frac{1}{5} & 2 \\
0 & \boxed{1/2} & 1 & -\frac{1}{10} & \frac{1}{5} & 2
\end{array}
\right]
\qquad
\begin{array}{l}
(2,\,0,\,2)\ \text{is a corner.}\\
2R_2 \to R_2
\end{array}
$$
$$\underset{\uparrow y}{}$$

$$
\left[
\begin{array}{ccccc|c}
1 & -\frac{1}{4} & 0 & \frac{3}{20} & -\frac{1}{5} & 2 \\
0 & 1 & 2 & -\frac{1}{5} & \frac{2}{5} & 4
\end{array}
\right]
\qquad
\tfrac{1}{4}R_2 + R_1 \to R_1
$$

$$
\left[
\begin{array}{ccccc|c}
1 & 0 & \frac{1}{2} & \frac{1}{10} & -\frac{1}{10} & 3 \\
0 & 1 & 2 & -\frac{1}{5} & \frac{2}{5} & 4
\end{array}
\right]
\qquad
(3,\,4,\,0)\ \text{is a corner.}
$$

21. Let $\begin{array}{l}2x+y+z+r=10\\2x+2y+3z+s=14\end{array}$. The augmented matrix is

$$
\begin{array}{ccccc}
x & y & z & r & s
\end{array}
$$
$$
\left[
\begin{array}{ccccc|c}
2 & 1 & 1 & 1 & 0 & 10 \\
2 & 2 & 3 & 0 & 1 & 14
\end{array}
\right],
\qquad
\text{and } (0,\,0,\,0)\ \text{is a corner.}
$$

$$
\begin{array}{c}
\leftarrow r
\end{array}
\left[
\begin{array}{ccccc|c}
\boxed{2} & 1 & 1 & 1 & 0 & 10 \\
2 & 2 & 3 & 0 & 1 & 14
\end{array}
\right]
\qquad
\tfrac{1}{2}R_1 \to R_1
$$
$$\underset{\uparrow x}{}$$

$$
\left[
\begin{array}{ccccc|c}
1 & \frac{1}{2} & \frac{1}{2} & \frac{1}{2} & 0 & 5 \\
2 & 2 & 3 & 0 & 1 & 14
\end{array}
\right]
\qquad
-2R_1 + R_2 \to R_2
$$

$$
\begin{array}{c}
\\
\leftarrow t
\end{array}
\left[
\begin{array}{ccccc|c}
1 & \frac{1}{2} & \frac{1}{2} & \frac{1}{2} & 0 & 5 \\
0 & \boxed{1} & 2 & -1 & 1 & 4
\end{array}
\right]
\qquad
\begin{array}{l}
(5,\,0,\,0)\ \text{is a corner.}\\
-\tfrac{1}{2}R_2 + R_1 \to R_1
\end{array}
$$
$$\underset{\uparrow y}{}$$

$$
\begin{array}{c}
\\
\leftarrow y
\end{array}
\left[
\begin{array}{ccccc|c}
1 & 0 & -\frac{1}{2} & 1 & -\frac{1}{2} & 3 \\
0 & 1 & \boxed{2} & -1 & 1 & 4
\end{array}
\right]
\qquad
\begin{array}{l}
(3,\,4,\,0)\ \text{is a corner.}\\
\tfrac{1}{2}R_2 \to R_2
\end{array}
$$
$$\underset{\uparrow z}{}$$

$$
\left[
\begin{array}{ccccc|c}
1 & 0 & -\frac{1}{2} & 1 & -\frac{1}{2} & 3 \\
0 & \frac{1}{2} & 1 & -\frac{1}{2} & \frac{1}{2} & 2
\end{array}
\right]
\qquad
\tfrac{1}{2}R_2 + R_1 \to R_1
$$

$$
\left[
\begin{array}{ccccc|c}
1 & \frac{1}{4} & 0 & \frac{3}{4} & -\frac{1}{4} & 4 \\
0 & \frac{1}{2} & 1 & -\frac{1}{2} & \frac{1}{2} & 2
\end{array}
\right]
\qquad
(4,\,0,\,2)\ \text{is a corner.}
$$

23. Let
$$x + y + z + r = 15$$
$$x + y \quad\quad + s = 10$$
$$4x - 3y \quad\quad + t = 12$$
$$-3x + 4y \quad\quad + w = 12.$$

The augmented matrix is

$$\begin{array}{ccccccc} x & y & z & r & s & t & w \end{array}$$
$$\left[\begin{array}{ccccccc|c} 1 & 1 & 1 & 1 & 0 & 0 & 0 & 15 \\ 1 & 1 & 0 & 0 & 1 & 0 & 0 & 10 \\ 4 & -3 & 0 & 0 & 0 & 1 & 0 & 12 \\ -3 & 4 & 0 & 0 & 0 & 0 & 1 & 12 \end{array}\right].$$

One corner is $(0, 0, 0)$ and another is $(0, 0, 15)$.

$$\leftarrow t \left[\begin{array}{ccccccc|c} 1 & 1 & 1 & 1 & 0 & 0 & 0 & 15 \\ 1 & 1 & 0 & 0 & 1 & 0 & 0 & 10 \\ \boxed{4} & -3 & 0 & 0 & 0 & 1 & 0 & 12 \\ -3 & 4 & 0 & 0 & 0 & 0 & 1 & 12 \end{array}\right] \qquad R_3/4 \to R_3$$
$$\uparrow x$$

$$\left[\begin{array}{ccccccc|c} 1 & 1 & 1 & 1 & 0 & 0 & 0 & 15 \\ 1 & 1 & 0 & 0 & 1 & 0 & 0 & 10 \\ 1 & -\frac{3}{4} & 0 & 0 & 0 & \frac{1}{4} & 0 & 3 \\ -3 & 4 & 0 & 0 & 0 & 0 & 1 & 12 \end{array}\right] \qquad \begin{array}{l} -R_3+R_1 \to R_1 \\ -R_3+R_2 \to R_2 \\ 3R_3+R_4 \to R_4 \end{array}$$

$$\leftarrow s \left[\begin{array}{ccccccc|c} 0 & \frac{7}{4} & 1 & 1 & 0 & -\frac{1}{4} & 0 & 12 \\ 0 & \boxed{7/4} & 0 & 0 & 1 & -\frac{1}{4} & 0 & 7 \\ 1 & -\frac{3}{4} & 0 & 0 & 0 & \frac{1}{4} & 0 & 3 \\ 0 & \frac{7}{4} & 0 & 0 & 0 & \frac{3}{4} & 1 & 21 \end{array}\right] \qquad \begin{array}{l} \text{Other corners are} \\ (3, 0, 12) \text{ and } (3, 0, 0). \\ \frac{4}{7}R_2 \to R_2 \end{array}$$
$$\uparrow y$$

$$\left[\begin{array}{ccccccc|c} 0 & \frac{7}{4} & 1 & 1 & 0 & -\frac{1}{4} & 0 & 12 \\ 0 & 1 & 0 & 0 & \frac{4}{7} & -\frac{1}{7} & 0 & 4 \\ 1 & -\frac{3}{4} & 0 & 0 & 0 & \frac{1}{4} & 0 & 3 \\ 0 & \frac{7}{4} & 0 & 0 & 0 & \frac{3}{4} & 1 & 21 \end{array}\right] \qquad \begin{array}{l} -\frac{7}{4}R_2+R_1 \to R_1 \\ \frac{3}{4}R_2+R_3 \to R_3 \\ -\frac{7}{4}R_2+R_4 \to R_4 \end{array}$$

$$\left[\begin{array}{ccccccc|c} 0 & 0 & 1 & 1 & 0 & 0 & 0 & 5 \\ 0 & 1 & 0 & 0 & \frac{4}{7} & -\frac{1}{7} & 0 & 4 \\ 1 & 0 & 0 & 0 & \frac{3}{7} & \frac{1}{7} & 0 & 6 \\ 0 & 0 & 0 & 0 & 1 & 1 & 1 & 14 \end{array}\right]$$

Other corners on $(6, 4, 5)$ and $(6, 4, 0)$.

25. Let $-3x + y + r = 3$
$$x - y + s = 5$$
$$2x - y + t = 15.$$

The augmented matrix is

	x	y	r	s	t		
$\leftarrow r$	-3	$\boxed{1}$	1	0	0		3
s	1	-1	0	1	0		5
t	2	-1	0	0	1		15
		$\uparrow$					
		y					

One corner is $(0, 0)$.
$R_1 + R_2 \to R_2$
$R_1 + R_3 \to R_3$

y	-3	1	1	0	0		3
s	-2	0	1	1	0		8
t	-1	0	1	0	1		18

Another corner is $(0, 3)$.

Since all the entries in column one are negative the region is unbounded.

27. Let $-4x + y + r = 16$
$$-3x + y + s = 7$$
$$x - 4y + t = 5$$
$$x - 3y + w = 6.$$

The augmented matrix is

	x	y	r	s	t	w		
$\leftarrow r$	-4	$\boxed{1}$	1	0	0	0		6
	-3	1	0	1	0	0		7
	1	-4	0	0	1	0		5
	1	-3	0	0	0	1		6
		$\uparrow$						
		y						

$(0, 0)$ is a corner.
$-R_1 + R_2 \to R_2$
$4R_1 + R_3 \to R_3$
$3R_1 + R_4 \to R_4$

	-4	1	1	0	0	0		6
$\leftarrow s$	$\boxed{1}$	0	-1	1	0	0		1
	-15	0	4	0	1	0		29
	-11	0	3	0	0	1		24
	$\uparrow$							
	x							

$(0, 6)$ is a corner.

$4R_2 + R_1 \to R_1,$ $\qquad$ $15R_2 + R_3 \to R_3,$ $\qquad$ $11R_2 + R_4 \to R_4$

0	1	-3	4	0	0		10
1	0	-1	1	0	0		1
0	0	-11	15	1	0		44
0	0	-8	11	0	1		35

$(1, 10)$ is a corner.

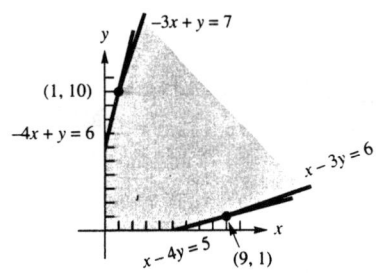

Since all the entries in column 3 are negative, and t and w are basic variables, it is not possible to move from $(1, 10)$ directly to a new corner of the region. The region is unbounded.

EXERCISE SET 4.5 THE SIMPLEX METHOD

1. Let
$$x + 5y + s_1 = 75$$
$$2x + 5y + s_2 = 90$$
$$x + y + s_3 = 33$$
$$z - 2x - 4y = 0. \quad \text{The augmented matrix is}$$

$$
\begin{bmatrix}
0 & 1 & 5 & 1 & 0 & 0 & 75 \\
0 & 2 & 5 & 0 & 1 & 0 & 90 \\
0 & 1 & 1 & 0 & 0 & 1 & 33 \\
1 & -2 & -4 & 0 & 0 & 0 & 0
\end{bmatrix}
$$

Tableau I

basis	z	x	y	s_1	s_2	s_3	values
s_1	0	1	[5]	1	0	0	75
s_2	0	2	5	0	1	0	90
s_3	0	1	1	0	0	1	33
z	1	-2	-4	0	0	0	0

↑

Tableau II

basis	z	x	y	s_1	s_2	s_3	values
y	0	$\frac{1}{5}$	1	$\frac{1}{5}$	0	0	15
s_2	0	[1]	0	-1	1	0	15
s_3	0	$\frac{4}{5}$	0	$-\frac{1}{5}$	0	1	18
z	1	$-\frac{6}{5}$	0	$\frac{4}{5}$	0	0	60

↑

Tableau III

basis	z	x	y	s_1	s_2	s_3	values
y	0	0	1	$\frac{2}{5}$	$-\frac{1}{5}$	0	12
x	0	1	0	-1	1	0	15
$\leftarrow$ s_3	0	0	0	$\frac{3}{5}$	$-\frac{4}{5}$	1	6
z	1	0	0	$-\frac{2}{5}$	$\frac{6}{5}$	0	78

$\uparrow$

Tableau IV

basis	z	x	y	s_1	s_2	s_3	values
y	0	0	1	0	$\frac{1}{3}$	$-\frac{2}{5}$	12
x	0	1	0	0	$-\frac{1}{3}$	$\frac{1}{3}$	25
s_1	0	0	0	1	$-\frac{4}{3}$	$\frac{1}{3}$	10
z	1	0	0	0	$\frac{2}{3}$	$\frac{2}{3}$	82

The maximum value of z is 82 which occurs when $x=25$ and $y=8$

3. Let
$$\begin{aligned}
-2x + 3y + r &= 3 \\
-x + 4y \quad + s &= 8 \\
z - 3x - 2y &= 0.
\end{aligned}$$

Tableau I

basis	x	y	r	s	values
r	-2	3	1	0	3
s	-1	4	0	1	8
z	-3	-2	0	0	0

Since all entries in the first column are negative the region of feasible solutions is unbounded and z does not reach a maximum value.

5. Let
$$\begin{aligned}
x - y + s_1 &= 60 \\
8x + 5y \quad + s_2 &= 600 \\
2x + y \quad + s_3 &= 90 \\
-x + y \quad + s_4 &= 60 \\
z - 5x - 5y &= 0.
\end{aligned}$$

Tableau I

basis	x	y	s_1	s_2	s_3	s_4	values
s_1	1	-1	1	0	0	0	60
s_2	8	5	0	1	0	0	600
s_3	2	1	0	0	1	0	90
$\leftarrow$ s_4	-1	$\boxed{1}$	0	0	0	1	60
z	-5	-5	0	0	0	0	0

$\uparrow$

Tableau II

basis	x	y	s_1	s_2	s_3	s_4	values
s_1	0	0	1	0	0	1	120
s_2	13	0	0	1	0	-5	300
$\leftarrow$ s_3	3	0	0	0	1	-1	30
y	-1	1	0	0	0	1	60
z	-10	0	0	0	0	5	300

$\uparrow$

Tableau III

basis	x	y	s_1	s_2	s_3	s_4	values
s_1	0	0	1	0	$-$	$-$	$-$
s_2	0	0	0	1	$-$	$-$	$-$
x	1	0	0	0	$\frac{1}{3}$	$-\frac{1}{3}$	10
y	0	1	0	0	$\frac{1}{3}$	$\frac{2}{3}$	70
z	0	0	0	0	$\frac{10}{3}$	$\frac{5}{3}$	400

The maximum value of z is 400 which occurs when $x=10$ and $y=70$.

7. Let
$$7x + 3y + s_1 \qquad\qquad = 210$$
$$x + 7y \qquad + s_2 \qquad = 140$$
$$2x + 5y \qquad\qquad + s_3 = 118$$
$$z - 14x - 24y \qquad\qquad\qquad = 0$$

Tableau I

basis	x	y	s_1	s_2	s_3	values
s_1	7	3	1	0	0	210
s_2	1	$\boxed{7}$	0	1	0	140
s_3	2	5	0	0	1	118
z	-14	-24	0	0	0	0

Tableau II

basis	x	y	s_1	s_2	s_3	values
s_1	$\frac{46}{7}$	0	1	$-\frac{3}{7}$	0	150
y	$\boxed{1/7}$	1	0	$\frac{1}{7}$	0	20
s_3	$\frac{9}{7}$	0	0	$-\frac{5}{7}$	1	18
z	$-\frac{74}{7}$	0	0	$\frac{24}{7}$	0	480

Tableau III

basis	x	y	s_1	s_2	s_3	values
s_1	0	0	1	$\boxed{29/9}$	$-\frac{46}{9}$	58
y	0	1	0	$\frac{2}{9}$	$-\frac{1}{9}$	18
x	1	0	0	$-\frac{5}{9}$	$\frac{7}{9}$	14
z	0	0	0	$-\frac{22}{9}$	$\frac{74}{9}$	628

Tableau IV

basis	x	y	s_1	s_2	s_3	values
s_2	0	0	$\frac{9}{29}$	1	$-\frac{46}{29}$	$\frac{2178}{29}$
y	0	1	—	0	—	14
x	1	0	—	0	—	24
z	0	0	$\frac{22}{29}$	0	$\frac{126}{29}$	672

The maximum value of z is 672 which occurs when $x=24$ and $y=14$.

9. Let

$$
\begin{aligned}
4x + \quad\quad 5z + s_1 \quad\quad\quad\quad &= 102 \\
x + y \quad\quad\quad + s_2 \quad\quad &= 48 \\
x \quad\quad\quad\quad\quad + s_3 &= 45 \\
w - 8x - 9y - 6z \quad\quad\quad\quad\quad &= 0.
\end{aligned}
$$

Tableau I

basis	x	y	z	s_1	s_2	s_3	values
s_1	4	0	5	1	0	0	210
s_2	1	☐1	0	0	1	0	48
s_3	1	0	0	0	0	1	45
w	−8	−9	−6	0	0	0	0

Tableau II

basis	x	y	z	s_1	s_2	s_3	values
s_1	4	0	☐5	1	0	0	210
y	1	1	0	0	1	0	48
s_3	1	0	0	0	0	1	45
w	1	0	−6	0	9	0	432

Tableau III

basis	x	y	z	s_1	s_2	s_3	values
z	$\frac{4}{5}$	0	1	$\frac{1}{5}$	0	0	42
y	1	1	0	0	1	0	48
s_3	1	0	0	0	0	1	45
w	$\frac{29}{5}$	0	0	$\frac{6}{5}$	9	0	684

The maximum value of w is 684, and occurs when $x=0$, $y=48$ and $z=42$.

11. Let
$$x + y + 2z + s_1 \qquad = 14$$
$$y + 4z \qquad + s_2 = 12$$
$$w - 2x - 4y - 12z \qquad = 0$$

Tableau I

basis	x	y	z	s_1	s_3	values
s_1	1	1	2	1	0	14
s_2	0	1	$\boxed{4}$	0	1	12
w	-2	-4	-12	0	0	0

Tableau II

basis	x	y	z	s_1	s_3	values
s_1	$\boxed{1}$	$\frac{1}{2}$	0	1	$-\frac{1}{2}$	8
z	0	$\frac{1}{4}$	1	0	$\frac{1}{4}$	3
w	-2	-1	0	0	3	36

Tableau III

basis	x	y	z	s_1	s_3	values
x	1	$\frac{1}{2}$	0	1	$-\frac{1}{2}$	8
z	0	$\boxed{1/4}$	1	0	$\frac{1}{4}$	3
w	0	0	0	2	2	52

The maximum value of w is 52 which occurs at $(8, 0, 3)$ but may occur elsewhere.

Tableau IV

basis	x	y	z	s_1	s_3	values
x	1	0	-2	1	-1	2
y	0	1	4	0	1	12
w	0	0	0	2	2	52

The maximum also occurs at the point $(2, 12, 0)$ and occurs for all (x, y, z) such that
$$x=(1-t)8+2t, \qquad y=12t, \qquad z=(1-t)3$$
for t a real number and $0 \leq t \leq 1$.

13. Let

$$
\begin{aligned}
x + 2y + z + s_1 &&&= 168 \\
2x + y + z &&+ s_2 &= 280 \\
x + y + 3z + &&+ s_3 &= 210 \\
w - x - y - z &&&= 0.
\end{aligned}
$$

Tableau I

basis	x	y	z	s_1	s_2	s_3	values
s_1	1	2	1	1	0	0	168
s_2	[2]	1	1	0	1	0	280
s_3	1	1	3	0	0	1	210
w	-1	-1	-1	0	0	0	0

Tableau II

basis	x	y	z	s_1	s_2	s_3	values
s_1	0	$\frac{3}{2}$	$\frac{1}{2}$	1	$-\frac{1}{2}$	0	28
x	1	$\frac{1}{2}$	$\frac{1}{2}$	0	$\frac{1}{2}$	0	140
s_3	0	$\frac{1}{2}$	$[5/2]$	0	$-\frac{1}{2}$	1	70
w	0	$-\frac{1}{2}$	$-\frac{1}{2}$	0	$\frac{1}{2}$	0	140

Tableau III

basis	x	y	z	s_1	s_2	s_3	values
s_1	0	$[7/5]$	0	1	$-\frac{2}{5}$	$-\frac{1}{5}$	14
x	1	$\frac{2}{5}$	0	0	$\frac{3}{5}$	$-\frac{1}{5}$	126
z	0	$\frac{1}{5}$	1	0	$-\frac{1}{5}$	$\frac{2}{5}$	28
w	0	$-\frac{2}{5}$	0	0	$\frac{2}{5}$	$\frac{1}{5}$	154

Tableau IV

basis	x	y	z	s_1	s_2	s_3	values
y	0	1	0	$\frac{5}{7}$	$-\frac{2}{7}$	$-\frac{1}{7}$	10
x	1	0	0	$-$	$-$	$-$	122
z	0	0	1	$-$	$-$	$-$	26
w	0	0	0	$\frac{2}{7}$	$\frac{2}{7}$	$\frac{1}{7}$	158

The maximum value of w is 158 and occurs when $x=122$, $y=10$ and $z=26$.

15. Let
$$x + 4y + 2z + s_1 \qquad = 200$$
$$4x + 6y + 5z \qquad + s_2 = 750$$
$$w - 10x - 25y - 15z \qquad = 0.$$

Tableau I

basis	x	y	z	s_1	s_2	values
s_1	1	$\boxed{4}$	2	1	0	200
s_2	4	6	5	0	1	750
w	-10	-25	-15	0	0	0

Tableau II

basis	x	y	z	s_1	s_2	values
y	$\frac{1}{4}$	1	$\frac{1}{2}$	$\frac{1}{4}$	0	50
s_2	$\boxed{5/2}$	0	2	$-\frac{3}{2}$	1	450
w	$-\frac{15}{4}$	0	$-\frac{5}{2}$	$\frac{25}{4}$	0	1250

$$\frac{3}{2}R_2 + R_3 \rightarrow R_3$$
$$-\frac{1}{10}R_2 + R_1 \rightarrow R_1$$

Tableau III

basis	x	y	z	s_1	s_2	values
y	0	1	$\frac{3}{10}$	$\frac{2}{5}$	$-\frac{1}{10}$	5
x	1	0	$\frac{4}{5}$	$-\frac{3}{5}$	$\frac{2}{5}$	180
w	0	0	$\frac{1}{2}$	4	$\frac{3}{2}$	1925

The maximum value of w is 1925 which occurs when $x=180$ and $y=5$, $z = 0$.

17. Let
$$8x + 10y - 9z + s_1 = 80$$
$$30x - 25y + 60z + s_2 = 300$$
$$-22x + 35y + 56z + s_3 = 280$$
$$w - 8x - 2y - 3z = 0.$$

Tableau I

basis	x	y	z	s_1	s_2	s_3	values
s_1	$\boxed{8}$	10	-9	1	0	0	80
s_2	30	-25	60	0	1	0	300
s_3	-22	35	56	0	0	1	280
w	-8	-2	-3	0	0	0	0

Tableau II

basis	x	y	z	s_1	s_2	s_3	values
x	1	$\frac{5}{4}$	$-\frac{9}{8}$	$\frac{1}{8}$	0	0	10
s_2	0	$-\frac{125}{2}$	$\boxed{\frac{375}{4}}$	$-\frac{15}{4}$	1	0	0
s_3	0	$\frac{195}{4}$	$\frac{125}{4}$	22	0	1	2040
w	0	8	-12	1	0	0	80

Tableau III

basis	x	y	z	s_1	s_2	s_3	values
x	1	$\boxed{1/2}$	0	$\frac{2}{25}$	$\frac{3}{250}$	0	10
z	0	$-\frac{2}{3}$	1	$-\frac{1}{25}$	$\frac{4}{375}$	0	0
s_3	0	$\frac{245}{4}$	0	$\frac{93}{4}$	$-\frac{1}{3}$	1	2040
w	0	0	0	$\frac{13}{25}$	$\frac{16}{125}$	0	80

The maximum value of w is 80 which occurs when $x=10$, $y=0$, and $z=0$.

Tableau IV

basis	x	y	z	s_1	s_2	s_3	values
y	2	1	0	$\frac{4}{25}$	$\frac{3}{125}$	0	20
z	$\frac{4}{3}$	0	1	-	-	0	$\frac{40}{3}$
s_3	-	0	0	-	-	1	3265
w	0	0	0	$\frac{13}{25}$	$\frac{16}{125}$	0	80

The maximum is also attained when $x=0$, $y=20$ and $z=\frac{40}{3}$ and at all points where

$$x=(1-t)10, \qquad y=20t \qquad \text{and} \qquad z=\frac{40}{3}t, \qquad \text{for } 0 \leq t \leq 1.$$

19. Let
$$
\begin{aligned}
x + y + z + s_1 &= 40 \\
2x + y - 3z + s_2 &= 20 \\
x - y + 2z + s_3 &= 30 \\
w - 3x - 6y + 3z &= 0.
\end{aligned}
$$

Tableau I

basis	x	y	z	s_1	s_2	s_3	values
s_1	1	1	1	1	0	0	40
s_2	2	☐1	−3	0	1	0	20
s_3	1	−1	2	0	0	1	30
w	−3	−6	3	0	0	0	0

Tableau II

basis	x	y	z	s_1	s_2	s_3	values
s_1	−1	0	☐4	1	−1	0	20
y	2	1	−3	0	1	0	20
s_3	3	0	−1	0	1	1	50
w	9	0	−15	0	6	0	120

Tableau III

basis	x	y	z	s_1	s_2	s_3	values
z	$-\frac{1}{4}$	0	1	$\frac{1}{4}$	$-\frac{1}{4}$	0	5
y	$\frac{5}{4}$	1	0	$\frac{3}{4}$	$\frac{1}{4}$	0	35
s_3	$\frac{11}{4}$	0	0	$\frac{1}{4}$	$\frac{3}{4}$	1	55
w	$\frac{21}{4}$	0	0	$\frac{15}{4}$	$\frac{9}{4}$	0	195

The maximum value of w is 195 which occurs when $x=0$, $y=35$ and $z=5$.

21. Let
$$
\begin{aligned}
3x + 2y \qquad\quad + s_1 \qquad\qquad &= 60 \\
3x + 2y + 3z \qquad + s_2 \qquad &= 72 \\
2y + 3z \qquad\qquad + s_3 &= 48 \\
w - 3x - 4y - 5z \qquad\qquad\qquad &= 0.
\end{aligned}
$$

Tableau I

basis	x	y	z	s_1	s_2	s_3	values
s_1	3	2	0	1	0	0	60
s_2	3	2	3	0	1	0	72
s_3	0	2	[3]	0	0	1	48
w	-3	-4	-5	0	0	0	0

Tableau II

basis	x	y	z	s_1	s_2	s_3	values
s_1	3	2	0	1	0	0	60
s_2	[3]	0	0	0	1	-1	24
z	0	$\frac{2}{3}$	1	0	0	$\frac{1}{3}$	16
w	-3	$-\frac{2}{3}$	0	0	0	$\frac{5}{3}$	80

Tableau III

basis	x	y	z	s_1	s_2	s_3	values
s_1	0	$\boxed{2}$	0	1	-1	1	36
x	1	0	0	0	$\frac{1}{3}$	$-\frac{1}{3}$	8
z	0	$\frac{2}{3}$	1	0	0	$\frac{1}{3}$	16
w	0	$-\frac{2}{3}$	0	0	1	$\frac{2}{3}$	104

Tableau IV

basis	x	y	z	s_1	s_2	s_3	values
y	0	1	0	$\frac{1}{2}$	$-\frac{1}{2}$	$\frac{1}{2}$	18
x	1	0	0	0	$\frac{1}{3}$	$-\frac{1}{3}$	8
z	0	0	1	$-\frac{1}{3}$	$\frac{1}{3}$	0	4
w	0	0	0	$\frac{1}{3}$	$\frac{2}{3}$	1	116

The maximum value of w is 116 which occurs when $x=8$, $y=18$ and $z=4$ and elsewhere.

23. Let
$$3x_1 + x_2 + 2x_3 + x_4 + s_1 = 10$$
$$x_1 + 2x_2 + x_3 + x_4 + s_2 = 5$$
$$z - 5x_1 - 10x_2 - 3x_3 - 25x_4 = 0.$$

Tableau I

basis	x_1	x_2	x_3	x_4	s_1	s_2	values
s_1	3	1	2	1	1	0	10
s_2	1	2	1	$\boxed{1}$	0	1	5
z	-5	-10	-5	-25	0	0	0

Tableau II

basis	x_1	x_2	x_3	x_4	s_1	s_2	values
s_1	2	-1	1	0	1	-1	5
x_4	1	2	1	1	0	1	5
z	20	40	20	0	0	25	125

The maximum value of z is 125 and this occurs when $x_1=x_2=x_3=0$ and $x_4=5$.

25. Let x be the number of acres of tomatoes and y the number of acres of peas.

$$x + \quad y \leq \quad 100$$
$$50x + 25y \leq 6250 \quad \text{or} \quad 2x+y \leq 250$$

the profit $z = 150x+100y$.

Let
$$x + \quad y + s_1 \qquad\qquad = 100$$
$$2x + \quad y \qquad + s_2 = 250$$
$$z - 150x - 100y \qquad\qquad = \quad 0.$$

Tableau I

basis	x	y	s_1	s_2	values
s_1	$\boxed{1}$	1	1	0	100
s_2	2	1	0	1	250
z	-150	-100	0	0	0

Tableau II

basis	x	y	s_1	s_2	values
x	1	1	1	0	100
s_2	0	-1	-2	1	50
z	0	50	150	0	15,000

The maximum profit is \$15,000 which occurs when the complete 100 acres is planted in tomatoes.

27. Let x be the number of acres planted in oranges and y the number of acres planted in lemons,

$$x+y \leq 100$$
$$0 \leq y \leq 50$$
$$0 \leq x$$
$$150x+300y \leq 18,000$$

The total profit is $z = 500x+250y$.

Let
$$y + s_1 \qquad\qquad\qquad = \quad 50$$
$$150x + 300y \qquad + s_2 \qquad = 18000$$
$$x + \quad y \qquad\qquad + s_3 = \quad 100$$
$$z - 500x - 250y \qquad\qquad\qquad = \quad 0.$$

Tableau I

basis	x	y	s_1	s_2	s_3	values
s_1	0	1	1	0	0	50
s_2	150	300	0	1	0	18,000
s_3	[1]	1	0	0	1	100
z	-500	-250	0	0	0	0

Tableau II

basis	x	y	s_1	s_2	s_3	values
s_1	0	1	1	0	0	50
s_2	0	150	0	1	-150	3000
x	1	1	0	0	1	100
z	0	250	0	0	500	50,000

The maximum profit is \$50,000 which occurs when all 100 acres are planted in oranges and none in lemons.

29. Let x, y and z be the number of units of products A, B and C, respectively, which are manufactured.

$$3x + 5y + 3z \le 450$$
$$3x + 2y + \le 60$$
$$x + 2y \le 120$$

and the total profit is $w = 70x + 100y + 40z$.

Let
$$3x + 5y + 3z + s_1 = 450$$
$$3x + 2y + s_2 = 60$$
$$x + 2y + s_3 = 120$$
$$w - 70x - 100y - 40z = 0.$$

Tableau I

basis	x	y	z	s_1	s_2	s_3	values
s_1	3	5	3	1	0	0	450
s_2	3	[2]	0	0	1	0	60
s_3	1	2	0	0	0	1	120
w	-70	-100	-40	0	0	0	0

Tableau II

basis	x	y	z	s_1	s_2	s_3	values
s_1	-3	0	$\boxed{3}$	1	$-\frac{5}{2}$	0	300
y	$\frac{3}{2}$	1	0	0	$\frac{1}{2}$	0	30
s_3	-2	0	0	0	-1	1	60
w	80	0	-40	0	50	0	3000

Tableau III

basis	x	y	z	s_1	s_2	s_3	values
z	-1	0	1	$\frac{1}{3}$	$-\frac{5}{6}$	0	100
y	$\frac{3}{2}$	1	0	0	$\boxed{1/2}$	0	30
s_3	-2	0	0	0	-1	1	60
w	40	0	0	$\frac{40}{3}$	$\frac{50}{3}$	0	7000

The maximum profit is $7000 which occurs when 100 units of product C is produced, 30 units of product B, and none of product A.

31. Let x, y and z be the number of acres in peanuts, wheat and barley respectively.

$$x + y + z \leq 40$$
$$100x + 50y + 100z \leq 3000$$
$$300x + 200y + 100z \leq 8000 \qquad \text{and the total profit is } w = 100x + 300y + 400z.$$

Tableau I

basis	x	y	z	s_1	s_2	s_3	values
s_1	1	1	1	1	0	0	40
s_2	10	5	$\boxed{10}$	0	1	0	300
s_3	3	2	1	0	0	1	80
w	-100	-300	-400	0	0	0	0

Tableau II

basis	x	y	z	s_1	s_2	s_3	values
s_1	0	$\boxed{1/2}$	0	1	$-.1$	0	10
z	1	$\frac{1}{2}$	1	0	$.1$	0	30
s_3	2	$\frac{3}{2}$	0	0	$-.1$	1	50
w	300	-100	0	0	40	0	12,000

Tableau III

basis	x	y	z	s_1	s_2	s_3	values
y	0	1	0	2	$-.2$	0	20
z	1	0	1	-1	1.1	0	20
s_3	2	0	0	$-\frac{3}{2}$	$-.2$	1	20
w	300	0	0	200	20	0	14,000

The maximum profit is \$14,000 which occurs when 20 acres are planted in wheat, 20 in barley and none in peanuts.

33. Let x, y and z be the pounds of varieties I, II and III, respectively. The total revenue is $w = 3x+4y+5z$.

$$\tfrac{1}{2}x + \tfrac{1}{3}y \qquad\quad \leq 1000$$
$$\tfrac{1}{2}x + \tfrac{1}{3}y + \tfrac{1}{2}z \leq 1200$$
$$\qquad\quad \tfrac{1}{3}y + \tfrac{1}{2}z \leq \; 800.$$

Tableau I

basis	x	y	z	s_1	s_2	s_3	values
s_1	$\frac{1}{2}$	$\frac{1}{3}$	0	1	0	0	1000
s_2	$\frac{1}{2}$	$\frac{1}{3}$	$\frac{1}{2}$	0	1	0	1200
s_3	0	$\frac{1}{3}$	$\boxed{1/2}$	0	0	1	800
w	-3	-4	-5	0	0	0	0

Tableau II

basis	x	y	z	s_1	s_2	s_3	values
s_1	$\frac{1}{2}$	$\frac{1}{3}$	0	1	0	0	1000
s_2	$\boxed{1/2}$	0	0	0	1	-1	400
z	0	$\frac{2}{3}$	1	0	0	2	1600
w	-3	$-\frac{2}{3}$	0	0	0	10	8000

Tableau III

basis	x	y	z	s_1	s_2	s_3	values
s_1	0	$\boxed{1/3}$	0	1	-1	1	600
x	1	0	0	0	2	-2	800
z	0	$\frac{2}{3}$	1	0	0	2	1600
w	0	$-\frac{2}{3}$	0	0	3	7	10,400

Tableau IV

basis	x	y	z	s_1	s_2	s_3	values
y	0	1	0	3	-3	3	1800
x	1	0	0	0	2	-2	800
z	0	0	1	-2	2	0	400
w	0	0	0	2	1	9	11,600

The maximum revenue is $11,600 and occurs for 800 pounds of variety I, 1800 pounds of variety II and 400 pounds of variety III.

35. Let x, y and z be the number of units of product A, B and C, respectively. The total profit is $w=10x+20y+40z$ and

$$x + 2y + 3z \leq 50$$
$$x + 3y + 5z \leq 60.$$

Tableau I

basis	x	y	z	s_1	s_3	values
s_1	1	2	3	1	0	50
s_2	1	3	$\boxed{5}$	0	1	60
w	-10	-20	-40	0	0	0

Tableau II

basis	x	y	z	s_1	s_3	values
s_1	$\boxed{2/5}$	$\frac{1}{5}$	0	1	$-\frac{3}{5}$	14
z	$\frac{1}{5}$	$\frac{3}{5}$	1	0	$\frac{1}{5}$	12
w	-2	4	0	0	8	480

Tableau III

basis	x	y	z	s_1	s_3	values
x	1	$\frac{1}{2}$	0	$\frac{5}{2}$	$-\frac{3}{2}$	35
z	0	$\frac{1}{2}$	1	$-\frac{1}{2}$	$-\frac{1}{10}$	5
w	0	5	0	5	5	550

The maximum profit is \$550 which occurs when 35 units of A, 0 units of B and 5 units of C are produced.

37. Let x, y and z be the number of type A, B and C houses, respectively. The profit is $w = 20,000x+40,000y+60,000z$ and

$$
\begin{array}{rrrcr}
.25x + & .25y + & z & \leq & 75 \\
50,000x + & 150,000y + & 100,000z & \leq & 30,000,000 \\
6000x + & 4000y + & 2000z & \leq & 80,000
\end{array}
$$

Tableau I

basis	x	y	z	s_1	s_2	s_3	values
s_1	$\frac{1}{4}$	$\frac{1}{4}$	$\boxed{1}$	1	0	0	75
s_2	50,000	150,000	100,000	0	1	0	30,000,000
s_3	6000	4000	2000	0	0	1	800,000
w	$-20,000$	$-40,000$	$-60,000$	0	0	0	0

Tableau II

basis	x	y	z	s_1	s_2	s_3	values
z	$\frac{1}{4}$	$\frac{1}{4}$	1	1	0	0	75
s_2	25,000	$\boxed{125,000}$	0	$-100,000$	1	0	22,500,000
s_3	5500	3500	0	-2000	0	1	650,000
w	-5000	$-25,000$	0	60,000	0	0	4,500,000

Tableau III

basis	x	y	z	s_1	s_2	s_3	values
z	.2	0	1	.8	$-.000002$	0	30
y	.2	1	0	$-.8$	.000008	0	180
s_3	$\boxed{500}$	0	0	800	$-.028$	1	20,000
w	0	0	0	40,000	.2	0	9,000,000

The maximum profit is $9,000,000 which occurs when 180 Type B houses, 30 Type C and 0 Type A houses are constructed.

Tableau IV

basis	x	y	z	s_1	s_2	s_3	values
z	0	0	1	$-$	$-$	$-$	22
y	0	1	0	$-$	$-$	$-$	172
x	1	0	0	$-$	$-$	$-$	40
w	0	0	0	$-$	$-$	$-$	9,000,000

The \$9,000,000 profit also occurs when 40, 172 and 22 Type A, B and C houses, respectively, are constructed and in general occurs when $(1-t)40$ type A houses, $(1-t)172+180t$ type B houses and $(1-t)22+30t$ type C houses are constructed for $0 \le t \le 1$.

39. Let $w = f(x, y, z) = mx + ny + pz$ and suppose $mx_1 + ny_1 + pz_1 = A = mx_2 + ny_2 + pz_2$. Then if $x = (1-t)x_1 + x_2$, $y = (1-t)y_1 + y_2$, $z = (1-t)z_1 + z_2$,

$$mx + ny + pz = f(x, y, z) = m((1-t)x_1 + tx_2) + n((1-t)y_1 + ty_2) + p((1-t)z_1 + tz_2)$$
$$= m(x_1 - x_1 t + tx_2) + n(y_1 - y_1 t + ty_2) + p(z_1 - z_1 t + tz_2)$$
$$= (mx_1 + ny_1 + pz_1) - t(mx_1 + ny_1 + pz_1) + t(mx_2 + ny_2 + pz_2)$$
$$= A - tA + tA = A$$

EXERCISE SET 4.6 MINIMIZATION AND DUALITY

1. The matrix of the primal problem is $\begin{bmatrix} 1 & 1 & 45 \\ 3 & 1 & 81 \\ 45 & 27 & z \end{bmatrix}$. Its transpose is $\begin{bmatrix} 1 & 3 & 45 \\ 1 & 1 & 27 \\ 45 & 81 & z \end{bmatrix}$.

Thus we maximize $Z = 45u + 81v$ subject to
$$u + 3v \le 45$$
$$\text{and } u + v \le 27, \qquad u \ge 0, v \ge 0.$$

Tableau I

basis	u	v	r	s	values
r	1	③	1	0	45
s	1	1	0	1	27
Z	-45	-81	0	0	0

Tableau II

basis	u	v	r	s	values
v	$\frac{1}{3}$	1	$\frac{1}{3}$	0	15
s	$\boxed{2/3}$	0	$-\frac{1}{3}$	1	12
Z	-18	0	27	0	1215

Tableau III

basis	u	v	r	s	values
v	0	1	$\frac{1}{2}$	$-\frac{1}{2}$	9
u	1	0	$-\frac{1}{2}$	$\frac{3}{2}$	18
Z	0	0	18	27	1539

The minimum value of z is 1539 and occurs when $x=18$ and $y=27$.

3. The primal matrix is $\begin{bmatrix} 2 & 5 & 210 \\ 5 & 3 & 240 \\ 57 & 133 & z \end{bmatrix}$. The dual matrix is $\begin{bmatrix} 2 & 5 & 57 \\ 5 & 3 & 133 \\ 210 & 240 & z \end{bmatrix}$.

We must maximize $Z=210u+240v$ subject to
$$2u+5v \leq 57$$
$$5u+3v \leq 133, \qquad u \geq 0, \ v \geq 0.$$

Tableau I

basis	u	v	r	s	values
r	2	$\boxed{5}$	1	0	57
s	5	3	0	1	133
Z	-210	-240	0	0	0

Tableau II

basis	u	v	r	s	values
v	.4	1	.2	0	11.4
s	$\boxed{3.8}$	0	$-.6$	1	98.8
Z	-114	0	48	0	2736

Tableau III

basis	u	v	r	s	values
v	0	1	.44	$-\frac{2}{19}$	1
u	1	0	$-\frac{3}{19}$	$\frac{5}{19}$	26
Z	0	0	30	30	5700

The minimum value of z is 5700 and this occurs when $x=y=30$.

5. The primal matrix is $\begin{bmatrix} 4 & 5 & 260 \\ 3 & 1 & 96 \\ 27 & 20 & z \end{bmatrix}$. The dual matrix is $\begin{bmatrix} 4 & 3 & 27 \\ 5 & 1 & 20 \\ 260 & 96 & z \end{bmatrix}$.

We must maximize $Z=260u+96v$ subject to
$$4u+3v \leq 27$$
$$5u+v \leq 20, \qquad u \geq 0, \ v \geq 0.$$

Tableau I

basis	u	v	r	s	values
r	4	3	1	0	27
s	⑤	1	0	1	20
Z	-260	-96	0	0	0

Tableau II

basis	u	v	r	s	values
r	0	$\boxed{11/5}$	1	$-\frac{4}{5}$	11
u	1	$\frac{1}{5}$	0	$\frac{1}{5}$	4
Z	0	-44	0	52	1040

Tableau III

basis	u	v	r	s	values
v	0	1	$\frac{5}{11}$	$-\frac{4}{11}$	5
u	1	0	$-\frac{1}{11}$	$\frac{3}{11}$	3
Z	0	0	20	36	1260

The minimum value of z is 1260 and this occurs when $x=20$ and $y=36$.

7. The primal matrix is $\begin{bmatrix} 2 & 1 & 7 \\ 1 & 1 & 5 \\ 2 & 5 & 16 \\ 16 & 25 & z \end{bmatrix}$. The dual matrix is $\begin{bmatrix} 2 & 1 & 2 & 16 \\ 1 & 1 & 5 & 25 \\ 7 & 5 & 16 & z \end{bmatrix}$.

We must maximize $Z=7u+5v+16w$ subject to
$$2u+v+2w \leq 16$$
$$u+v+5w \leq 25, \qquad u \geq 0, \ v \geq 0, \ w \geq 0.$$

Tableau I

basis	u	v	w	r	s	values
r	2	1	2	1	0	16
s	1	1	$\boxed{5}$	0	1	25
Z	-7	-5	-16	0	0	0

Tableau II

basis	u	v	w	r	s	values
r	$\boxed{1.6}$	.6	0	1	$-.4$	6
w	.2	.2	1	0	.2	5
Z	-3.8	-1.8	0	0	3.2	80

Tableau III

basis	u	v	w	r	s	values
u	1	$\boxed{.375}$	0	.625	$-.25$	3.75
w	0	.125	1	$-.125$	.25	4.25
Z	0	$-.375$	0	2.375	2.25	94.25

Tableau IV

basis	u	v	w	r	s	values
v	$\frac{8}{3}$	1	0	$\frac{5}{3}$	$\frac{2}{3}$	10
w	$-\frac{1}{3}$	0	1	$-\frac{1}{3}$	$\frac{1}{6}$	3
Z	1	0	0	3	2	98

The minimum value of z is 98 and occurs when $x=3$ and $y=2$.

9. The primal matrix is $\begin{bmatrix} 4 & 1 & 1 & 8 \\ 0 & 1 & 0 & 9 \\ 5 & 0 & 0 & 6 \\ 210 & 48 & 45 & z \end{bmatrix}$. The dual matrix is $\begin{bmatrix} 4 & 0 & 5 & 210 \\ 1 & 1 & 0 & 48 \\ 1 & 0 & 0 & 45 \\ 8 & 9 & 6 & z \end{bmatrix}$.

Thus we must maximize $W=8u+9v+6w$ subject to
$$
\begin{aligned}
4u \quad + 5w &\le 210 \\
u + v \quad &\le 48 \\
u \quad &\le 45 \\
u \ge 0, \ v \ge 0, \ q &\ge 0.
\end{aligned}
$$

Tableau I

basis	u	v	q	r	s	t	values
r	4	0	5	1	0	0	210
s	1	$\boxed{1}$	0	0	1	0	48
t	1	0	0	0	0	1	45
W	-8	-9	-6	0	0	0	0

Tableau II

basis	u	v	q	r	s	t	values
r	4	0	$\boxed{5}$	1	0	0	210
v	1	1	0	0	1	0	48
t	1	0	0	0	0	1	45
W	1	0	-6	0	9	0	432

Tableau III

basis	u	v	q	r	s	t	values
q	$\frac{4}{5}$	0	1	$\frac{1}{5}$	0	0	42
v	1	1	0	0	1	0	48
t	1	0	0	0	0	1	45
W	$\frac{29}{5}$	0	0	$\frac{6}{5}$	9	0	684

The minimum value of w is 684 and this occurs when $x=1.2$, $y=9$ and $z=0$.

11. The primal matrix is $\begin{bmatrix} 1 & -1 & 2 & 5 \\ 1 & 3 & 1 & 1 \\ 1 & 2 & 3 & 1 \\ 66 & 72 & 108 & w \end{bmatrix}$. The dual matrix is $\begin{bmatrix} 1 & 1 & 1 & 66 \\ -1 & 3 & 2 & 72 \\ 2 & 1 & 3 & 108 \\ 5 & 1 & 1 & w \end{bmatrix}$.

Thus we must maximize $W = 5u + v + q$ subject to

$$
\begin{aligned}
u + v + q &\le 66 \\
u + 3v + 2q &\le 72 \\
2u + v + 3q &\le 108 \\
u \ge 0,\ v \ge 0,\ q &\ge 0.
\end{aligned}
$$

Tableau I

basis	u	v	q	r	s	t	values
r	1	1	1	1	0	0	66
s	-1	3	2	0	1	0	72
t	$\boxed{2}$	1	3	0	0	1	108
W	-5	-1	-1	0	0	0	0

Tableau II

basis	u	v	q	r	s	t	values
r	0	$\frac{1}{2}$	$-\frac{1}{2}$	1	0	$-\frac{1}{2}$	12
s	0	$\frac{7}{2}$	$\frac{7}{2}$	0	1	$\frac{1}{2}$	126
u	1	$\frac{1}{2}$	$\frac{3}{2}$	0	0	$\frac{1}{2}$	54
W	0	$\frac{3}{2}$	$\frac{13}{2}$	0	0	$\frac{5}{2}$	270

The minimum of w is 270 and occurs when $x=0$, $y=0$ and $z=\frac{5}{2}$.

13. The primal matrix is $\begin{bmatrix} 1 & 4 & 20 \\ 4 & 6 & 50 \\ 2 & 5 & 30 \\ 200 & 750 & z \end{bmatrix}$. The dual matrix is $\begin{bmatrix} 1 & 4 & 2 & 200 \\ 4 & 6 & 5 & 750 \\ 20 & 50 & 30 & z \end{bmatrix}$.

We must maximize $Z=20u+50v+30w$ subject to
$$u+4v+2w \le 200$$
$$4u+6v+5w \le 750, \qquad u \ge 0,\ v \ge 0,\ w \ge 0.$$

Tableau I

basis	u	v	w	r	s	values
r	1	$\boxed{4}$	2	1	0	200
s	4	6	5	0	1	750
Z	-20	-50	-30	0	0	0

Tableau II

basis	u	v	w	r	s	values
v	$\frac{1}{4}$	1	$\frac{1}{2}$	$\frac{1}{4}$	0	50
s	$\boxed{5/2}$	0	2	$-\frac{3}{2}$	1	450
Z	$-\frac{15}{2}$	0	-5	$\frac{25}{2}$	0	2500

Tableau III

basis	u	v	w	r	s	values
v	0	1	$\frac{3}{10}$	$\frac{2}{5}$	$-\frac{1}{10}$	5
u	1	0	$\frac{4}{5}$	$-\frac{3}{5}$	$\frac{2}{5}$	180
Z	0	0	1	8	3	3850

The minimum value of z is 3850 and this occurs when $x=8$ and $y=3$.

15. The primal matrix is $\begin{bmatrix} 1 & 1 & 10 \\ 5 & 3 & 20 \\ 2 & 4 & 15 \\ 80 & 60 & z \end{bmatrix}$. The dual matrix is $\begin{bmatrix} 1 & 5 & 2 & 80 \\ 1 & 3 & 4 & 60 \\ 10 & 20 & 15 & z \end{bmatrix}$.

We must maximize $Z=10u+20v+15w$ subject to
$$u+5v+2w \le 80$$
$$u+3v+4w \le 60, \qquad u \ge 0,\ v \ge 0,\ w \ge 0.$$

Tableau I

basis	u	v	w	r	s	values
r	1	$\boxed{5}$	2	1	0	80
s	1	3	4	0	1	60
Z	-10	-20	-15	0	0	0

Tableau II

basis	u	v	w	r	s	values
v	$\frac{1}{5}$	1	$\frac{2}{5}$	$\frac{1}{5}$	0	16
s	$\frac{2}{5}$	0	$\boxed{14/5}$	$-\frac{3}{5}$	1	12
Z	-6	0	-7	4	0	320

Tableau III

basis	u	v	w	r	s	values
v	$\frac{1}{7}$	1	0	$\frac{2}{7}$	$-\frac{1}{7}$	$\frac{100}{7}$
w	$\boxed{1/7}$	0	1	$-\frac{3}{14}$	$\frac{5}{14}$	$\frac{30}{7}$
Z	-5	0	0	$\frac{5}{2}$	$\frac{5}{2}$	350

Tableau IV

basis	u	v	w	r	s	values
v	0	1	-1	$\boxed{1/2}$	$-\frac{1}{2}$	10
u	1	0	7	$-\frac{3}{2}$	$\frac{5}{2}$	30
Z	0	0	35	-5	15	500

Tableau V

basis	u	v	w	r	s	values
r	0	2	-2	1	-1	20
u	1	3	4	0	1	60
Z	0	10	25	0	10	600

The minimum value of z is 600 and this occurs when $x=0$ and $y=10$.

17. The primal matrix is
$$\begin{bmatrix} 1 & 2 & 1 & 2 \\ 5 & 1 & 7 & 24 \\ 2 & 3 & 0 & 4 \\ 420 & 910 & 560 & w \end{bmatrix}.$$

The dual matrix is
$$\begin{bmatrix} 1 & 5 & 2 & 420 \\ 2 & 1 & 3 & 910 \\ 1 & 7 & 0 & 560 \\ 2 & 24 & 4 & w \end{bmatrix}.$$

Thus we must maximize $W=2u+24v+4q$ subject to
$$\begin{aligned} u + 5v + 2q &\le 420 \\ 2u + v + 3q &\le 910 \\ u + 7v &\le 560 \\ u \ge 0,\ v \ge 0,\ q &\ge 0. \end{aligned}$$

Tableau I

basis	u	v	q	r	s	t	values
r	1	5	2	1	0	0	420
s	2	1	3	0	1	0	910
t	1	$\boxed{7}$	0	0	0	1	560
W	-2	-24	-4	0	0	0	0

Tableau II

basis	u	v	q	r	s	t	values
r	$\frac{2}{7}$	0	$\boxed{2}$	1	0	$-\frac{5}{7}$	20
s	$\frac{3}{7}$	0	3	0	1	$-\frac{1}{7}$	830
v	$\frac{1}{7}$	1	0	0	0	$\frac{1}{7}$	80
W	$\frac{10}{7}$	0	-4	0	0	$\frac{24}{7}$	1920

Tableau III

basis	u	v	q	r	s	t	values
q	$\frac{1}{7}$	0	1	$\frac{1}{2}$	0	$-\frac{5}{14}$	10
s	$\frac{10}{7}$	0	0	$-\frac{3}{2}$	1	$\frac{13}{14}$	800
v	$\frac{1}{7}$	1	0	0	0	$\frac{1}{7}$	80
W	2	0	0	2	0	2	1960

The minimum value of w is 1960 and this occurs when $x=2$, $y=0$ and $z=2$.

19. The primal matrix is
$$\begin{bmatrix} 1 & 1 & 0 & 0 & 3 \\ 1 & -2 & 0 & 0 & 0 \\ 0 & 0 & 1 & 1 & 2 \\ 0 & 0 & -2 & 1 & 0 \\ 16 & 28 & 8 & 20 & w \end{bmatrix}.$$

The dual matrix is
$$\begin{bmatrix} 1 & 1 & 0 & 0 & 16 \\ 1 & -2 & 0 & 0 & 28 \\ 0 & 0 & 1 & -2 & 8 \\ 0 & 0 & 1 & 1 & 20 \\ 3 & 0 & 2 & 0 & w \end{bmatrix}.$$

Thus we must maximize $W=3u+2p$ subject to

$$\begin{aligned} u + v &\le 16 \\ u - 2v &\le 28 \\ p - 2q &\le 8 \\ p + q &\le 20, \end{aligned}$$

$u \ge 0,\ v \ge 0,\ p \ge 0,\ q \ge 0.$

Tableau I

basis	u	v	p	q	r	s	t	y	values
r	$\boxed{1}$	1	0	0	1	0	0	0	16
s	1	-2	0	0	0	1	0	0	28
t	0	0	1	-2	0	0	1	0	8
v	0	0	1	1	0	0	0	1	20
W	-3	0	-2	0	0	0	0	0	0

Tableau II

basis	u	v	p	q	r	s	t	y	values
u	1	1	0	0	1	0	0	0	16
s	0	-3	0	0	-1	1	0	0	12
t	0	0	$\boxed{1}$	-2	0	0	1	0	8
v	0	0	1	1	0	0	0	1	20
W	0	3	-2	0	3	0	0	0	48

Tableau III

basis	u	v	p	q	r	s	t	y	values
u	1	1	0	0	1	0	0	0	16
s	0	-3	0	0	-1	1	0	0	12
t	0	0	1	-2	0	0	1	0	8
p	0	0	0	$\boxed{3}$	0	0	-1	1	12
W	0	3	0	-4	3	0	2	0	64

Tableau IV

basis	u	v	p	q	r	s	t	y	values
u	1	1	0	0	1	0	0	0	16
s	0	-3	0	0	-1	1	0	0	12
t	0	0	1	0	0	0	$\frac{1}{3}$	$\frac{2}{3}$	16
q	0	0	0	1	0	0	$-\frac{1}{3}$	$\frac{1}{3}$	4
W	0	3	0	0	3	0	$\frac{2}{3}$	$\frac{4}{3}$	80

The minimum value of w is 80 and occurs when $x_1=3$, $x_2=0$, $x_3=\frac{2}{3}$, and $x_4=\frac{4}{3}$.

21. The primal matrix is
$$\begin{bmatrix} 1 & 2 & 3 & 2 \\ 1 & -1 & -1 & 3 \\ -1 & 1 & 2 & -1 \\ 2 & 1 & 1 & 5 \\ 8 & 7 & 9 & w \end{bmatrix}.$$

The dual matrix is
$$\begin{bmatrix} 1 & 1 & -1 & 2 & 8 \\ 2 & -1 & 1 & 1 & 7 \\ 3 & -1 & 2 & 1 & 9 \\ 2 & 3 & -1 & 5 & w \end{bmatrix}.$$

Thus we must maximize $W=2u+3v-p+5q$ subject to
$$\begin{aligned}
u + v - p + 2q &\le 8 \\
2u - v + p + q &\le 7 \\
3u - v + 2p + q &\le 9 \\
u \ge 0,\ v \ge 0,\ p \ge 0,\ q &\ge 0.
\end{aligned}$$

Tableau I

basis	u	v	p	q	r	s	t	values
r	1	1	-1	$\boxed{2}$	1	0	0	8
s	2	-1	1	1	0	1	0	7
t	3	-1	2	1	0	0	1	9
W	-2	-3	1	-5	0	0	0	0

Tableau II

basis	u	v	p	q	r	s	t	values
q	$\frac{1}{2}$	$\frac{1}{2}$	$-\frac{1}{2}$	1	$\frac{1}{2}$	0	0	4
s	$\frac{3}{2}$	$-\frac{3}{2}$	$\boxed{3/2}$	0	$-\frac{1}{2}$	1	0	3
t	$\frac{5}{2}$	$-\frac{3}{2}$	$\frac{5}{2}$	0	$-\frac{1}{2}$	0	1	5
W	$\frac{1}{2}$	$-\frac{1}{2}$	$-\frac{3}{2}$	0	$\frac{5}{2}$	0	0	20

Tableau III

basis	u	v	p	q	r	s	t	values
q	1	0	0	1	$\frac{1}{3}$	$\frac{1}{3}$	0	5
p	1	−1	1	0	$-\frac{1}{3}$	$\frac{2}{3}$	0	2
t	0	$\boxed{1}$	0	0	$\frac{1}{3}$	$-\frac{5}{3}$	1	0
W	2	−2	0	0	2	1	0	23

Tableau IV

basis	u	v	p	q	r	s	t	values
q	1	0	0	1	$\frac{1}{3}$	$\boxed{1/3}$	0	5
p	1	0	1	0	0	−1	1	2
v	0	1	0	0	$\frac{1}{3}$	$-\frac{5}{3}$	1	0
W	2	0	0	0	$\frac{8}{3}$	$-\frac{7}{3}$	2	23

Tableau V

basis	u	v	p	q	r	s	t	values
s	3	0	0	3	1	1	0	15
p	4	0	1	3	1	0	1	17
v	5	1	0	5	2	0	1	25
W	9	0	0	7	5	0	2	58

The minimum value of w is 58 and this occurs when $x=5$, $y=0$ and $z=2$.

23. Let x be the dollars spent on radio advertising and y the dollars spent on TV advertising.

$$10x + 20y \geq 700,000$$
$$30x + 20y \geq 1,200,000$$

Minimize $w = x+y$.

The primal matrix is $\begin{bmatrix} 10 & 20 & 700,000 \\ 30 & 20 & 1,200,000 \\ 1 & 1 & w \end{bmatrix}$.

The dual matrix is $\begin{bmatrix} 10 & 30 & 1 \\ 20 & 20 & 1 \\ 700,000 & 1,200,000 & w \end{bmatrix}$.

We must maximize $W=700,000u+1,200,000v$ subject to
$$10u + 30v \leq 1$$
$$20u + 20v \leq 1, \quad u \geq 0, v \geq 0$$

Tableau I

basis	u	v	r	s	values
r	10	$\boxed{30}$	1	0	1
s	20	20	0	1	1
W	$-700,000$	$-1,200,000$	0	0	0

Tableau II

basis	u	v	r	s	values
v	$\frac{1}{3}$	1	$\frac{1}{30}$	0	$\frac{1}{30}$
s	$\boxed{40/3}$	0	$-\frac{2}{3}$	1	$\frac{1}{3}$
W	$-300,000$	0	$40,000$	0	$40,000$

Tableau III

basis	u	v	r	s	values
v	0	1	$\frac{1}{60}$	$-\frac{1}{40}$	$\frac{1}{40}$
u	1	0	$-\frac{1}{20}$	$\frac{3}{40}$	$\frac{1}{40}$
W	0	0	$25,000$	$22,500$	$47,500$

The minimum cost is $47,500 which occurs when $25,000 is spent on radio advertising and $22,500 is spent on TV advertising.

25. Let x, y and z be the number of days sites A, B and C are operated, respectively.

Minimize $30,000x+36,000y+27,000z$ subject to

$$6x + 3y + 3z \geq 270$$
$$5x + 5y + 5z \geq 300$$
$$4x + 8y + 4z \geq 288, \quad x \geq 0, y \geq 0, z \geq 0.$$

The primal matrix is
$$\begin{bmatrix} 6 & 3 & 3 & 270 \\ 5 & 5 & 5 & 300 \\ 4 & 8 & 4 & 288 \\ 30,000 & 36,000 & 27,000 & w \end{bmatrix}.$$

The dual matrix is
$$\begin{bmatrix} 6 & 5 & 4 & 30,000 \\ 3 & 5 & 8 & 36,000 \\ 3 & 5 & 4 & 27,000 \\ 270 & 300 & 288 & w \end{bmatrix}.$$

We must maximize $W=270u+300v+288q$ subject to

$$6u + 5v + 4q \leq 30{,}000$$
$$3u + 5v + 8q \leq 36{,}000$$
$$3u + 5v + 4q \leq 27{,}000, \qquad u \geq 0,\ v \geq 0,\ q \geq 0.$$

Tableau I

basis	u	v	q	r	s	t	values
r	6	5	4	1	0	0	30,000
s	3	5	8	0	1	0	36,000
t	3	[5]	4	0	0	1	27,000
W	-270	-300	-288	0	0	0	0

Tableau II

basis	u	v	q	r	s	t	values
r	[3]	0	0	1	0	-1	3000
s	0	0	4	0	1	-1	9000
v	$\frac{3}{5}$	1	$\frac{4}{5}$	0	0	$\frac{1}{5}$	5400
W	-90	0	-48	0	0	60	1,620,000

Tableau III

basis	u	v	q	r	s	t	values
u	1	0	0	$\frac{1}{3}$	0	$-\frac{1}{3}$	1000
s	0	0	[4]	0	1	-1	9000
v	0	1	$\frac{4}{5}$	$-\frac{1}{5}$	0	$\frac{2}{5}$	4800
W	0	0	-48	30	0	30	1,710,000

Tableau IV

basis	u	v	q	r	s	t	values
u	1	0	0	$\frac{1}{3}$	0	$-\frac{1}{3}$	1000
q	0	0	1	0	$\frac{1}{4}$	$-\frac{1}{4}$	2250
v	0	1	0	$-\frac{1}{5}$	$-\frac{1}{5}$	$\frac{3}{5}$	3000
W	0	0	0	30	12	18	1,818,000

The minimum cost is $1,818,000. This occurs when plants A, B and C are operated for 30, 12 and 18 days, respectively.

27. Let x, y and z be the ounces of foods X, Y and Z, respectively. We must minimize $16x+8y+4z$ subject to

$$
\begin{array}{rcrcrcl}
15x & + & 10y & + & 10z & \geq & 160 \\
8x & + & 6y & + & 2z & \geq & 56 \\
10x & + & 6y & + & 2z & \geq & 48, \qquad x \geq 0,\ y \geq 0,\ z \geq 0.
\end{array}
$$

The primal matrix is
$$
\begin{bmatrix}
15 & 10 & 10 & 160 \\
8 & 6 & 2 & 56 \\
10 & 6 & 2 & 48 \\
16 & 8 & 4 & w
\end{bmatrix}.
$$

$$
\begin{bmatrix}
15 & 8 & 10 & 16 \\
10 & 6 & 6 & 8 \\
10 & 2 & 2 & 4 \\
160 & 56 & 48 & w
\end{bmatrix}
$$
is the dual matrix.

We maximize $W=160u+56v+48q$ subject to

$$
\begin{array}{rcrcrcl}
15u & + & 8v & + & 10q & \leq & 16 \\
10u & + & 6v & + & 6q & \leq & 8 \\
10u & + & 2v & + & 2q & \leq & 4 \qquad u \geq 0,\ v \geq 0,\ q \geq 0.
\end{array}
$$

Tableau I

basis	u	v	q	r	s	t	values
r	15	8	10	1	0	0	16
s	10	6	6	0	1	0	8
t	[10]	2	2	0	0	1	4
W	-160	-56	-48	0	0	0	0

Tableau II

basis	u	v	q	r	s	t	values
r	0	5	7	1	1	$-\frac{3}{2}$	10
s	0	[4]	4	0	1	-1	4
u	1	$\frac{1}{5}$	$\frac{1}{5}$	0	0	$\frac{1}{10}$	$\frac{2}{5}$
W	0	-24	-16	0	0	16	64

Tableau III

basis	u	v	q	r	s	t	values
r	0	0	2	1	$-\frac{1}{4}$	$-\frac{1}{4}$	5
v	0	1	1	0	$\frac{1}{4}$	$-\frac{1}{4}$	1
u	1	0	0	0	$-\frac{1}{20}$	$\frac{1}{20}$	$\frac{1}{5}$
W	0	0	8	0	6	10	88

The minimum cholesterol is 88 units which occurs for 0, 6 and 10 ounces of foods A, B and C respectively.

29. Let x, y and z be the days operating the plants at Arlington, Bellingham, and Mount Vernon, respectively. We must minimize $w = 20{,}000x + 14{,}000y + 24{,}000z$ subject to

$$10x + 10y + 20z \geq 630$$
$$20x + 10y + 10z \geq 360, \qquad x \geq 0, \; y \geq 0, \; z \geq 0.$$

The primal matrix is
$$\begin{bmatrix} 10 & 10 & 20 & 630 \\ 20 & 10 & 10 & 360 \\ 20{,}000 & 14{,}000 & 24{,}000 & w \end{bmatrix}.$$

The dual matrix is
$$\begin{bmatrix} 10 & 20 & 20{,}000 \\ 10 & 10 & 14{,}000 \\ 20 & 10 & 24{,}000 \\ 630 & 360 & w \end{bmatrix}.$$

We will maximize $W = 630u + 360v$ subject to

$$10u + 20v \leq 20{,}000$$
$$10u + 10v \leq 14{,}000$$
$$20u + 10v \leq 24{,}000 \qquad u \geq 0, \; v \geq 0.$$

Tableau I

basis	u	v	r	s	t	values
r	10	20	1	0	0	20,000
s	10	10	0	1	0	14,000
t	$\boxed{20}$	10	0	0	1	24,000
W	-630	-360	0	0	0	0

Tableau II

basis	u	v	r	s	t	values
r	0	15	1	0	$-\frac{1}{2}$	8000
s	0	$\boxed{5}$	0	1	$-\frac{1}{2}$	2000
u	1	$\frac{1}{2}$	0	0	$\frac{1}{20}$	1200
W	0	-45	0	0	$\frac{63}{2}$	756,000

Tableau III

basis	u	v	r	s	t	values
r	0	0	1	$-\frac{3}{5}$	$-\frac{1}{5}$	2000
v	0	1	0	$\frac{1}{5}$	$-\frac{1}{10}$	400
u	1	0	0	$-\frac{1}{10}$	$\frac{1}{20}$	1000
W	0	0	0	9	27	774,000

The minimum cost is \$774,000 which occurs when the plants at Arlington, Bellingham and Mount Vernon operate for 0, 9 and 27 days respectively.

31. Let x, y, z and w be the number of VCRs sent from Site A to Distributors I and II and from Site B to Distributors I and II respectively. We must minimize $p=6x+12y+15z+9w$ subject to

$$x+y \leq 750$$
$$z+w \leq 600$$
$$x+z \geq 450$$
$$y+w \geq 750, \qquad x \geq 0,\ y \geq 0,\ z \geq 0,\ w \geq 0.$$

The primal matrix is
$$\begin{bmatrix} -1 & -1 & 0 & 0 & -750 \\ 0 & 0 & -1 & -1 & -600 \\ 1 & 0 & 1 & 0 & 450 \\ 0 & 1 & 0 & 1 & 750 \\ 6 & 12 & 15 & 9 & p \end{bmatrix}.$$

The dual matrix is
$$\begin{bmatrix} -1 & 0 & 1 & 0 & 6 \\ -1 & 0 & 0 & 1 & 12 \\ 0 & -1 & 1 & 0 & 15 \\ 0 & -1 & 0 & 1 & 9 \\ -750 & -600 & 450 & 750 & p \end{bmatrix}.$$

We will maximize $P=-750a-600b+450c+750d$ subject to

$$\begin{aligned}
-a \quad\ +c \quad\ &\le 6 \\
-a \quad\quad\ +d &\le 12 \\
-b+c \quad\ &\le 15 \\
-b \quad\ +d &\le 9, \qquad a \ge 0,\ b \ge 0,\ c \ge 0,\ d \ge 0.
\end{aligned}$$

Tableau I

basis	a	b	c	d	r	s	t	u	values
r	−1	0	1	0	1	0	0	0	6
s	−1	0	0	1	0	1	0	0	12
t	0	−1	1	0	0	0	1	0	15
u	0	−1	0	☐1	0	0	0	1	9
P	750	600	−450	−750	0	0	0	0	0

Tableau II

basis	a	b	c	d	r	s	t	u	values
r	−1	0	☐1	0	1	0	0	0	6
s	−1	1	0	0	0	1	0	−1	3
t	0	−1	1	0	0	0	1	0	15
d	0	−1	0	1	0	0	0	1	9
P	750	−150	−450	0	0	0	0	750	6750

Tableau III

basis	a	b	c	d	r	s	t	u	values
c	−1	0	1	0	1	0	0	0	6
s	−1	☐1	0	0	0	1	0	−1	3
t	1	−1	0	0	−1	0	1	0	9
d	0	−1	0	1	0	0	0	1	9
P	300	−150	0	0	450	0	0	750	9450

Tableau IV

basis	a	b	c	d	r	s	t	u	values
c	−1	0	1	0	1	0	0	0	6
b	−1	1	0	0	0	1	0	−1	3
t	0	0	0	0	−1	1	1	−1	12
d	−1	0	0	1	0	1	0	0	12
P	150	0	0	0	450	150	0	600	9,900

The minimum cost is \$9,900 and this occurs when 450, 150, 0, and 600 are the values of x, y, z and w respectively.

33. Let x, y, z and w be the number of motorhomes shipped from Factory A to Distributors I and II and from Factory B to Distributors I and II respectively. We must minimize $p=200x+250y+225z+175w$ subject to

$$x+y \leq 25$$
$$z+w \leq 25$$
$$x+z \geq 20$$
$$y+w \geq 30, \qquad x \geq 0, \ y \geq 0, \ z \geq 0, \ w \geq 0.$$

The primal matrix is
$$\begin{bmatrix} -1 & -1 & 0 & 0 & -25 \\ 0 & 0 & -1 & -1 & -25 \\ 1 & 0 & 1 & 0 & 20 \\ 0 & 1 & 0 & 1 & 30 \\ 200 & 250 & 225 & 175 & p \end{bmatrix}.$$

The dual matrix is
$$\begin{bmatrix} -1 & 0 & 1 & 0 & 200 \\ -1 & 0 & 0 & 1 & 250 \\ 0 & -1 & 1 & 0 & 225 \\ 0 & -1 & 0 & 1 & 175 \\ -25 & -25 & 20 & 30 & p \end{bmatrix}.$$

We will maximize $P=-25a-25b+20c+30d$ subject to

$$-a \qquad +c \qquad \leq 200$$
$$-a \qquad \qquad +d \leq 250$$
$$\qquad -b+c \qquad \leq 225$$
$$\qquad -b \qquad +d \leq 175, \qquad a \geq 0, \ b \geq 0, \ c \geq 0, \ d \geq 0.$$

Tableau I

basis	a	b	c	d	r	s	t	u	values
r	-1	0	1	0	1	0	0	0	200
s	-1	0	0	1	0	1	0	0	250
t	0	-1	1	0	0	0	1	0	225
u	0	-1	0	$\boxed{1}$	0	0	0	1	175
P	25	25	-20	-30	0	0	0	0	0

Tableau II

basis	a	b	c	d	r	s	t	u	values
r	-1	0	$\boxed{1}$	0	1	0	0	0	200
s	-1	1	0	0	0	1	0	-1	75
t	0	-1	1	0	0	0	1	0	225
d	0	-1	0	1	0	0	0	1	175
P	25	-5	-20	0	0	0	0	30	5250

Tableau III

basis	a	b	c	d	r	s	t	u	values
c	-1	0	1	0	1	0	0	0	200
s	-1	$\boxed{1}$	0	0	0	1	0	-1	75
t	1	-1	0	0	-1	0	1	0	25
d	0	-1	0	1	0	0	0	1	175
P	5	-5	0	0	20	0	0	30	9250

Tableau IV

basis	a	b	c	d	r	s	t	u	values
c	−1	0	1	0	1	0	0	0	200
b	−1	1	0	0	0	1	0	−1	75
t	0	0	0	0	−1	1	1	−1	100
d	−1	0	0	1	0	1	0	0	250
P	0	0	0	0	20	5	0	25	9,625

The minimum cost is \$9,625 which occurs when x, y, z and w equal 20, 5, 0 and 25, respectively.

EXERCISE SET 4.7 THE BIG M METHOD

1. $\quad 2x + 3y + s_1 \qquad\qquad = 36$
    $\quad -x + \quad y - s_2 + a_1 = \quad 2, \qquad x \geq 0,\ y \geq 0,\ s_1 \geq 0,\ s_2 \geq 0,\ a_1 \geq 0$

    Maximize $P_m = 3x + 2y - a_1 M$.

Tableau I

basis	x	y	s_1	s_2	a_1	values
s_1	2	3	1	0	0	36
a_1	−1	1	0	−1	1	2
P_m	−3	−2	0	0	M	0

Tableau II

basis	x	y	s_1	s_2	a_1	values
s_1	2	3	1	0	0	36
a_1	−1	$\boxed{1}$	0	−1	1	2
P_m	$M-3$	$-M-2$	0	M	0	$-2M$

Tableau III

basis	x	y	s_1	s_2	a_1	values
s_1	$\boxed{5}$	0	1	3	-3	30
y	-1	1	0	-1	1	2
P_m	-5	0	0	-2	$M+2$	4

Tableau IV

basis	x	y	s_1	s_2	values
x	1	0	$\frac{1}{5}$	$\frac{3}{5}$	6
y	0	1	$\frac{1}{5}$	$-\frac{2}{5}$	8
P_m	0	0	1	1	34

The procedure terminates with $x=3$, $y=8$ and $s_1=s_2=0$. The solution to the original problem is that the maximum value of 34 occurs when $x=6$ and $y=8$.

3.
$$2x + 3y - s_1 + a_1 = 36$$
$$x + 4y + s_2 = 8$$

Maximize $P_m=3x+2y-a_1M$.

Tableau I

basis	x	y	s_1	s_2	a_1	values
a_1	2	3	-1	0	1	36
s_2	1	4	0	1	0	8
P_m	-3	-2	0	0	M	0

Tableau II

basis	x	y	s_1	s_2	a_1	values
a_1	2	3	-1	0	1	36
s_2	1	$\boxed{4}$	0	1	0	8
P_m	$-2M-3$	$-3M-2$	M	0	0	$-36M$

Tableau III

basis	x	y	s_1	s_2	a_1	values
a_1	$\frac{5}{4}$	0	-1	$-\frac{3}{4}$	1	30
y	$\boxed{1/4}$	1	0	$\frac{1}{4}$	0	2
P_m	$-\frac{5}{4}M-\frac{5}{2}$	0	M	$\frac{3}{4}M+\frac{1}{2}$	0	$-30M+4$

Tableau IV

basis	x	y	s_1	s_2	a_1	values
a_1	0	-5	-1	-2	1	20
x	1	4	0	1	0	8
P_m	0	$5M+10$	M	$2M+3$	0	$-20M+24$

The procedure terminates with the artificial variable a_1 having a positive value. Thus the original problem has no solution.

5. $\quad\begin{aligned} 2x + 4y + 3z - s_1 + a_1 &= 48 \\ 2x + y + 3z + s_2 &= 45 \\ 2x + 6y + z + s_3 &= 56 \\ x + 2y + 3z + a_4 &= 30 \end{aligned}$

We maximize $P_m = 4x+4y+3z-a_1 M-a_4 M$.

Tableau I

basis	x	y	z	s_1	s_2	s_3	a_1	a_4	values
a_1	2	4	3	-1	0	0	1	0	48
s_2	2	1	3	0	1	0	0	0	45
s_3	2	6	1	0	0	1	0	0	56
a_4	1	2	3	0	0	0	0	1	30
P_m	-4	-4	-3	0	0	0	M	M	0

Tableau II

basis	x	y	z	s_1	s_2	s_3	a_1	a_4	values
a_1	2	4	3	-1	0	0	1	0	48
s_2	2	1	3	0	1	0	0	0	45
s_3	2	$\boxed{6}$	1	0	0	1	0	0	56
a_4	1	2	3	0	0	0	0	1	30
P_m	$-3M-4$	$-6M-4$	$-6M-3$	M	0	0	0	0	$-78M$

Tableau III

basis	x	y	z	s_1	s_2	s_3	a_1	a_4	values
a_1	$\frac{2}{3}$	0	$\frac{7}{3}$	-1	0	$-\frac{2}{3}$	1	0	$\frac{32}{3}$
s_2	$\frac{5}{3}$	0	$\frac{17}{6}$	0	1	$-\frac{1}{6}$	0	0	$\frac{107}{3}$
y	$\frac{1}{3}$	1	$\frac{1}{6}$	0	0	$\frac{1}{6}$	0	0	$\frac{28}{3}$
a_4	$\frac{1}{3}$	0	$\boxed{8/3}$	0	0	$-\frac{1}{3}$	0	1	$\frac{34}{3}$
P_m	$-M-\frac{8}{3}$	0	$-5M-\frac{7}{3}$	M	0	$M+\frac{2}{3}$	0	0	$-22M+\frac{112}{3}$

Tableau IV

basis	x	y	z	s_1	s_2	s_3	a_1	a_4	values
a_1	$\boxed{3/8}$	0	0	-1	0	$-\frac{3}{8}$	1	$-\frac{7}{8}$	$\frac{3}{4}$
s_2	$\frac{21}{16}$	0	0	0	1	$\frac{3}{16}$	0	$-\frac{17}{16}$	$\frac{189}{8}$
y	$\frac{5}{16}$	1	0	0	0	$\frac{3}{16}$	0	$-\frac{1}{16}$	$\frac{69}{8}$
z	$\frac{1}{8}$	0	1	0	0	$-\frac{1}{8}$	0	$\frac{3}{8}$	$\frac{17}{4}$
P_m	$-\frac{3}{8}M-\frac{19}{8}$	0	0	M	0	$\frac{3}{8}M+\frac{3}{8}$	0	$\frac{15}{8}M+\frac{7}{8}$	$-\frac{3}{4}M+\frac{567}{12}$

Tableau V

basis	x	y	z	s_1	s_2	s_3	a_1	a_4	values
x	1	0	0	$-\frac{8}{3}$	0	-1	$\frac{8}{3}$	$-\frac{7}{3}$	2
s_2	0	0	0	$\boxed{7/2}$	1	$\frac{3}{2}$	$-\frac{7}{2}$	2	21
y	0	1	0	$\frac{5}{6}$	0	$\frac{1}{2}$	$-\frac{5}{6}$	$\frac{2}{3}$	8
z	0	0	1	$\frac{1}{3}$	0	0	$-\frac{1}{3}$	$\frac{2}{3}$	4
P_m	0	0	0	$-\frac{19}{3}$	0	-2	$M+\frac{19}{3}$	$M-\frac{14}{3}$	52

Simplify and pivot,

Tableau VI

basis	x	y	z	s_1	s_2	s_3	values
x	1	0	0	0	$\frac{16}{21}$	$\frac{1}{7}$	18
s_1	0	0	0	1	$\frac{2}{7}$	$\frac{3}{7}$	6
y	0	1	0	0	$-\frac{5}{21}$	$\frac{1}{7}$	3
z	0	0	1	0	$-\frac{2}{21}$	$-\frac{1}{7}$	2
P_m	0	0	0	0	$\frac{38}{21}$	$\frac{5}{7}$	90

The maximum value is 90. It occurs when $x = 18$, $y = 3$ and $z = 2$.

7. $4x - y + 3z - s_1 + a_1 = 29$
   $6x + 3y + s_2 = 39$
   $5x + y + 6z + s_3 = 61$
   $x + y + z + a_4 = 12$

We maximize $P_m = 3x + 2y + 5z - Ma_1 - Ma_4$.

Tableau I

basis	x	y	z	s_1	s_2	s_3	a_1	a_4	values
a_1	4	-1	3	-1	0	0	1	0	29
s_2	6	3	0	0	1	0	0	0	39
s_3	5	1	6	0	0	1	0	0	61
a_4	1	1	1	0	0	0	0	1	12
P_m	-3	-2	-5	0	0	0	M	M	0

Tableau II

basis	x	y	z	s_1	s_2	s_3	a_1	a_4	values
a_1	4	-1	3	-1	0	0	1	0	29
s_2	$\boxed{6}$	3	0	0	1	0	0	0	39
s_3	5	1	6	0	0	1	0	0	61
a_4	1	1	1	0	0	0	0	1	12
P_m	$-5M-3$	-2	$-4M-5$	M	0	0	0	0	$-41M$

Tableau III

basis	x	y	z	s_1	s_2	s_3	a_1	a_4	values
a_1	0	-3	$\boxed{3}$	-1	$-\frac{2}{3}$	0	1	0	3
x	1	$\frac{1}{2}$	0	0	$\frac{1}{6}$	0	0	0	$\frac{13}{2}$
s_3	0	$-\frac{3}{2}$	6	0	$-\frac{5}{6}$	1	0	0	$\frac{57}{2}$
a_4	0	$\frac{1}{2}$	1	0	$-\frac{1}{6}$	0	0	1	$\frac{11}{2}$
P_m	0	$\frac{5M-1}{2}$	$-4M-5$	M	$\frac{5M+3}{6}$	0	0	0	$\frac{-17M+39}{2}$

Tableau IV

basis	x	y	z	s_1	s_2	s_3	a_1	a_4	values
z	0	-1	1	$-\frac{1}{3}$	$-\frac{2}{9}$	0	$\frac{1}{3}$	0	1
x	1	$\frac{1}{2}$	0	0	$\frac{1}{6}$	0	0	0	$\frac{13}{2}$
s_3	0	$-\frac{9}{2}$	0	2	$\frac{1}{2}$	1	-2	0	$\frac{45}{2}$
a_4	0	$\boxed{3/2}$	0	$\frac{1}{3}$	$\frac{1}{18}$	0	$-\frac{1}{3}$	1	$\frac{9}{2}$
P_m	0	$-\frac{3M}{2}-\frac{11}{2}$	0	$-\frac{M}{3}-\frac{5}{3}$	$\frac{-M-11}{18}$	0	$\frac{4M+5}{3}$	0	$\frac{-9M+49}{2}$

Tableau V

basis	x	y	z	s_1	s_2	s_3	a_1	a_4	values
z	0	0	1	$-\frac{1}{9}$	$-\frac{5}{27}$	0	$-\frac{1}{9}$	$\frac{2}{3}$	4
x	1	0	0	$-\frac{1}{9}$	$\frac{4}{27}$	0	$\frac{1}{9}$	$-\frac{1}{3}$	5
s_3	0	0	0	$\boxed{1}$	$\frac{1}{3}$	1	-1	-3	9
y	0	1	0	$\frac{2}{9}$	$\frac{1}{27}$	0	$-\frac{2}{9}$	$\frac{2}{3}$	3
P_m	0	0	0	$-\frac{4}{9}$	$-\frac{11}{27}$	0	$M+\frac{4}{9}$	$M+\frac{11}{3}$	41

Tableau VI

basis	x	y	z	s_1	s_2	s_3	values
z	0	0	1	0	$-\frac{4}{27}$	$\frac{1}{9}$	5
x	1	0	0	0	$\frac{5}{27}$	$\frac{1}{9}$	6
s_1	0	0	0	1	$\boxed{1/3}$	1	9
y	0	1	0	0	$-\frac{1}{27}$	$-\frac{2}{9}$	1
P_m	0	0	0	0	$-\frac{7}{27}$	$\frac{4}{9}$	45

Tableau VII

basis	x	y	z	s_1	s_2	s_3	values
z	0	0	1	$\frac{4}{9}$	0	$\frac{5}{9}$	9
x	1	0	0	$-\frac{5}{9}$	0	$-\frac{4}{9}$	1
s_2	0	0	0	3	1	3	27
y	0	1	0	$\frac{1}{9}$	0	$-\frac{1}{9}$	2
P_m	0	0	0	$\frac{14}{27}$	0	$\frac{13}{9}$	52

The maximum value for the original problem is 52.

9. We minimize $P_m = -2x - 3y - 5z - a_1M - a_4M$

where
$$\begin{aligned}
4x - y + 3z - s_1 + a_1 &= 29 \\
6x + 3y + s_2 &= 39 \\
5x + y + 6z + s_3 &= 61 \\
x + y + z + a_4 &= 12
\end{aligned}$$

Tableau I

basis	x	y	z	s_1	s_2	s_3	a_1	a_4	values
a_1	4	-1	3	-1	0	0	1	0	29
s_2	6	3	0	0	1	0	0	0	39
s_3	5	1	6	0	0	1	0	0	61
a_4	1	1	1	0	0	0	0	1	12
P_m	$-5M+2$	3	$-4M+5$	M	0	0	0	0	$-41M$

Tableau II

basis	x	y	z	s_1	s_2	s_3	a_1	a_4	values
a_1	0	-3	3	-1	$-\frac{2}{3}$	0	1	0	3
x	1	$\frac{1}{2}$	0	0	$\frac{1}{6}$	0	0	0	$\frac{13}{2}$
s_3	0	$-\frac{3}{2}$	6	0	$-\frac{5}{6}$	1	0	0	$\frac{57}{2}$
a_4	0	$\frac{1}{2}$	1	0	$-\frac{1}{6}$	0	0	1	$\frac{11}{12}$
P_m	0	$\frac{5}{2}M+2$	$-4M+5$	M	$\frac{5M}{6}-\frac{1}{3}$	0	0	0	$\frac{17M}{2}-13$

Tableau III

basis	x	y	z	s_1	s_2	s_3	a_1	a_4	values
z	0	-1	1	$-\frac{1}{3}$	$-\frac{2}{9}$	0	$\frac{1}{3}$	0	1
x	1	$\frac{1}{2}$	0	0	$\frac{1}{6}$	0	0	0	$\frac{13}{2}$
s_3	0	$\frac{9}{2}$	0	2	$\frac{1}{2}$	1	-2	0	$\frac{45}{2}$
a_4	0	$\boxed{3/2}$	0	$\frac{1}{3}$	$\frac{1}{18}$	0	$-\frac{1}{3}$	1	$\frac{9}{2}$
P_m	0	$-\frac{3}{2}M+7$	0	$-\frac{M}{3}+\frac{5}{3}$	$-\frac{M}{18}+\frac{7}{9}$	0	$\frac{4M}{3}-\frac{5}{3}$	0	$-\frac{9M}{2}-18$

Tableau IV

basis	x	y	z	s_1	s_2	s_3	a_1	a_4	values
z	0	0	1	$-\frac{1}{9}$	$-\frac{5}{27}$	0	$\frac{1}{9}$	$\frac{2}{3}$	4
x	1	0	0	$-\frac{1}{9}$	$\frac{4}{27}$	0	$\frac{1}{9}$	$\frac{1}{3}$	5
s_3	0	0	0	1	$\frac{1}{3}$	1	-1	-3	9
y	0	1	0	$\frac{2}{9}$	$\frac{1}{27}$	0	$-\frac{2}{9}$	$\frac{2}{3}$	3
P_m	0	0	0	$\frac{1}{9}$	$\frac{14}{27}$	0	$M-\frac{1}{9}$	$M-\frac{14}{3}$	-39

The minimum value for the original problem is 39.

11. Maximize $P_m=-9x-4y-4z-a_2M-a_3M$ where

$$
\begin{aligned}
2x + 3y + 4z + s_1 \qquad\qquad &= 34 \\
5x + 5y + 8z - s_2 + a_2 \;\;\; &= 64 \\
x + 2y + 4z \qquad\quad + a_3 &= 26;
\end{aligned}
$$
x, y, z, s_1, s_2, a_2, a_3 all non-negative.

Tableau I

basis	x	y	z	s_1	s_2	a_2	a_3	values
s_1	2	3	4	1	0	0	0	34
a_2	5	5	8	0	-1	1	0	64
a_3	1	2	$\boxed{4}$	0	0	0	1	26
P_m	$-6M+9$	$-7M+4$	$-12M+4$	0	M	0	0	$-90M$

Tableau II

basis	x	y	z	s_1	s_2	a_2	a_3	values
s_1	1	1	0	1	0	0	-1	8
a_2	③	1	0	0	-1	1	-2	12
z	$\frac{1}{4}$	$\frac{1}{2}$	1	0	0	0	$\frac{1}{4}$	$\frac{13}{2}$
P_m	$-3M+8$	$-M+2$	0	0	M	0	$3M-1$	$-12M-26$

Tableau III

basis	x	y	z	s_1	s_2	a_2	a_3	values
s_1	0	$\boxed{2/3}$	0	1	$\frac{1}{3}$	$-\frac{1}{3}$	$-\frac{1}{3}$	4
x	1	$\frac{1}{3}$	0	0	$-\frac{1}{3}$	$\frac{1}{3}$	$-\frac{2}{3}$	4
z	0	$\frac{5}{12}$	1	0	$\frac{1}{12}$	$-\frac{1}{12}$	$\frac{5}{12}$	$\frac{11}{2}$
P_m	0	$-\frac{2}{3}$	0	0	$\frac{8}{3}$	$M-\frac{8}{3}$	$M+\frac{13}{3}$	-58

Tableau IV

basis	x	y	z	s_1	s_2	values
y	0	1	0	$\frac{3}{2}$	$\frac{1}{2}$	6
x	1	0	0	$-\frac{1}{2}$	$-\frac{1}{2}$	2
z	0	0	1	$-\frac{5}{8}$	$-\frac{1}{8}$	8
P_m	0	0	0	1	3	-54

The maximum value of P_m is -54 and the minimum value for the original problem is 54. The minimum occurs when $x = 2$, $y = 6$ and $z = 3$.

13. Maximize $P_m = -9x - 2y - 3z - a_2M - a_3M$ where

$$3x + 2y + 3z + s_1 \qquad\qquad = 33$$
$$7x + 3y + 6z - s_2 + a_2 = 62$$
$$2x + y + 3z \qquad\quad + a_3 = 25; \qquad x,\ y,\ z,\ s_1,\ s_2,\ a_2,\ a_3 \text{ all non-negative.}$$

Tableau I

basis	x	y	z	s_1	s_2	a_2	a_3	values
s_1	3	2	3	1	0	0	0	33
a_2	7	3	6	0	-1	1	0	62
a_3	2	1	$\boxed{3}$	0	0	0	1	25
P_m	$-9M+9$	$-4M+2$	$-9M+3$	0	M	0	0	$-87M$

Tableau II

basis	x	y	z	s_1	s_2	a_2	a_3	values
s_1	1	1	0	1	0	0	-1	8
a_2	$\boxed{3}$	1	0	0	-1	1	-2	12
z	$\frac{2}{3}$	$\frac{1}{3}$	1	0	0	0	$\frac{1}{3}$	$\frac{25}{3}$
P_m	$-3M+7$	$-M+1$	0	0	M	0	$3M-1$	$-12M-25$

Tableau III

basis	x	y	z	s_1	s_2	a_2	a_3	values
s_1	0	$\boxed{2/3}$	0	1	$\frac{1}{3}$	$-\frac{1}{3}$	$-\frac{1}{3}$	4
x	1	$\frac{1}{3}$	0	0	$-\frac{1}{3}$	$\frac{1}{3}$	$-\frac{2}{3}$	4
z	0	$\frac{1}{9}$	1	0	$\frac{2}{9}$	$-\frac{2}{9}$	$\frac{7}{9}$	$\frac{17}{3}$
P_m	0	$-\frac{4}{3}$	0	0	$\frac{7}{3}$	$M-\frac{7}{3}$	$M+\frac{11}{3}$	-53

Tableau IV

basis	x	y	z	s_1	s_2	values
y	0	1	0	$\frac{3}{2}$	$\frac{1}{2}$	6
x	1	0	0	$-\frac{1}{2}$	$-\frac{1}{2}$	2
z	0	0	1	$-\frac{1}{6}$	$-\frac{1}{18}$	5
P_m	0	0	0	2	3	-45

The maximum value of P_m is -45. The minimum value of C is 45.

15. Maximize $P_m = -4x - 3y - 5z - a_1 M$ subject to

$$
\begin{aligned}
-13x + 33y + 4z + s_1 &= 183 \\
16x + 24y - 13z + s_2 &= 219 \\
9x - 4y + 8z + s_3 &= 121 \\
2x + 3y + z - s_4 + a_4 &= 30, \qquad \text{all variables non-negative.}
\end{aligned}
$$

Tableau I

basis	x	y	z	s_1	s_2	s_3	s_4	a_4	values
s_1	-13	$\boxed{33}$	4	1	0	0	0	0	183
s_2	16	24	-13	0	1	0	0	0	219
s_3	9	-4	8	0	0	1	0	0	121
a_4	2	3	1	0	0	0	-1	1	30
P_m	$-2M+4$	$-3M+3$	$-M+5$	0	0	0	M	0	$-30M$

Tableau II

basis	x	y	z	s_1	s_2	s_3	s_4	a_4	values
y	$-\frac{13}{33}$	1	$\frac{4}{33}$	$\frac{1}{33}$	0	0	0	0	$\frac{61}{11}$
s_2	$\boxed{\frac{280}{11}}$	0	$-\frac{175}{11}$	$-\frac{8}{11}$	1	0	0	0	$\frac{945}{11}$
s_3	$\frac{245}{33}$	0	$\frac{280}{33}$	$\frac{4}{33}$	0	1	0	0	$\frac{1575}{11}$
a_4	$\frac{35}{11}$	0	$\frac{7}{11}$	$-\frac{1}{11}$	0	0	-1	1	$\frac{147}{11}$
P_m	$\frac{-35M}{11}+\frac{57}{11}$	0	$\frac{-7M}{11}+\frac{51}{11}$	$\frac{M}{11}-\frac{1}{11}$	0	0	M	0	$-\frac{147}{11}M-\frac{183}{11}$

Tableau III

basis	x	y	z	s_1	s_2	s_3	s_4	a_4	values
y	0	1	$-\frac{1}{8}$	$\frac{2}{105}$	$\frac{13}{840}$	0	0	0	$\frac{55}{8}$
x	1	0	$-\frac{5}{8}$	$-\frac{1}{35}$	$\frac{11}{280}$	0	0	0	$\frac{27}{8}$
s_3	0	0	$\frac{105}{8}$	$\frac{1}{3}$	$-\frac{7}{24}$	1	0	0	$\frac{945}{8}$
a_4	0	0	$\boxed{21/8}$	0	$-\frac{1}{8}$	0	-1	1	$\frac{21}{8}$
P_m	0	0	$\frac{-21M}{8}+\frac{63}{8}$	$\frac{2}{35}$	$\frac{M}{8}-\frac{57}{280}$	0	M	0	$-\frac{21M}{8}-\frac{273}{8}$

Tableau IV

basis	x	y	z	s_1	s_2	s_3	s_4	a_4	values
y	0	1	0	$\frac{2}{105}$	$\frac{1}{105}$	0	$-\frac{1}{21}$	$\frac{1}{21}$	7
x	1	0	0	$-\frac{1}{35}$	$\frac{1}{35}$	0	$-\frac{5}{21}$	$\frac{5}{21}$	4
s_3	0	0	0	$\frac{1}{3}$	$\frac{1}{3}$	1	5	-5	30
z	0	0	1	0	$-\frac{1}{21}$	0	$-\frac{8}{21}$	$\frac{8}{21}$	1
P_m	0	0	0	$\frac{2}{35}$	$\frac{16}{35}$	0	3	$M-3$	-42

The maximum value of P_m is -42 and the minimum value of C is 42.

17. x is the acreage allocated for peanuts and y is the acreage allocated for corn.

$$x+y \le 18$$
$$y-2x \ge 3$$
$$2y-x \ge 12, \qquad x \ge 0, \, y \ge 0$$

$P=400x+300y$. We maximize P_m.
$P_m=400x+300y-a_1M-a_2M$ where

$$
\begin{aligned}
x + y + s_1 \quad\quad\quad &= 18 \\
-2x + y - s_2 + a_2 \quad &= 3 \\
-x + 2y - s_3 + a_3 &= 12;
\end{aligned}
$$
$x, y, s_1, s_2, s_3, a_2, a_3$ all non-negative.

Tableau I

basis	x	y	s_1	s_2	s_3	a_2	a_3	values
s_1	1	1	1	0	0	0	0	18
a_2	-2	$\boxed{1}$	0	-1	0	1	0	3
a_3	-1	2	0	0	-1	0	1	12
P_m	$3M-400$	$-3M-300$	0	M	M	0	0	$-15M$

Tableau II

basis	x	y	s_1	s_2	s_3	a_1	a_2	values
s_1	3	0	1	1	0	-1	0	15
y	-2	1	0	-1	0	1	0	3
a_2	$\boxed{3}$	0	0	2	-1	-2	1	6
P_m	$-3M-1000$	0	0	$-2M-300$	M	$3M+300$	0	$-6M+900$

Tableau III

basis	x	y	s_1	s_2	s_3	a_1	a_2	values
s_1	0	0	1	-1	$\boxed{1}$	1	-1	9
y	0	1	0	$\frac{1}{3}$	$-\frac{2}{3}$	$-\frac{1}{3}$	$\frac{2}{3}$	7
x	1	0	0	$\frac{2}{3}$	$-\frac{1}{3}$	$-\frac{2}{3}$	$\frac{1}{3}$	2
P_m	0	0	0	$\frac{1100}{3}$	$-\frac{1000}{3}$	$M-\frac{1100}{3}$	$M+\frac{1000}{3}$	2900

Tableau IV

basis	x	y	s_1	s_2	s_3	values
s_3	0	0	1	-1	1	9
y	0	1	$\frac{2}{3}$	$-\frac{1}{3}$	0	13
x	1	0	$\frac{1}{3}$	$\frac{1}{3}$	0	5
P_m	0	0	$\frac{1000}{3}$	$\frac{100}{3}$	0	5900

This maximum profit occurs for 5 acres of peanuts and 13 acres of corn.

19. $P=120x+95y$ where

$$2x+3y \le 60,000$$
$$\tfrac{1}{4}x+\tfrac{1}{2}y \le 8,000 \qquad \text{(or } x+2y \le 32,000)$$
$$15y-x \ge 3,000$$

We maximize $P_m=120x+95y-a_3M$ subject to

$$
\begin{aligned}
2x + 3y + s_1 \quad\quad\quad\quad &= 60,000 \\
x + 2y + s_2 \quad\quad\quad &= 32,000 \\
-\ x + 15y - s_3 + a_3 &= 3,000, \quad\quad \text{all variables nonnegative.}
\end{aligned}
$$

Tableau I

basis	x	y	s_1	s_2	s_3	a_3	values
s_1	2	3	1	0	0	0	60,000
s_2	1	2	0	1	0	0	32,000
a_3	-1	$\boxed{15}$	0	0	-1	1	3,000
P_m	$M-120$	$-15M-95$	0	0	M	0	$-3000M$

Tableau II

basis	x	y	s_1	s_2	s_3	a_3	values
s_1	$\boxed{11/5}$	0	1	0	$\frac{1}{5}$	$-\frac{1}{5}$	59,400
s_2	$\frac{17}{15}$	0	0	1	$\frac{2}{15}$	$-\frac{2}{15}$	31,600
y	$-\frac{1}{15}$	1	0	0	$-\frac{1}{15}$	$\frac{1}{15}$	200
P_m	$-\frac{379}{3}$	0	0	0	$-\frac{19}{3}$	$M+\frac{19}{3}$	19,000

Tableau III

basis	x	y	s_1	s_2	s_3	values
x	1	0	$\frac{5}{11}$	0	$\frac{1}{11}$	27,000
s_2	0	0	$-\frac{17}{33}$	1	$-\frac{1}{33}$	1,000
y	0	1	$\frac{1}{33}$	0	$-\frac{2}{33}$	2,000
P_m	0	0	$\frac{379}{3}$	0	$\frac{170}{33}$	3,430,000

The profit is a maximum when 27,000 units of model A and 2,000 units of model B are produced.

21. $R=1.05x+1.15y$ subject to

$$x+y \le 1600$$
$$x \ge 500$$
$$y \ge 200$$
$$5y-x \ge 2000, \qquad x \ge 0,\ y \ge 0$$

We maximize $R_m=1.05x+1.15y-a_2M-a_3M-a_4M$ subject to

$$
\begin{aligned}
x + y + s_1 &= 1600 \\
x - s_2 + a_2 &= 500 \\
y - s_3 + a_3 &= 200 \\
- x + 5y - s_4 + a_4 &= 2000
\end{aligned}
$$
all variables nonnegative.

Tableau I

basis	x	y	s_1	s_2	s_3	s_4	a_2	a_3	a_4	values
s_1	1	1	1	0	0	0	0	0	0	1600
a_2	1	0	0	-1	0	0	1	0	0	500
a_3	0	$\boxed{1}$	0	0	-1	0	0	1	0	200
a_4	-1	5	0	0	0	-1	0	0	1	2000
P_m	-1.05	$-6M-1.15$	0	M	M	M	0	0	0	$-2700M$

Tableau II

basis	x	y	s_1	s_2	s_3	s_4	a_2	a_3	a_4	values
s_1	1	0	1	0	1	0	0	-1	0	1400
a_2	1	0	0	-1	0	0	1	0	0	500
y	0	1	0	0	-1	0	0	1	0	200
a_4	-1	0	0	0	$\boxed{5}$	-1	0	-5	1	1000
P_m	-1.05	0	0	M	$-5M-1.15$	M	0	$6M+1.15$	0	$-1500M+230$

Tableau III

basis	x	y	s_1	s_2	s_3	s_4	a_2	a_3	a_4	values
s_1	$\frac{6}{5}$	0	1	0	0	$\frac{1}{5}$	0	0	$-\frac{1}{5}$	1200
a_2	$\boxed{1}$	0	0	-1	0	0	1	0	0	500
y	$-\frac{1}{5}$	1	0	0	0	$-\frac{1}{5}$	0	0	$\frac{1}{5}$	400
s_3	$-\frac{1}{5}$	0	0	0	1	$-\frac{1}{5}$	0	-1	$\frac{1}{5}$	200
P_m	$-M-1.28$	0	0	M	0	$-.23$	0	M	$M+.23$	$-500M$

Tableau IV

basis	x	y	s_1	s_2	s_3	s_4	a_2	a_3	a_4	values
s_1	0	0	1	$\boxed{6/5}$	0	$\frac{1}{5}$	$-\frac{6}{5}$	0	$-\frac{1}{5}$	600
x	1	0	0	-1	0	0	1	0	0	500
y	0	1	0	$-\frac{1}{5}$	0	$-\frac{1}{5}$	$\frac{1}{5}$	0	$\frac{1}{5}$	500
s_3	0	0	0	$-\frac{1}{5}$	1	$-\frac{1}{5}$	$\frac{1}{5}$	-1	$\frac{1}{5}$	300
P_m	0	0	0	-1.28	0	$-.23$	$M+1.28$	M	$M+.23$	1100

Tableau V

basis	x	y	s_1	s_2	s_3	s_4	values
s_2	0	0	$\frac{5}{6}$	1	0	$\boxed{1/6}$	500
x	1	0	$\frac{5}{6}$	0	0	$\frac{1}{6}$	1000
y	0	1	$\frac{1}{6}$	0	0	$-\frac{1}{6}$	600
s_3	0	0	$\frac{1}{6}$	0	1	$-\frac{1}{6}$	400
P_m	0	0	$\frac{16}{15}$	0	0	$-\frac{1}{60}$	1740

Tableau VI

basis	x	y	s_1	s_2	s_3	s_4	values
s_4	0	0	5	6	0	1	3000
x	1	0	0	-1	0	0	500
y	0	1	1	1	0	0	1100
s_3	0	0	1	1	1	0	900
P_m	0	0	$\frac{23}{20}$	$\frac{1}{10}$	0	0	1790

The maximum revenue occurs when 500 gallons of regular and 1100 gallons of premium are sold.

23. $C = 20x + 35y + 25z$ subject to

$$
\begin{aligned}
x + y + z &= 500 \\
x + y &\leq 400 \\
x - y &\leq 100 \\
2x \qquad - z &\geq 100, \qquad x \geq 0,\ y \geq 0,\ z \geq 0.
\end{aligned}
$$

We maximize
$$-C_m = -20x - 35y - 25z - a_1 M - a_4 M \text{ subject to}$$

$$
\begin{aligned}
x + y + z \qquad\qquad + a_1 &= 500 \\
x + y \qquad + s_2 \qquad\qquad &= 400 \\
x - y \qquad\qquad + s_3 \qquad &= 100 \\
2x \qquad - z \qquad\qquad - s_4 + a_4 &= 100, \qquad \text{all variables nonnegative.}
\end{aligned}
$$

Tableau I

basis	x	y	z	s_2	s_3	s_4	a_1	a_4	values
a_1	1	1	1	0	0	0	1	0	500
s_2	1	1	0	1	0	0	0	0	400
s_3	1	-1	0	0	1	0	0	0	100
a_4	$\boxed{2}$	0	-1	0	0	-1	0	1	100
$-C_m$	$-3M+20$	$-M+35$	25	0	0	M	0	0	$-600M$

Tableau II

basis	x	y	z	s_2	s_3	s_4	a_1	a_4	values
a_1	0	1	$\frac{3}{2}$	0	0	$\frac{1}{2}$	1	$-\frac{1}{2}$	450
s_2	0	1	$\frac{1}{2}$	1	0	$\frac{1}{2}$	0	$-\frac{1}{2}$	350
s_3	0	-1	$\boxed{1/2}$	0	1	$\frac{1}{2}$	0	$-\frac{1}{2}$	50
x	1	0	$-\frac{1}{2}$	0	0	$-\frac{1}{2}$	0	$\frac{1}{2}$	50
$-C_m$	0	$-M+35$	$-\frac{3}{2}M+35$	0	0	$\frac{-M}{2}+10$	0	$\frac{3M}{2}+10$	$-450M-1000$

Tableau III

basis	x	y	z	s_2	s_3	s_4	a_1	a_4	values
a_1	0	$\boxed{4}$	0	0	-3	-1	1	1	300
s_2	0	2	0	1	-1	0	0	0	300
z	0	-2	1	0	2	1	0	-1	100
x	1	-1	0	0	1	0	0	0	100
$-C_m$	0	$-4M+105$	0	0	$3M-70$	$M-25$	0	45	$-300M-4500$

Tableau IV

basis	x	y	z	s_2	s_3	s_4	a_1	a_4	values
y	0	1	0	0	$-\frac{3}{4}$	$-\frac{1}{4}$	$\frac{1}{4}$	$\frac{1}{4}$	75
s_2	0	0	0	1	$\frac{1}{2}$	$\frac{1}{2}$	$-\frac{1}{2}$	$-\frac{1}{2}$	150
z	0	0	1	0	$\frac{1}{2}$	$\frac{1}{2}$	$\frac{1}{2}$	$-\frac{1}{2}$	250
x	1	0	0	0	$\frac{1}{4}$	$-\frac{1}{4}$	$\frac{1}{4}$	$\frac{1}{4}$	175
$-C_m$	0	0	0	0	$\frac{35}{4}$	$\frac{5}{4}$	$M+\frac{105}{4}$	$\frac{5M}{2}+\frac{145}{4}$	$-12,375$

The minimum cost occurs when 175 shares of stock A, 75 shares of stock B and 250 shares of stock C are purchased.

25. $C=3x+3.75y$ subject to

$$3x + 12y \geq 120 \quad \text{or} \quad x + 4y \geq 30,$$
$$2x + 4y \geq 60 \quad \text{or} \quad x + 2y \geq 30,$$
$$6x + 3y \geq 90 \quad \text{or} \quad 2x + y \geq 30$$
$$2x + 3y \leq 70$$

We maximize $-C_M=-3x-3.75y-a_1M-a_2M-a_3M$ subject to

$$x + 4y - s_1 + a_1 = 30$$
$$x + 2y - s_2 + a_2 = 30$$
$$2x + y - s_3 + a_3 = 30$$
$$2x + 3y + s_4 = 70 \qquad \text{all variables nonnegative.}$$

Tableau I

basis	x	y	s_1	s_2	s_3	s_4	a_1	a_1	a_3	values
a_1	1	$\boxed{4}$	-1	0	0	0	1	0	0	30
a_2	1	2	0	-1	0	0	0	1	0	30
a_3	2	1	0	0	-1	0	0	0	1	30
s_4	2	3	0	0	0	1	0	0	0	70
$-C_m$	$-4M+3$	$-7M+3.75$	M	M	M	0	0	0	0	$-90M$

Tableau II

basis	x	y	s_1	s_2	s_3	s_4	a_1	a_2	a_3	values
y	$\frac{1}{4}$	1	$-\frac{1}{4}$	0	0	0	$\frac{1}{4}$	0	0	$\frac{15}{2}$
a_2	$\frac{1}{2}$	0	$\frac{1}{2}$	-1	0	0	$-\frac{1}{2}$	1	0	15
a_3	$\boxed{7/4}$	0	$\frac{1}{4}$	0	-1	0	$-\frac{1}{4}$	0	1	$\frac{45}{2}$
s_4	$\frac{5}{4}$	0	$\frac{3}{4}$	0	0	1	$-\frac{3}{4}$	0	0	$\frac{95}{2}$
$-C_m$	$\frac{-9M}{4}+\frac{33}{16}$	0	$\frac{-3M}{4}+\frac{15}{16}$	M	M	0	$\frac{7M}{4}-\frac{15}{16}$	0	0	$-\frac{75M}{2}-\frac{225}{8}$

Tableau III

basis	x	y	s_1	s_2	s_3	s_4	a_1	a_2	a_3	values
y	0	1	$-\frac{2}{7}$	0	$\frac{1}{7}$	0	$\frac{2}{7}$	0	$-\frac{1}{7}$	$\frac{30}{7}$
a_2	0	0	$\boxed{3/7}$	-1	$\frac{2}{7}$	0	$-\frac{3}{7}$	1	$-\frac{2}{7}$	$\frac{60}{7}$
x	1	0	$\frac{1}{7}$	0	$-\frac{4}{7}$	0	$-\frac{1}{7}$	0	$\frac{4}{7}$	$\frac{90}{7}$
s_4	0	0	$\frac{4}{7}$	0	$\frac{5}{7}$	1	$-\frac{4}{7}$	0	$-\frac{5}{7}$	$\frac{220}{7}$
$-C_m$	0	0	$\frac{-3M}{7}+\frac{9}{14}$	M	$\frac{-2M}{7}+\frac{33}{28}$	0	$\frac{10M}{7}-\frac{9}{28}$	0	$\frac{9M}{7}-\frac{33}{28}$	$\frac{-60M}{7}-\frac{765}{14}$

Tableau IV

basis	x	y	s_1	s_2	s_3	s_4	a_1	a_2	a_3	values
y	0	1	0	$-\frac{2}{3}$	$\frac{1}{3}$	0	0	$\frac{2}{3}$	$-\frac{1}{3}$	10
s_1	0	0	1	$-\frac{7}{3}$	$\frac{2}{3}$	0	-1	$\frac{7}{3}$	$-\frac{2}{3}$	20
x	1	0	0	$\frac{1}{3}$	$-\frac{2}{3}$	0	0	$-\frac{1}{3}$	$\frac{2}{3}$	10
s_4	0	0	0	$\frac{4}{3}$	$\frac{1}{3}$	1	0	$-\frac{4}{3}$	$-\frac{1}{3}$	20
$-C_m$	0	0	0	$\frac{3}{2}$	$\frac{3}{4}$	0	$M+\frac{9}{28}$	$M-\frac{3}{2}$	$M-\frac{5}{28}$	$\frac{135}{2}$

The cost is minimized when 10 pounds of Type I and 10 pounds of Type II are used.

27. $C=8x_1+6x_2+7x_3+5x_4$ subject to

$$x_1 + x_2 \le 900,$$
$$x_3 + x_4 \le 500,$$
$$x_1 + x_3 \ge 600,$$
$$x_2 + x_4 \ge 700, \qquad x_1,\ x_2,\ x_3,\ x_4 \text{ nonnegative.}$$

$-C_M=-8x_1-6x_2-7x_3-5x_4-a_3M-a_4M$ subject to

$$x_1 + x_2 + s_1 \qquad\qquad = 900$$
$$x_3 + x_4 + s_2 \qquad\qquad = 500$$
$$x_1 + x_3 - s_3 + a_3 = 600$$
$$x_2 + x_4 - s_4 + a_4 = 700, \qquad \text{all variables nonnegative.}$$

Tableau I

basis	x_1	x_2	x_3	x_4	s_1	s_2	s_3	s_4	a_3	a_4	values
s_1	1	1	0	0	1	0	0	0	0	0	900
s_2	0	0	1	$\boxed{1}$	0	1	0	0	0	0	500
a_3	1	0	1	0	0	0	-1	0	1	0	600
a_4	0	1	0	1	0	0	0	-1	0	1	700
$-C_m$	$-M+8$	$-M+6$	$-M+7$	$-M+5$	0	0	M	M	0	0	$-1300M$

Tableau II

basis	x_1	x_2	x_3	x_4	s_1	s_2	s_3	s_4	a_3	a_4	values
s_1	1	1	0	0	1	0	0	0	0	0	900
x_4	0	0	1	1	0	1	0	0	0	0	500
a_3	1	0	1	0	0	0	-1	0	1	0	600
a_4	0	$\boxed{1}$	-1	0	0	-1	0	-1	0	1	200
$-C_m$	$-M+8$	$-M+6$	2	0	0	$M-5$	M	M	0	0	$-800M-2500$

Tableau III

basis	x_1	x_2	x_3	x_4	s_1	s_2	s_3	s_4	a_3	a_4	values
s_1	1	0	1	0	1	1	0	1	0	-1	700
x_4	0	0	$\boxed{1}$	1	0	1	0	0	0	0	500
a_3	1	0	1	0	0	0	-1	0	1	0	600
x_2	0	1	-1	0	0	-1	0	-1	0	1	200
$-C_m$	$-M+8$	0	$-M+8$	0	0	1	M	6	0	$M-6$	$-600M-3700$

Tableau IV

basis	x_1	x_2	x_3	x_4	s_1	s_2	s_3	s_4	a_3	a_4	values
s_1	1	0	0	-1	1	0	0	1	0	-1	200
x_3	0	0	1	1	0	1	0	0	0	0	500
a_3	$\boxed{1}$	0	0	-1	0	-1	-1	0	1	0	100
x_2	0	1	0	1	0	0	0	-1	0	1	700
$-C_m$	$-M+8$	0	0	$M-8$	0	$M-7$	M	6	0	$M-6$	$-100M-7700$

Tableau V

basis	x_1	x_2	x_3	x_4	s_1	s_2	s_3	s_4	a_3	a_4	values
s_1	0	0	0	0	1	1	1	1	-1	-1	100
x_3	0	0	1	1	0	1	0	0	0	0	500
x_1	1	0	0	-1	0	-1	-1	0	1	0	100
x_2	0	1	0	1	0	0	0	-1	0	1	700
$-C_m$	0	0	0	0	0	1	8	6	$M-8$	$M-6$	-8500

The minimum cost occurs when
> 100 motors are shipped from site A to plant I,
> 700 motors are shipped from site A to plant II,
> 500 motors are shipped from site B to plant I,
and 0 motors are shipped from site B to plant II.

EXERCISE SET 4.8 CHAPTER REVIEW

1.

point	$z = 3x+2y$
(1, 5)	13
(3, 8)	25
(5, 4)	23
(2, 1)	8

The maximum value of z is 25.
The minimum value is 8.

3.

point	$w = 2x+5y+3z$
(1, 1, 1)	10
(4, 3, 5)	38
(5, 2, 3)	29
(7, 4, 6)	52

The maximum value of w is 52.
The minimum value is 10.

5.

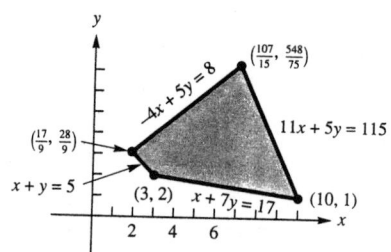

point	$z = 3x+4y$
(17/9, 28/9)	$18.\overline{1}$
(107/15, 548/75)	$50.62\overline{6}$
(10, 1)	34
(3, 2)	17

The maximum value of z is $50.62\overline{6}$.
The minimum value is 17.

7. (a)

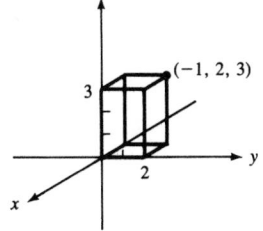

(b)

(c)

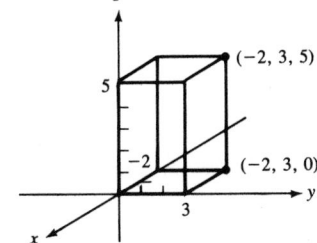

9. $\dfrac{5!}{2!(5-2)!}=\dfrac{5\cdot 4}{2}=10$

11. (a)

$$\begin{bmatrix} -10 & 12 & 9 & | & 72 \\ 18 & 15 & -4 & | & 90 \\ 16 & -7 & 10 & | & 80 \end{bmatrix}$$
$\begin{array}{l} 4E_1+9E_2\rightarrow E_1 \\ 5E_2+2E_3\rightarrow E_2 \end{array}$

$$\begin{bmatrix} 122 & 183 & 0 & | & 1098 \\ 122 & 61 & 0 & | & 610 \\ 16 & -7 & 10 & | & 80 \end{bmatrix}$$
$\begin{array}{l} E_1-E_2\rightarrow E_2 \\ 7E_1+183E_3\rightarrow E_3 \end{array}$

$$\begin{bmatrix} 122 & 183 & 0 & | & 1098 \\ 0 & 122 & 0 & | & 488 \\ 3782 & 0 & 1830 & | & 22{,}326 \end{bmatrix}$$
$\begin{array}{l} E_2/122\rightarrow E_2 \\ E_3/2\rightarrow E_3 \end{array}$

$$\begin{bmatrix} 122 & 183 & 0 & | & 1098 \\ 0 & 1 & 0 & | & 4 \\ 1891 & 0 & 915 & | & 11{,}163 \end{bmatrix}$$
$E_1-183E_2\rightarrow E_1$

$$\begin{bmatrix} 122 & 0 & 0 & | & 366 \\ 0 & 1 & 0 & | & 2 \\ 1891 & 0 & 915 & | & 11{,}163 \end{bmatrix}$$
$E_1/122\rightarrow E_1$

$$\begin{bmatrix} 1 & 0 & 0 & 3 \\ 0 & 1 & 0 & 4 \\ 1891 & 0 & 915 & 11,163 \end{bmatrix} \qquad -1891E_1 + E_3 \to E_3$$

$$\begin{bmatrix} 1 & 0 & 0 & 3 \\ 0 & 1 & 0 & 4 \\ 0 & 0 & 915 & 5490 \end{bmatrix} \qquad E_3/915 \to E_3$$

$$\begin{bmatrix} 1 & 0 & 0 & 3 \\ 0 & 1 & 0 & 4 \\ 0 & 0 & 1 & 6 \end{bmatrix} \qquad \text{The point of intersection is } (3, 4, 6).$$

(b)

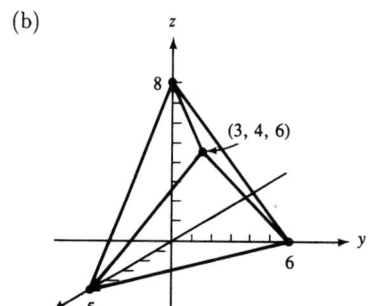

(c)

point	$w = 2x+5y+3z$
(0, 0, 8)	24
(5, 0, 0)	10
(0, 6, 0)	30
(0, 0, 0)	0
(3, 4, 6)	44

The maximum value of w is 44.
The minimum value is 0.

13. (a) Let $\begin{matrix} 2x+y+r=24 \\ 2x+5y+s=40 \end{matrix}$. The augmented matrix is

$$\begin{bmatrix} 2 & 1 & 1 & 0 & 24 \\ 2 & 5 & 0 & 1 & 40 \end{bmatrix} \qquad \text{and one corner is at } (0, 0).$$

$$\leftarrow r \begin{bmatrix} \boxed{2} & 1 & 1 & 0 & 24 \\ 2 & 5 & 0 & 1 & 40 \end{bmatrix} \qquad R_1/2 \to R_2$$
$$\quad\; \underset{x}{\uparrow}$$

$$\begin{bmatrix} 1 & \frac{1}{2} & \frac{1}{2} & 0 & 12 \\ 2 & 5 & 0 & 1 & 40 \end{bmatrix} \qquad -2R_1 + R_2 \to R_2$$

$$\leftarrow s \begin{bmatrix} 1 & \frac{1}{2} & \frac{1}{2} & 0 & 12 \\ 0 & \boxed{4} & -1 & 1 & 16 \end{bmatrix}. \qquad (12, 0) \text{ is a corner.}$$
$$\qquad\quad \underset{y}{\uparrow}$$

$$R_2/4 \to R_2$$

$$\begin{bmatrix} 1 & \frac{1}{2} & \frac{1}{2} & 0 & \bigm| & 12 \\ 0 & 1 & -\frac{1}{4} & \frac{1}{4} & \bigm| & 4 \end{bmatrix} \qquad -R_2/2 + R_1 \to R_1$$

$$\begin{bmatrix} 1 & 0 & \frac{5}{8} & -\frac{1}{8} & \bigm| & 10 \\ 0 & 1 & -\frac{1}{4} & \frac{1}{4} & \bigm| & 4 \end{bmatrix} \qquad (10,\,4) \text{ is also a corner.}$$

13. (b) Let $x + y + r = 15$
$5x + 3y + s = 55$
and $5x + 2y + t = 50$.

The augmented matrix is $\begin{bmatrix} 1 & 1 & 1 & 0 & 0 & \bigm| & 15 \\ 5 & 3 & 0 & 1 & 0 & \bigm| & 55 \\ 5 & 2 & 0 & 0 & 1 & \bigm| & 50 \end{bmatrix}$.

$(0,\,0)$ is a corner.

$\leftarrow r \begin{bmatrix} 1 & \boxed{1} & 1 & 0 & 0 & \bigm| & 15 \\ 5 & 3 & 0 & 1 & 0 & \bigm| & 55 \\ 5 & 2 & 0 & 0 & 1 & \bigm| & 50 \end{bmatrix}$ $\begin{array}{l} -3R_1 + R_2 \to R_2 \\ -2R_1 + R_3 \to R_3 \end{array}$

$\underset{y}{\uparrow}$

$\leftarrow s \begin{bmatrix} 1 & 1 & 1 & 0 & 0 & \bigm| & 15 \\ \boxed{2} & 0 & -3 & 1 & 0 & \bigm| & 10 \\ 3 & 0 & -2 & 0 & 1 & \bigm| & 20 \end{bmatrix}$ $\begin{array}{l} (0,\,15) \text{ is a corner.} \\ R_2/2 \to R_2 \end{array}$

$\underset{x}{\uparrow}$

$\begin{bmatrix} 1 & 1 & 1 & 0 & 0 & \bigm| & 15 \\ 1 & 0 & -\frac{3}{2} & \frac{1}{2} & 0 & \bigm| & 5 \\ 3 & 0 & -2 & 0 & 1 & \bigm| & 20 \end{bmatrix}$ $\begin{array}{l} -R_2 + R_1 \to R_1 \\ -3R_2 + R_3 \to R_3 \end{array}$

$\begin{bmatrix} 0 & 1 & \frac{5}{2} & -\frac{1}{2} & 0 & \bigm| & 10 \\ 1 & 0 & -\frac{3}{2} & \frac{1}{2} & 0 & \bigm| & 5 \\ 0 & 0 & \boxed{5/2} & -\frac{3}{2} & 1 & \bigm| & 5 \end{bmatrix}$ $\begin{array}{l} (5,\,10) \text{ is a corner.} \\ -R_3 + R_1 \to R_1 \\ \frac{2}{5} R_3 \to R_3 \end{array}$

$\leftarrow t \underset{r}{\uparrow}$

$\begin{bmatrix} 0 & 1 & 0 & 1 & -1 & \bigm| & 5 \\ 1 & 0 & -\frac{3}{2} & \frac{1}{2} & 0 & \bigm| & 5 \\ 0 & 0 & 1 & -\frac{3}{5} & \frac{2}{5} & \bigm| & 2 \end{bmatrix}$ $\frac{3}{2} R_3 + R_2 \to R_2$

$$\begin{bmatrix} 0 & 1 & 0 & 1 & -1 & \bigm| & 5 \\ 1 & 0 & 0 & -\frac{2}{5} & \frac{3}{5} & \bigm| & 8 \\ 0 & 0 & 1 & -\frac{3}{5} & \frac{2}{5} & \bigm| & 2 \end{bmatrix}$$

$(8, 5)$ is a corner. (It may be shown that $(10, 0)$ is also a corner.)

13. **(c)** Let $\ -10x + 12y + \ 9z + r = 72$
$\qquad 18x + 15y - \ 4z + s = 90$
$\qquad 16x - \ 7y + 10z + t = 80$

The augmented matrix is
$$\begin{bmatrix} -10 & 12 & 9 & 1 & 0 & 0 & \bigm| & 72 \\ 18 & 15 & -4 & 0 & 1 & 0 & \bigm| & 90 \\ 16 & -7 & 10 & 0 & 0 & 1 & \bigm| & 80 \end{bmatrix}$$

and $(0, 0, 0)$ is a corner.

$\leftarrow s$
$$\begin{bmatrix} -10 & 12 & 9 & 1 & 0 & 0 & \bigm| & 72 \\ \boxed{18} & 15 & -4 & 0 & 1 & 0 & \bigm| & 90 \\ 16 & -7 & 10 & 0 & 0 & 1 & \bigm| & 80 \end{bmatrix}$$
$\underset{x}{\uparrow}$
$\qquad \frac{1}{18}R_2 \rightarrow R_2$

$$\begin{bmatrix} -10 & 12 & 9 & 1 & 0 & 0 & \bigm| & 72 \\ \boxed{1} & \frac{5}{6} & -\frac{2}{9} & 0 & \frac{1}{18} & 0 & \bigm| & 5 \\ 16 & -7 & 10 & 0 & 0 & 1 & \bigm| & 80 \end{bmatrix}$$
$\qquad \begin{array}{l} 10R_2 + R_1 \rightarrow R_1 \\ -16R_2 + R_3 \rightarrow R_3 \end{array}$

$\leftarrow x$
$$\begin{bmatrix} 0 & \frac{61}{3} & \frac{61}{9} & 1 & \frac{5}{9} & 0 & \bigm| & 122 \\ 1 & \boxed{5/6} & -\frac{2}{9} & 0 & \frac{1}{18} & 0 & \bigm| & 5 \\ 0 & -\frac{61}{3} & \frac{122}{9} & 0 & -\frac{8}{9} & 1 & \bigm| & 0 \end{bmatrix}$$
$\underset{y}{\uparrow}$
$\qquad (5, 0, 0)$ is a corner.
$\qquad \frac{6}{5}R_2 \rightarrow R_2$

$$\begin{bmatrix} 0 & \frac{61}{3} & \frac{61}{9} & 1 & \frac{5}{9} & 0 & \bigm| & 122 \\ \frac{6}{5} & 1 & -\frac{4}{15} & 0 & \frac{1}{15} & 0 & \bigm| & 6 \\ 0 & -\frac{61}{3} & \frac{122}{9} & 0 & -\frac{8}{9} & 1 & \bigm| & 0 \end{bmatrix}$$
$\qquad \begin{array}{l} -\frac{61}{3}R_2 + R_1 \rightarrow R_1 \\ \frac{61}{3}R_2 + R_3 \rightarrow R_3 \end{array}$

$\leftarrow t$
$$\begin{bmatrix} -\frac{122}{5} & 0 & \frac{61}{5} & 1 & -\frac{4}{5} & 0 & \bigm| & 0 \\ \frac{6}{5} & 1 & -\frac{4}{15} & 0 & \frac{1}{15} & 0 & \bigm| & 6 \\ \boxed{122/5} & 0 & \frac{122}{15} & 0 & \frac{7}{15} & 1 & \bigm| & 122 \end{bmatrix}$$
$\underset{x}{\uparrow}$
$\qquad (0, 6, 0)$ is a corner.
$\qquad \frac{5}{122}R_3 \rightarrow R_3$

$$\begin{bmatrix} -\frac{122}{5} & 0 & \frac{61}{5} & 1 & -\frac{4}{5} & 0 & \Big| & 0 \\ \frac{6}{5} & 1 & -\frac{4}{15} & 0 & \frac{1}{15} & 0 & \Big| & 6 \\ 1 & 0 & \frac{1}{3} & 0 & \frac{127}{2196} & \frac{5}{122} & \Big| & 5 \end{bmatrix}$$

$(122/5)R_3 + R_1 \to R_1$
$-\frac{6}{5}R_3 + R_2 \to R_2$

(irrelevant entries are not entered.)

$\leftarrow x \quad \begin{bmatrix} 0 & 0 & \frac{61}{3} & 1 & - & 1 & \Big| & 122 \\ 0 & 1 & -\frac{2}{3} & 0 & - & -\frac{3}{61} & \Big| & 6 \\ 1 & 0 & \boxed{1/3} & 0 & - & - & \Big| & 5 \end{bmatrix}$

$\uparrow$
z

giving $(5, 0, 0)$ again as a corner.
$3R_3 \to R_3$

$$\begin{bmatrix} 0 & 0 & \frac{61}{3} & 1 & - & 1 & \Big| & 122 \\ 0 & 1 & -\frac{2}{3} & 0 & - & -\frac{3}{61} & \Big| & 0 \\ 3 & 0 & 1 & 0 & - & - & \Big| & 15 \end{bmatrix}$$

$-\frac{61}{3}R_3 + R_1 \to R_1$
$\frac{2}{3}R_3 + R_2 \to R_2$

$$\begin{bmatrix} -61 & 0 & 0 & 1 & - & - & \Big| & - \\ - & 1 & 0 & 0 & - & - & \Big| & 10 \\ 3 & 0 & 1 & 0 & - & - & \Big| & 15 \end{bmatrix}$$

$(0, 10, 15)$ is a corner.

15. Let
$$-2x + y + s_1 = 2$$
$$-2x + 3y + s_2 = 14$$
and $\quad z - 3x - 8y = 0$

Tableau I

basis	x	y	s_1	s_2	values
s_1	-2	$\boxed{1}$	1	0	2
s_2	-2	3	0	1	14
z	-3	-8	0	0	0

Tableau II

basis	x	y	s_1	s_2	values
y	-2	1	1	0	2
s_2	$\boxed{4}$	0	-3	1	8
z	-19	0	8	0	16

Tableau III

basis	x	y	s_1	s_2	values
y	0	1	$-\frac{1}{2}$	$\frac{1}{2}$	6
x	1	0	$-\frac{3}{4}$	$\frac{1}{4}$	2
z	0	0	$-\frac{25}{4}$	$\frac{19}{4}$	54

Since the s_1 column contains only negative values the region is unbounded and z does not attain a maximum.

17. Let
$$
\begin{aligned}
-\ 10\,x + 12\,y +\ 9\,z + s_1 \qquad\qquad\qquad &= 72 \\
18\,x + 15\,y -\ 4\,z \qquad + s_2 \qquad\quad &= 90 \\
16\,x -\ 7\,y + 10\,z \qquad\qquad + s_3 &= 80 \\
\text{and } w -\ 2\,x -\ 5\,y -\ 3\,z \qquad\qquad\qquad &= 0
\end{aligned}
$$

Tableau I

basis	x	y	z	s_1	s_2	s_3	values
s_1	-10	$\boxed{12}$	9	1	0	0	72
s_2	18	15	-4	0	1	0	90
s_3	16	-7	10	0	0	1	80
w	-2	-5	-3	0	0	0	0

Tableau II

basis	x	y	z	s_1	s_2	s_3	values
y	$-\frac{5}{6}$	1	$\frac{3}{4}$	$\frac{1}{12}$	0	0	6
s_2	$\boxed{61/2}$	0	$-\frac{61}{4}$	$-\frac{5}{4}$	1	0	0
s_3	$\frac{61}{6}$	0	$\frac{61}{4}$	$\frac{7}{12}$	0	1	122
w	$-\frac{37}{6}$	0	$\frac{3}{4}$	$\frac{5}{12}$	0	0	30

Tableau III

basis	x	y	z	s_1	s_2	s_3	values
y	0	1	$\frac{1}{3}$	$\frac{3}{61}$	$\frac{5}{183}$	0	6
x	1	0	$-\frac{1}{2}$	$-\frac{5}{122}$	$\frac{2}{61}$	0	0
s_3	0	0	$\boxed{61/3}$	1	$-\frac{1}{3}$	1	122
w	0	0	$-\frac{7}{3}$	$\frac{10}{61}$	$\frac{74}{366}$	0	30

Tableau IV

basis	x	y	z	s_1	s_2	s_3	values
y	0	1	0	–	–	–	4
x	1	0	0	–	–	–	3
z	0	0	1	$\frac{3}{61}$	$-\frac{1}{61}$	$\frac{3}{61}$	6
w	0	0	0	$\frac{17}{61}$	$\frac{10}{61}$	$\frac{7}{61}$	44

The maximum value of w is 44 and this occurs when $x=3$, $y=4$ and $z=6$.

19. Let
$$
\begin{aligned}
-x + 3y + 2z + s_1 &= 216 \\
x + y - z + s_2 &= 198 \\
2x + y + 3z + s_3 &= 324 \\
\text{and } w - 15x - 3y - 3z &= 0
\end{aligned}
$$

Tableau I

basis	x	y	z	s_1	s_2	s_3	values
s_1	-1	3	2	1	0	0	216
s_2	1	1	1	0	1	0	198
s_3	$\boxed{2}$	1	3	0	0	1	324
w	-15	-3	-3	0	0	0	0

Tableau II

basis	x	y	z	s_1	s_2	s_3	values
s_1	0	$\frac{7}{2}$	$\frac{9}{2}$	1	0	$\frac{1}{2}$	378
s_2	0	$\frac{1}{2}$	$-\frac{1}{2}$	0	1	$-\frac{1}{2}$	36
x	1	$\frac{1}{2}$	$\frac{3}{2}$	0	0	$\frac{1}{2}$	162
w	0	$\frac{9}{2}$	$\frac{39}{2}$	0	0	$\frac{15}{2}$	2430

The maximum value of w is 2430 and occurs when $x=162$ and $y=z=0$.

21. The primal matrix is $\begin{bmatrix} 3 & 1 & 50 \\ 1 & 1 & 30 \\ 1 & 4 & 60 \\ 5 & 2 & w \end{bmatrix}$. The dual matrix is $\begin{bmatrix} 3 & 1 & 1 & 5 \\ 1 & 1 & 4 & 2 \\ 50 & 30 & 60 & w \end{bmatrix}$.

We maximize $W=50u+30v+60q$ subject to
$$3u+v+q \le 5$$
$$u+v+4q \le 2, \qquad u \ge 0, \ v \ge 0.$$

Tableau I

basis	u	v	q	r	s	values
r	3	1	1	1	0	5
s	1	1	$\boxed{4}$	0	1	2
W	−50	−30	−60	0	0	0

Tableau II

basis	u	v	q	r	s	values
r	$\boxed{11/4}$	$\frac{3}{4}$	0	1	$-\frac{1}{4}$	$\frac{9}{2}$
q	$\frac{1}{4}$	$\frac{1}{4}$	1	0	$\frac{1}{4}$	$\frac{1}{2}$
W	−35	−15	0	0	15	30

Tableau III

basis	u	v	q	r	s	values
u	1	$\frac{3}{11}$	0	$\frac{4}{11}$	$-\frac{1}{11}$	$\frac{18}{11}$
q	0	$\boxed{2/11}$	1	$-\frac{1}{11}$	$\frac{3}{11}$	$\frac{1}{11}$
W	0	$-\frac{60}{11}$	0	$\frac{140}{11}$	$\frac{130}{11}$	$\frac{960}{11}$

Tableau IV

basis	u	v	q	r	s	values
u	1	0	$-\frac{3}{2}$	$\frac{1}{2}$	$-\frac{1}{2}$	$\frac{3}{2}$
v	0	1	$\frac{11}{2}$	$-\frac{1}{2}$	$\frac{3}{2}$	$\frac{1}{2}$
W	0	0	30	10	20	90

The minimum value of w is 90 which occurs when $x=10$ and $y=20$.

23. Let x be the number of gadgets A and y the number of gadgets B.

$$2x + y \le 200$$
$$x + y \le 140$$
$$x + 2y \le 240 \qquad \text{The total profit is } z=90x+120y.$$

Tableau I

basis	x	y	s_1	s_2	s_3	values
s_1	2	1	1	0	0	200
s_2	1	1	0	1	0	140
s_3	1	$\boxed{2}$	0	0	1	240
z	-90	-120	0	0	0	0

Tableau II

basis	x	y	s_1	s_2	s_3	values
s_1	$\frac{3}{2}$	0	1	0	$-\frac{1}{2}$	80
s_2	$\boxed{1/2}$	0	0	1	$-\frac{1}{2}$	20
y	$\frac{1}{2}$	1	0	0	$\frac{1}{2}$	120
z	-30	0	0	0	60	14,400

Tableau III

basis	x	y	s_1	s_2	s_3	values
s_1	0	0	1	-3	1	20
x	1	0	0	2	-1	40
y	0	1	0	-1	1	100
z	0	0	0	60	30	15,600

The maximum profit is \$15,600 which occurs for 40 gadgets A and gadgets B.

25. Let x and y be the number of hours at the Arlington and Marysville plants, respectively.

We must minimize $w=168x+288y$ subject to

$$50x + 100y \geq 1200$$
$$100x + 150y \geq 2250, \qquad x \geq 0, \ y \geq 0.$$

The primal matrix is $\begin{bmatrix} 50 & 100 & 1200 \\ 100 & 150 & 2250 \\ 168 & 288 & w \end{bmatrix}$. The dual matrix is $\begin{bmatrix} 50 & 100 & 168 \\ 100 & 150 & 288 \\ 1200 & 2250 & w \end{bmatrix}$.

We must maximize $W=1200u+2250v$ subject to

$$50u + 100v \leq 168$$
$$100u + 150v \leq 288, \qquad u \geq 0, \ v \geq 0.$$

Tableau I

basis	u	v	r	s	values
r	50	[100]	1	0	168
s	100	150	0	1	288
W	−1200	−2250	0	0	0

Tableau II

basis	u	v	r	s	values
v	.5	1	.01	0	1.68
s	[25]	0	−1.5	1	36
W	−75	0	22.5	0	3780

Tableau III

basis	u	v	r	s	values
v	0	1	.04	−.02	.96
u	1	0	−.06	.04	1.44
W	0	0	18	3	3888

The minimum daily cost is $3,888 which occurs when the Arlington plant operates for 18 hours and the Marysville plant operates for 3 hours.

27. Maximize $P_m = 2x + 5y + 3z - a_1 M - a_2 M$ where

$$
\begin{aligned}
4x - y + 3z - s_1 + a_1 &= 29 \\
x + y + z \quad\quad\quad + a_2 &= 12 \\
2x + y \quad\quad + s_3 \quad\quad\quad &= 13 \\
5x + y + 6z + s_4 \quad\quad\quad &= 61,
\end{aligned}
$$

$x, y, z, s_1, s_3, s_4, a_1, a_2$ all nonnegative.

Tableau I

basis	x	y	z	s_1	s_3	s_4	a_1	a_2	values
a_1	4	−1	3	−1	0	0	1	0	29
a_2	1	1	1	0	0	0	0	1	12
s_3	[2]	1	0	0	1	0	0	0	13
s_4	5	1	6	0	0	1	0	0	61
P_m	−5M−2	−5	−4M−3	M	0	0	0	0	−41M

— 244 —

Tableau II

basis	x	y	z	s_1	s_3	s_4	a_1	a_2	values
a_1	0	-3	$\boxed{3}$	-1	-2	0	1	0	3
a_2	0	$\frac{1}{2}$	1	0	$-\frac{1}{2}$	0	0	1	$\frac{11}{2}$
x	1.	$\frac{1}{2}$	0	0	$\frac{1}{2}$	0	0	0	$\frac{13}{2}$
s_4	0	$-\frac{3}{2}$	6	0	$-\frac{5}{2}$	1	0	0	$\frac{57}{2}$
P_m	0	$\frac{5M}{2}-4$	$-4M-3$	M	$\frac{5M}{2}+1$	0	0	0	$-\frac{17M}{2}+13$

Tableau III

basis	x	y	z	s_1	s_3	s_4	a_1	a_2	values
z	0	-1	1	$-\frac{1}{3}$	$-\frac{2}{3}$	0	$\frac{1}{3}$	0	1
a_2	0	$\boxed{3/2}$	0	$\frac{1}{3}$	$\frac{1}{6}$	0	$-\frac{1}{3}$	1	$\frac{9}{2}$
x	1	$\frac{1}{2}$	0	0	$\frac{1}{2}$	0	0	0	$\frac{13}{2}$
s_4	0	$\frac{9}{2}$	0	2	$\frac{3}{2}$	1	-2	0	$\frac{45}{2}$
P_m	0	$-\frac{3M}{2}-7$	0	$-\frac{M}{3}-1$	$-\frac{M}{6}-1$	0	$\frac{4M}{3}+1$	0	$-\frac{9M}{2}+16$

Tableau IV

basis	x	y	z	s_1	s_3	s_4	a_1	a_2	values
z	0	0	1	$-\frac{1}{9}$	$-\frac{5}{9}$	0	$-\frac{1}{9}$	$\frac{2}{3}$	4
y	0	1	0	$\frac{2}{9}$	$\frac{1}{9}$	0	$-\frac{2}{9}$	$\frac{2}{3}$	3
x	1	0	0	$-\frac{1}{9}$	$\frac{4}{9}$	0	$\frac{1}{9}$	$-\frac{1}{3}$	5
s_4	0	0	0	1	$\boxed{1}$	1	-1	-3	9
P_m	0	0	0	$\frac{5}{9}$	$-\frac{2}{9}$	0	$M-\frac{5}{9}$	$M+\frac{14}{3}$	37

Tableau V

basis	x	y	z	s_1	s_3	s_4	values
z	0	0	1	$\frac{4}{9}$	0	$\frac{5}{9}$	9
y	0	1	0	$\frac{1}{9}$	0	$-\frac{1}{9}$	2
x	1	0	0	$-\frac{5}{9}$	0	$-\frac{4}{9}$	1
s_3	0	0	0	1	1	1	9
P_m	0	0	0	$\frac{7}{9}$	0	$\frac{2}{9}$	39

The maximum is 39.

29. We maximize $W_m = 7x + 4y + z - a_1 M - a_2 M$ under the conditions of Problem 27.

Tableau I

basis	x	y	z	s_1	s_3	s_4	a_1	a_2	values
a_1	4	-1	3	-1	0	0	1	0	29
a_2	1	1	1	0	0	0	0	1	12
s_3	$\boxed{2}$	1	0	0	1	0	0	0	13
s_4	5	1	6	0	0	1	0	0	61
W_m	$-5M-7$	-4	$-4M-1$	M	0	0	0	0	$-41M$

Tableau II

basis	x	y	z	s_1	s_3	s_4	a_1	a_2	values
a_1	0	-3	$\boxed{3}$	-1	-2	0	1	0	3
a_2	0	$\frac{1}{2}$	1	0	$-\frac{1}{2}$	0	0	1	$\frac{11}{2}$
x	1	$\frac{1}{2}$	0	0	$\frac{1}{2}$	0	0	0	$\frac{13}{2}$
s_4	0	$-\frac{3}{2}$	6	0	$-\frac{5}{2}$	1	0	0	$\frac{57}{2}$
W_m	0	$\frac{5M}{2}-\frac{1}{2}$	$-4M-1$	M	$\frac{5M}{2}+\frac{7}{2}$	0	0	0	$\frac{17M}{2}+\frac{91}{2}$

Tableau III

basis	x	y	z	s_1	s_3	s_4	a_1	a_2	values
z	0	-1	1	$-\frac{1}{3}$	$-\frac{2}{3}$	0	$\frac{1}{3}$	0	1
a_2	0	$\boxed{3/2}$	0	$\frac{1}{3}$	$\frac{1}{6}$	0	$-\frac{1}{3}$	1	$\frac{9}{2}$
x	1	$\frac{1}{2}$	0	0	$\frac{1}{2}$	0	0	0	$\frac{13}{2}$
s_4	0	$\frac{9}{2}$	0	2	$\frac{3}{2}$	1	-2	0	$\frac{45}{2}$
W_m	0	$-\frac{3M}{2}-\frac{3}{2}$	0	$-\frac{M}{3}-\frac{1}{3}$	$-\frac{M}{6}+\frac{17}{6}$	0	$\frac{4M+1}{3}$	0	$-\frac{9M}{2}+\frac{93}{2}$

Tableau IV

basis	x	y	z	s_1	s_3	s_4	a_1	a_2	values
z	0	0	1	$-\frac{1}{9}$	$-\frac{5}{9}$	0	$\frac{1}{9}$	$\frac{2}{3}$	4
y	0	1	0	$\frac{2}{9}$	$\frac{1}{9}$	0	$-\frac{2}{9}$	$\frac{2}{3}$	3
x	1	0	0	$-\frac{1}{9}$	$\frac{4}{9}$	0	$\frac{1}{9}$	$-\frac{1}{3}$	5
s_4	0	0	0	1	1	1	-1	-3	9
W_m	0	0	0	0	3	0	$M+\frac{2}{3}$	$M+1$	51

The maximum value is 51.

CHAPTER 5

SEQUENCES AND MATHEMATICS OF FINANCE

EXERCISE 5.1 SEQUENCES, ARITHMETIC AND GEOMETRIC PROGRESSIONS

1. $s_1 = \dfrac{2}{2-1} = \dfrac{2}{1} = 2$ $\qquad$ $s_2 = \dfrac{2}{4-1} = 2/3$ $\qquad$ $s_3 = \dfrac{2}{6-1} = 2/5$

   $s_4 = \dfrac{2}{8-1} = 2/7$ $\qquad$ $s_5 = \dfrac{2}{10-1} = 2/9$

3. $a_1 = \dfrac{-2}{1+1} = -1$ $\qquad$ $a_2 = \dfrac{-2}{4+1} = -2/5$ $\qquad$ $a_3 = \dfrac{-2}{9+1} = -2/10$

   $a_4 = \dfrac{-2}{16+1} = -2/17$ $\qquad$ $a_5 = \dfrac{-2}{25+1} = -2/26$

5. $t_1 = \dfrac{1}{1+1} = 1/2$ $\qquad$ $t_2 = \dfrac{-2}{2+1} = -2/3$ $\qquad$ $t_3 = \dfrac{3}{3+1} = 3/4$

   $t_4 = \dfrac{-4}{4+1} = -4/5$ $\qquad$ $t_5 = \dfrac{5}{5+1} = 5/6$

7. $s_1 = \dfrac{1+3}{3-1} = 2$ $\qquad$ $s_2 = \dfrac{2+3}{6-1} = 1$ $\qquad$ $s_3 = \dfrac{-(3+3)}{9-1} = -3/4$

   $s_4 = \dfrac{-(4+3)}{12-1} = -7/11$ $\qquad$ $s_5 = \dfrac{-(5+3)}{15-1} = -8/14$

9. $a = -4$, adding three to successive terms we obtain

a_1	$=-4$	a_6	$=11$
a_2	$=-1$	a_7	$=14$
a_3	$=2$	a_8	$=17$
a_4	$=5$	a_9	$=20$
a_5	$=8$	a_{10}	$=23$

11. $a = -5$ and $d = -2$. Adding -2 to successive terms we obtain

$a_1 = -5$	$a_3 = -9$	$a_5 = -13$	$a_7 = -17$
$a_2 = -7$	$a_4 = -11$	$a_6 = -15$	$a_8 = -19$

13. $a_{20} = 73$, $a_{30} = 113$ and $a_k = a + (k-1)d$. $a_{20} = a + 19d = 73$ and $a_{30} = a + 29d = 113$.
$a_{30} - a_{20} = a + 29d - a - 19d = 113 - 73$, $10d = 40$, $d=4$ and $a + 19(4) = 73$ so $a = -3$.

The first five terms are $a_1 = -3$, $a_2 = 1$, $a_3 = 5$, $a_4 = 9$ and $a_5 = 13$.

15. $a_{16} = a + 15d = -20$
$a_{36} = a + 35d = -60$
$a_{36} - a_{16} = 20d = -40$, $d = -2$ and $a-30 = -20$ so $a = 10$.

The first five terms are 10, 8, 6, 4 and 2.

17. $a_{19} = a + 18d = 46$
$a_{63} = a + 62d = 178$
$a_{32} - a_{19} = 44d = 132$, $d = 3$, $a + 54 = 46$, $a = -8$.

The first 8 terms are -8, -5, -2, 1, 4, 7, 10, 13.

19. $a = 12$, $d = 5$, $a_{70} = 12 + 69(5) = 357$. $S_{70} = 70(12 + 357)/2 = 12915$.

21. $a = 0$, $d = 4$, $a_{90} = 0 + 4(89) = 356$. $S_{90} = (0 + 356)90/2 = 16020$.

23. $a = 2$, $d = 3$, $a_{100} = 2 + 3(99) = 299$. $S_{100} = \dfrac{100(2 + 299)}{2} = 15050$.

25. $a = -4$, $d = 5$, $a_{90} = -4 + 89(5) = 441$. $S_{90} = \dfrac{90(441 - 4)}{2} = 19665$

27. The change in value is $\dfrac{25000 - 5000}{8} = \$2500/\text{year}$,

$V_0 = \$25,000$
$V_1 = \$25,000 - 2,500 = \$22,500$
$V_2 = \$22,500 - 2,500 = \$20,000$
$V_3 = \$20,000 - 2,500 = \$17,500$
$V_4 = \$17,500 - 2,500 = \$15,000$
$V_5 = \$12,500$
$V_6 = \$10,000$
$V_7 = \$7,500$
$V_8 = \$5,000$

29. $V_0 = \$300,000$
$V_1 = \$300,000 - .05(300,000) = \$285,000$
$V_2 = \$285,000 - .05(285,000) = \$270,750$
$V_3 = \$270,750 - .05(270,750) = \$257,212.50$
$V_4 = \$257,212.50 - .05(257,212.50) = \$244,351.875$
$V_5 = \$244,351.875(.95) = \$232,134.2813$

31. $A_n = 500(1 + .005)^n$. $A_1 = \$502.50$, $A_2 = \$505.01$, $A_3 = \$507.54$, $A_4 = \$510.08$,
$A_5 = \$512.62$, $A_6 = \$515.19$

33. $A_n = 800(1 + .08/12)^n$, $A_1 = 805.33$, $A_2 = 810.70$, $A_3 = 816.11$, $A_4 = 821.55$, $A_5 = 827.02$,
$A_6 = 832.54$

35. $A = 600(1 + .07/12)^n$

$n = 1, A = \$603.50$ $n = 4, A = \$614.12$
$n = 2, A = \$607.02$ $n = 5, A = \$617.71$
$n = 3, A = \$610.56$ $n = 6, A = \$621.31$

37. $b_1 = 2$ and $r = 5$.
$b_2 = 2.5 = 10$, $b_3 = 10(5) = 50$, $b_4 = 50(5) = 250$, $b_5 = 250(5) = 1250$

39. $t_1 = 512$ and $r = \frac{1}{2}$.

$t_2 = 512(\frac{1}{2}) = 256$, $t_3 = 256(\frac{1}{2}) = 128$, $t_4 = 128/2 = 64$, $t_5 = 64/2 = 32$

41. $a_5 = ar^4 = 324$ and $a_7 = ar^6 = 2916$.
$a_7/a_5 = ar^6/ar^4 = r^2 = 2916/324 = 9$. Hence $r = \pm 3$.
If $r = 3$, $a = 324/81 = 4$. If $r = -3$, $a = 4$.
If $r = 3$, $a = 4$, $a_2 = 4(3) = 12$, $a_3 = 4(9) = 36$.
If $r = -3$, $a = 4$, $a_2 = -12$, $a_3 = 36$.

43. $a_7 = ar^6 = -1458$ and $a_{10} = ar^9 = -39366$
$ar^9/ar^6 = r^3 = 39366/1458 = 27$, $r = 3$, and $a = -1458/3^6 = -2$. The first 3 terms are
$a = -2$, $a_2 = -2(3) = -6$, $a_3 = -6(3) = -18$.

45. $ar^4 = 9$ and $ar^7 = \frac{1}{3}$. $ar^7/ar^4 = r^3 = (\frac{1}{3})/9 = 1/27$. Hence $r = \frac{1}{3}$ and
$a = 9/(\frac{1}{3})^4 = 9 \cdot 81 = 729$

The first 3 terms are 729, $729/3 = 243$, $243/3 = 81$.

47. $b_1 = 2$ and $r = 5$. $S_{20} = \dfrac{2(1 - 5^{20})}{1 - 5} = \dfrac{5^{20} - 1}{2}$.

49. $t_1 = 512$ and $r = \frac{1}{2}$. $S_{30} = \dfrac{512\,(1 - .5^{30})}{1 - .5} = 1024(1 - .5^{30}) = 1023.9999$

51. If $a = 4$ and $r = 3$, $S_{25} = 4(1 - 3^{25})/(1 - 3) = 2(3^{25} - 1)$.
If $a = 4$ and $r = -3$, $S_{25} = 4(1 + 3^{25})/(1 + 3) = 1 + 3^{25}$.

53. $a = -2$ and $r = 3$. $S_{27} = \dfrac{-2(1 - 3^{27})}{1 - 3} = 1 - 3^{27}$.

55. $a = 729$ and $r = \frac{1}{3}$. $S_{28} = \dfrac{729(1 - 3^{-28})}{2/3} = \dfrac{3^7(1 - 3^{-28})}{2} = (3^7 - 3^{-21})/2$.

57. $250(1.005)^{14} + 250(1.005)^{13} + \cdots + 250 = \dfrac{250(1 - 1.005^{15})}{1 - 1.005} = \$3884.14$

59. $400(1.005)^9 + 400(1.005)^8 + \cdots + 400(1.005) = 400(1.005)(1 - 1.005^9)/(1 - 1.005)$
$= \$3691.21$

61. $450(1.005)^{22} + 450(1.005)^{21} + \cdots + 450(1.005)$
$\quad = 450(1.005)(1 - 1.005^{22})/(1 - 1.005) = \$10,489.68$

63. $64 + 32(2) + 16(2) + 8(2) + 4(2) + 2(2) + 1(2) + (\frac{1}{2})2 + (\frac{1}{4})2 + (\frac{1}{8})2$
$\qquad\qquad = 64 + 2(32 + 16 + 8 + 4 + 2 + 1 + .5 + .25 + .125)$
$\qquad\qquad = 64 + 2(32)(1 - .5^9)/(1 - .5) = 191.75 \text{ inches}$

65. $30000 + 30000(1.06) + 30000(1.06)^2 + \cdots + 30000(1.06)^{11} = 30000(1 - 1.06^{12})/(1 - 1.06)$
$\quad = \$506,098.24$

67. $.01 + .02 + .04 + .08 + \cdots + (.01)2^{20} = (.01)(1 - 2^{20})/(1 - 2) = .01(2^{21} - 1) = \$20,971.51$
The second option is best.

69. Lose first hand $\Rightarrow$ lose $1
Lose 2nd hand $\Rightarrow$ lose $2 or a total of $3
Lose 3rd hand $\Rightarrow$ lose $4 or a total of $7
Lose 4th hand $\Rightarrow$ lose $8 or a total of $15
Lose 5th hand $\Rightarrow$ lose $16 or a total of $31
Lose 6th hand $\Rightarrow$ lose $32 or a total of $63
Lose 7th hand $\Rightarrow$ lose $64 or a total of $127
Lose 8th hand $\Rightarrow$ lose $128 or a total of $255
Lose 9th hand $\Rightarrow$ lose $256 or a total of $511
Lose 10th hand $\Rightarrow$ lose $500 or a total of $1011

If one loses the first k hands, where $k > 9$, and wins the next, they will come out behind, as they can't win more than $500 in one game.

EXERCISE SET 5.2 ANNUITIES

1. 1.576899264	**3.** 2.772469785	**5.** .591457366
7. 10.113249	**9.** 15.813679	**11.** 22.562977
13. 11.618933	**15.** 15.562251	

17. $300 + 300(1 + \frac{.06}{12}) + 300(1 + \frac{.06}{12})^2 + \cdots + 300(1 + \frac{.06}{12})^{39} = \dfrac{300(1 - 1.005^{40})}{1 - 1.005} = \$13,247.65.$
Interest earned is $1247.65.

19. Let D be the size of the deposit. 58 deposits are made.
$\quad D + D(1.01) + D(1.01)^2 + \cdots + D(1.01)^{57} = 9000$

$\quad \dfrac{D(1 - 1.01^{58})}{1 - 1.01} = D(78.09005966) = 9000. \qquad D = \$115.25$

21. Let $S = A_1 + A_2 + A_3 + A_4$ where
$\qquad A_1 = (1 + .08)^{18} = 9,500$
$\qquad A_2 = (1 + .08)^{19} = 9,500$
$\qquad A_3 = (1 + .08)^{20} = 9,500$
$\qquad A_4 = (1 + .08)^{21} = 9,500$

$$S = 9500(1+.08)^{-18} + 9500(1+.08)^{-19} + 9500(1+.08)^{-20} + 9500(1+.08)^{-21}$$
$$= \frac{9500(1+.08)^{-18}(1-(1+.08)^{-4})}{1-(1+.08)^{-1}} = \frac{9500}{.08}(1+.08)^{-17}(1-(1+.08)^{-4}) = \$8504.07$$

23. $250 + 250(1.005) + \cdots + 250(1.005)^{71} = \dfrac{250(1-1.005^{72})}{-.005} = \$21,602.21$

Interest earned is $21,602.21 - 250(72) = \$3,602.21$

25. $350(1.02) + \cdots + 350(1.02)^{20} = \dfrac{350(1.02)(1-1.02^{20})}{1-1.02} = \$8,674.16$

Interest earned is $8,674.16 - 350(20) = \$1674.16$

27. Using formula (3), $A = 950 \cdot \dfrac{1-(1.02)^{-24}}{.02} = \$17,968.23$

29. $500 + 500 \cdot \dfrac{1-(1+\frac{.09}{12})^{-47}}{.09/12} = 500 + 19,743.08 = \$20,243.08$

31. The amount paid in is
$$350 + 350(1.0075) + \cdots + 350(1.0075)^{359} = \frac{350(1-1.0075^{360})}{1-1.0075} = \$640,760.22.$$
Thus if R is the monthly amount withdrawn for 120 months,
$$640,760.22 = \frac{R[1-(1.0075)^{-120}]}{.0075} = R(78.94169267) \text{ and } R = \$8,116.88$$

33. Let R be the amount of the monthly payment. Then
$$500,000 = R + R(1.035) + R(1.035)^2 + \cdots + R(1.035)^9 = \frac{R(1-1.035^{10})}{1-1.035} = R(11.73139316).$$
$R = \$42,620.68$

35. In four years the account will be worth
$$100 + 100(1.01) + \cdots + 100(1.01)^{47} = 100s_{\overline{48}|}.01 = 100(61.222608) = \$6,122.26$$
In the next eight years this will grow to $6,122.26(1.01)^{96} = \$15,913.42$

The remaining payments must yield $150,000 - 15,913.42 = \$134,086.58.$

If R is the size of the remaining monthly payments,
$$R + R(1.01) + \cdots + R(1.01)^{95} = \frac{R(1-1.01^{96})}{1-1.01} = R(159.9272926) = 134,086.58.$$
$R = \$838.42$

37. If P is the size of the payment,
$$200,000 = \frac{P}{1.0075} + \frac{P}{1.0075^2} + \cdots + \frac{P}{1.0075^{156}} = \frac{P(1-1.0075^{-156})}{.0075}$$
$P = \$2,179.36$

39. At the end of 8 years the account contained
$$250(1.005) + 250(1.005)^2 + \cdots + 250(1.005)^{96} = \frac{250(1.005)(1-1.005^{96})}{-.005} = \$30,860.67.$$

This has grown to $30,860.67(1.0075)^{203} = \$140,653.80$.

The other deposits amount to $250 + 250(1.0075) + 250(1.0075)^2 + \cdots + 250(1.0075)^{203}$

$$= \frac{250(1 - 1.0075^{204})}{-.0075} = \$119,729.56.$$

Thus the total in the account is $119,729.56 + 140,653.80 = \$260,383.36$

41.   Amount deposited $= 9500 + P$ where $P(.08) = 9500$

$$P = \frac{9500}{.08} = \$118,750$$

The total deposit is $128,250.

43.   Since the payments are made at the beginning of the period, the amount of the annuity is

$R(1+i) + R(1+i)^2 + \cdots + R(1+i)^n$ which is the sum of a G.P. with the first term

$R(1+i)$ and common ratio $1+i$. The sum is $\dfrac{R(1+i)(1-(1+i)^n)}{1-(1+i)} = \dfrac{R(1+i-(1+i)^{n+1})}{-i}$

$= \dfrac{R((1+i)^{n+1} - 1 - i)}{i} = R\left(\dfrac{(1+i)^{n+1} - 1}{i} - \dfrac{i}{i}\right) = R(S_{\overline{n+1}|i} - 1)$.

45.   In example 5, $R = 300$, $i = .01$ and $n = 48$. The amount in the account will be

$$\frac{300(1.01^{49} - 1 - .01)}{.01} = \$18,550.45$$

47. $R = 250$, $n = 36$, $i = .0075$. The amount of the account will be

$$\frac{250((1.0075)^{37} - 1 - .0075)}{.0075} = \$10,365.34$$

49. $R = 15,000$, $n = 15$, $i = .07$. The present value is $\dfrac{15,000[1.07 - (1.07)^{-14}]}{.07} = \$146,182.02$

EXERCISE SET 5.3 MORTGAGES

1.   The downpayment is $36,500(.2) = \$7,300$. Thus the amount of the mortgage is $29,200.

$A = \$29,200$, $i = \dfrac{.0725}{12}$, $n = 300$

(a) $R = \dfrac{29,200(.0725/12)}{1 - (1 + .0725/12)^{-300}} = \$211.06$

(b) $I_{150} = 211.06[1 - (1 + .0725/12)^{150-1-300}] = \$126.06$

(c) $B_{150} = 211.06(1 + .0725/12)^{150-1-300} = \$85.00$

(d) $A_{144} = \dfrac{211.06[1 - (1 + .0725/12)^{144-300}]}{.0725/12} = \$21,283.27$

The amount that has been paid is $7,916.73

(e) $300(211.06) - 29,200 = \$34,118$.

3. The downpayment is $12,500(.2) = \$2,500$. $A = 12,500 - 2,500 = \$10,000$, $i = .0075$, $n = 60$

(a) $R = \dfrac{10,000(.0075)}{1 - (1.0075)^{-60}} = \$207.58$

(b) $I_{25} = R[1 - (1.0075)^{24-60}] = \$48.96$

(c) $B_{25} = R(1.0075)^{24-60} = \$158.62$

(d) $A_{36} = R\,\dfrac{1 - (1.0075)^{36-60}}{.0075} = \$4543.75$

     The amount that has been paid on the principal is $10,000 - 4543.75 = \$5456.25$.

(e) $60(207.58) - 10,000 = \$2,454.80$

5. The downpayment is $120,000(.25) = \$30,000$. $A = \$90,000$, $i = .01$, $n = 240$

(a) $R = \dfrac{90,000(.01)}{1 - (1.01)^{-240}} = \$990.98$

(b) $I_{125} = R[1 - (1.01)^{124-240}] = \$678.52$

(c) $B_{125} = R(1.01)^{124-240} = \$312.45$

(d) $A_{156} = R\,\dfrac{1 - (1.01)^{156-240}}{.01} = \$56,137.48$

     The amount paid is $\$90,000 - 56,137.48 = \$33,862.52$

(e) $240R - 90,000 = \$147,835.20$

7. The size of the loan is $210,000(.75) = \$157,500$ and $i = .09/12 = .0075$

(a) $n = 40(12) = 480$. Using formula 1, $157,500 = R \cdot \dfrac{1 - (1 + .0075)^{-480}}{.0075}$.

    $R = 157,500(.0077136) = \$1214.89$.

Similarly,

(b) $n = 30(12) = 360$ and $R = \$1267.28$
(c) $n = 25(12) = 300$ and $R = \$1321.73$
(d) $n = 20(12) = 240$ and $R = \$1417.07$
(e) $n = 15(12) = 180$ and $R = \$1597.47$
(f) $n = 10(12) = 120$ and $R = \$1995.14$

     As the number of payments decreases, the size of the monthly payments increases.

9. The monthly payment is $.25(5,212) = \$1,303$.
$i = .115/12$, $n = 240$. The downpayment is $.15A$.

$$A = 1,303\,\frac{1 - (1 + .115/12)^{-240}}{.115/12} = \$122,183.40. \qquad \text{(value of house)}(.85) = A$$

The value of the house is $\$143,745.18$.

11. The monthly payment is $\dfrac{45{,}228(.25)}{12} = \dfrac{\$11{,}307}{12} = \$942.25.$

$$A = 942.25 \,\frac{1 - (1 + .1/12)^{-420}}{.1/12} = \$109{,}605.70.$$

Value of the house is $\dfrac{109{,}605.7}{.85} = \$128{,}947.89$

13. $A = \$8{,}000,\ n = 36,\ i = .01$

    (a) $R = 8000(.01)/(1 - (1.01)^{-36}) = \$265.71$

    (b) $I_{19} = R \cdot [1 - 1.01^{-18}] = \$43.57$

    (c) $B_{19} = R \cdot (1.01)^{-18} = \$222.14$

    (d) $A_{25} = R \cdot \dfrac{1 - 1.01^{-11}}{.01} = \$2{,}754.78$

15. $A = \$9{,}800,\ i = .0075,\ R = \$250.$

The number of payments is $n = \dfrac{\ln(250) - \ln(250 - 9800(.0075))}{\ln(1.0075)} \doteq \dfrac{5.5215 - 5.1733}{.0075} = 46.4.$

The loan will be repaid in 47 months.

17. $A = \$57{,}600,\ i = .01,\ R = \$1{,}200.$

The number of payments is $n = \dfrac{\ln(1200) - \ln(1200 - 57600(.01))}{\ln(1.01)} \doteq \dfrac{7.0901 - 6.4362}{.00995} = 65.72$

The loan will be repaid in 66 months.

19. $A - (B_1 + B_2 + \cdots + B_k)$

$$= R \cdot \frac{1 - (1 + i)^{-n}}{i} - [R(1 + i)^{-n} + R(1 + i)^{1-n} + R(1 + i)^{2-n} + \cdots + R(1 + i)^{k-1-n}].$$

The sum within the bracket is the sum of terms of a G.P. with first term $R(1 + i)^{-n}$ and common ratio $1 + i.$ Thus

$$A - (B_1 + B_2 + \cdots + B_k) = R \cdot \frac{1 - (1 + i)^{-n}}{i} - \frac{R(1 + i)^{-n}(1 - (1 + i)^k}{1 - (1 + i)}$$

$$= R \cdot \frac{1 - (1 + i)^{-n}}{i} - \frac{R(1 + i)^{-n}}{-i} + \frac{R(1 + i)^{k-n}}{-i} = R \cdot \frac{1 - (1 + i)^{-n} + (1 + i)^{-n} - (1 + i)^{k-n}}{i}$$

$$= R \cdot \frac{1 - (1 + i)^{k-n}}{i} = A_k$$

21. By formula (4), $I_1 + I_2 + \cdots + I_n =$

$$R[1 - (1 + i)^{-n}] + R[1 - (1 + i)^{1-n}] + R[1 - (1 + i)^{2-n}] + \cdots + R[1 - (1 + i)^{-1}] =$$

$$Rn - R[(1 + i)^{-n} + (i + 1)^{1-n} + \cdots + (1 + i)^{-1}]$$

The sum in the brackets is the sum of n terms of a geometric progression with first term $(1 + i)^{-n}$ and common ratio $1 + i.$

$$I_1 + I_2 + \cdots + I_n = Rn - R\frac{(1 + i)^{-n}(1 - (1 + i)^n)}{1 - (1 + i)} = Rn - \frac{R[(1 + i)^{-n} - 1]}{-i} =$$

$$Rn - \frac{R(1 - (1 + i)^{-n})}{i} = nR - A, \text{ by formula (1).}$$

1. (a) $s_1 = \dfrac{4}{3+1} = 1$ $\qquad\qquad$ $s_2 = \dfrac{4}{6+1} = 4/7$ $\qquad\qquad$ $s_3 = \dfrac{4}{9+1} = 4/10$

$\qquad$ $s_4 = \dfrac{4}{12+1} = 4/13$ $\qquad\qquad$ $s_5 = \dfrac{4}{15+1} = 4/16$

$\quad$ (b) $s_1 = \dfrac{-2}{1+3} = -2/4$ $\qquad\qquad$ $s_2 = \dfrac{4}{2+3} = 4/5$ $\qquad\qquad$ $s_3 = \dfrac{-6}{3+3} = -6/6$

$\qquad$ $s_4 = \dfrac{8}{4+3} = 8/7$ $\qquad\qquad$ $s_5 = \dfrac{-10}{5+3} = -10/8$

$\quad$ (c) $s_1 = \dfrac{1+2+3}{2+1-2} = 6$ $\qquad\quad$ $s_2 = \dfrac{4+4+3}{8+2-2} = 11/8$ $\qquad\quad$ $s_3 = \dfrac{9+6+3}{18+3-2} = 18/19$

$\qquad$ $s_4 = \dfrac{16+8+3}{32+4-2} = 27/34$ $\qquad$ $s_5 = \dfrac{25+10+3}{50+5-2} = 38/53$

3. $a = -3$ and $d = 3$

$\quad$ (a) The first four terms are -3, $-3+3 = 0$, $0+3 = 3$ and $3+3 = 6$.

$\quad$ (b) $a_{200} = -3 + 199(3) = 594$ $\qquad$ $S_{200} = \dfrac{200(-3+594)}{2} = 59{,}100$

5. $s_1 = 4$ and $r = -2$

$\quad$ (a) $s_2 = 4(-2) = -8$ $\qquad\qquad$ $s_3 = -8(-2) = 16$ $\qquad\qquad$ $s_4 = 16(-2) = -32$

$\quad$ (b) $S_{20} = \dfrac{4(1-2^{20})}{1+2} = -1{,}398{,}100$

7. $325(1.005)^{53} + 325(1.005)^{52} + \cdots + 325 = \dfrac{325(1-1.005^{54})}{1-1.005} = \$20{,}090.42$

9. $20000(.83)^{10} = \$3{,}103.21$

11. $300 + 300(1 + \tfrac{.065}{12}) + 300(1 + \tfrac{.065}{12})^2 + \cdots + 300(1 + \tfrac{.065}{12})^{29} = \dfrac{300\left(1 - (1 + \tfrac{.065}{12})^{30}\right)}{1 - (1 + \tfrac{.065}{12})}$

$\qquad = \$9{,}743.96$

13. $A_1 + A_2 + A_3 + A_4 = S$ where

$\qquad A_1(1.09)^{18} = 12000$
$\qquad A_2(1.09)^{19} = 12000$
$\qquad A_3(1.09)^{20} = 12000$
$\qquad A_4(1.09)^{21} = 12{,}000$

$S = \dfrac{12000}{(1.09)^{18}} + \dfrac{12000}{(1.09)^{19}} + \dfrac{12000}{(1.09)^{20}} + \dfrac{12000}{(1.09)^{21}} = \dfrac{12000}{(1.09)^{18}} \dfrac{\left(1 - (\tfrac{1}{1.09})^4\right)}{1 - \tfrac{1}{1.09}}$

$\quad = \dfrac{12000(1.09)^{-17}(1 - 1.09^{-4})}{.09} = \$8{,}983.35$

15. $350 + 350(1.02) + 350(1.02)^2 + \cdots + 350(1.02)^{19} = \dfrac{350(1 - 1.02^{20})}{1 - 1.02} = \$8{,}504.08$

17. $15000 + \dfrac{15000}{1.09} + \dfrac{15000}{1.09^2} + \cdots + \dfrac{15000}{1.09^{19}} = \dfrac{15000\left(1 - (\tfrac{1}{1.09})^{20}\right)}{1 - \tfrac{1}{1.09}} = \$149{,}251.72$

19. In 40 years, John will have accumulated $300 + 300(1 + .08/12) + \cdots + 300(1 + .08/12)^{479}$

$$= \frac{300\left(1 - (1+.08/12)^{480}\right)}{1 - (1+.08/12)} = \$1,047,302.35. \text{ If the size of a withdrawal is } D,$$

$$1,047,302.35 = A_1 + A_2 + \cdots + A_{120} = \frac{D}{1.00\overline{6}} + \frac{D}{(1.00\overline{6})^2} + \cdots + \frac{D}{(1.00\overline{6})^{120}} =$$

$$\frac{D(1 - 1.00\overline{6}^{-120})}{.00\overline{6}} = D(82.42148)$$

Hence $D = \$12,706.66$

21. In five years, Judy will have $250 + 250(1.0075) + \cdots + 250(1.0075)^{59} = \dfrac{250(1 - 1.0075^{60})}{1 - 1.0075}$

$= \$18,856.03.$ In 15 years, this will grow to $18,856.03(1.0075)^{180} = \$72,370.26.$ The remainder needed is $200,000 - 72,370.26 = \$127,629.74.$ Thus if R is the size of the remaining monthly payments

$$R + R(1.0075) + \cdots + R(1.0075)^{179} = \frac{R(1 - 1.0075^{180})}{1 - 1.0075} = R(378.41) = 127,629.74.$$

$R = \$337.28$

23. $R = .25(5,540) = \$1385,\ i = .01,\ n = 240.$ $A = \dfrac{1385\left(1 - (1.01)^{-240}\right)}{.01} = \$125,784.89.$

The value of the house is $\dfrac{A}{.8} = \$157,231.11.$

25. $A = \$12,500,\ i = .01,\ R = \$250.$ The number of monthly payments is

$$n = \frac{\ln(250) - \ln\left(250 - 12500(.01)\right)}{\ln(1.01)} = 69.66. \text{ The loan will be repaid in 70 months.}$$

CHAPTER 6

PROBABILITY

EXERCISE SET 6.1 INTRODUCTION

1. (a)

Number of tubes tested	Number of defective tubes	relative frequency
n	s	s/n
100	3	.03
200	5	.025
300	8	$.02\overline{6}$
400	11	.0275
500	15	.03
600	17	$.028\overline{3}$
700	22	.031428571
800	25	.03125
900	29	$.03\overline{2}$
1000	32	.032
1100	35	$.031\overline{8}$
1200	38	$.031\overline{6}$
1300	40	.03076923
1400	44	.03142857
1500	47	$.031\overline{3}$

(b) The probability is approximately $.031\overline{3}$ since this is the relative frequency for 1500 tests, the largest number of tests.

3. (a)

Number of dryers tested n	Number of defective dryers s	Relative frequency s/n
200	4	.02
400	9	.0225
600	11	.018$\overline{3}$
800	17	.02125
1000	21	.021
1200	26	.021$\overline{6}$
1400	30	.021428571
1600	34	.02125
1800	35	.019$\overline{4}$
2000	39	.0195

(b) The approximate probability, based on 2000 tests is .0195.

5. Let $E_6 = \{(1,5), (2,4), (3,6), (4,2), (5,1)\}$ $P(E_6) = 5/36$ since there are 5 members of E_6 and 36 equally likely members of the sample space.

7. $E_8 = \{(2,6), (3,5), (4,4), (5,3), (6,2)\}$ $P(E_8) = 5/36$

9. $E_5 = \{(1,1,3), (1,3,1), (3,1,1), (1,2,2), (2,1,2), (2,2,1)\}$ $P(E_5) = \frac{6}{216} = \frac{1}{36}$ since the event E_5 has 6 members and there are $6 \cdot 6 \cdot 6 = 216$ equiprobable outcomes in the sample space which consists of ordered triples (x,y,z) where x, y and z are members of the set $\{1,2,3,4,5,6\}$.

11. (a) $E_8 = \{(2, 6), (3, 5), (4, 4), (5, 3), (6, 2)\}$, $P(E_8) = 5/36$

(b)   Let $E_9 = \{(3,6), (4,5), (5,4), (6,3)\}$. $P(E_9) = 4/36 = 1/9$

(c) $E_8 \cup E_9 = \{(2,6), (3,5), (4,4), (5,3), (6,2), (3,6), (4,5), (5,4), (6,3)\}$.
$P(E_8 \cup E_9) = 9/36 = 1/4$.

(d) $P(E_8 \cap E_9) = P(\emptyset) = 0$.

(e) $P(E_8') = 1 - P(E_8) = 1 - 5/36 = 31/36$.

13. (a) $S = \{(T,T,T), (T,T,H), (T,H,T), (H,T,T), (H,H,T), (H,T,H), (T,H,H), (H,H,H)\}$

(b) $E_1 = \{(T,T,H), (T,H,T), (H,T,T)\}$. $P(E_1) = 3/8$
$E_2 = \{(T,T,H), (T,H,T), (H,T,T), (T,H,H), (H,T,H), (H,H,T), (H,H,H)\}$.
$P(E_2) = 7/8$
$E_3 = \{(H,H,H)\}$. $P(E_3) = 1/8$
$E_4 = \{(H,H,T), (H,T,H), (T,H,H), (H,H,H)\}$. $P(E_4) = 4/8 = 1/2$

15. The probability the Celtics win the championship is $\frac{4}{4+20} = 1/6$. The probability the Celtics will not win the championship is $1 - \frac{1}{6} = \frac{5}{6}$.

17. Let E be the event the Seahawks win the next Super Bowl. $P(E) = .08$,
$P(E') = 1 - .08 = .92$
$P(E)/P(E') = \frac{.08}{.92} = \frac{8}{92} = \frac{2}{23}$. The odds the Seahawks win the next Super Bowl are 2 to 23.

19. If E is the event that John wins the championship, $P(E) = .35$ and $P(E') = 1-.35 = .65$. $P(E)/P(E') = .35/.65 = 35/65 = 7/13$. The odds John wins the championship are 7 to 13.

21. If E is the event that player E wins the tournament, $.23+.13+.10+.35+P(E) = 1$ and $P(E) = .19$

EXERCISE SET 6.2 COUNTING TECHNIQUES

1. $7! = 7 \cdot 6 \cdot 5 \cdot 4 \cdot 3 \cdot 2 \cdot 1 = 5,040$

3. $\dfrac{15!}{9!3!3!} = \dfrac{15 \cdot 14 \cdot 13 \cdot 12 \cdot 11 \cdot 10}{2 \cdot 3 \cdot 6} = 100,100$

5. $P(6,6) = 6! = 720$

7. $C(5,3) = \dfrac{5!}{(5-3)!3!} = \dfrac{5!}{2!3!} = \dfrac{5 \cdot 4}{2} = 10$

9. $C(13,7) = \dfrac{13!}{(13-7)!7!} = \dfrac{13!}{6!7!} = \dfrac{13 \cdot 12 \cdot 11 \cdot 10 \cdot 9 \cdot 8}{6 \cdot 5 \cdot 4 \cdot 3 \cdot 2} = 1,716$

11. $C(10,3) = C(10,10-3) = C(10,7) = \dfrac{10!}{7!3!} = \dfrac{10 \cdot 9 \cdot 8}{6} = 120$

13. $P(10; 6,4) = \dfrac{10!}{6!4!} = \dfrac{10 \cdot 9 \cdot 8 \cdot 7}{24} = 210$

15.

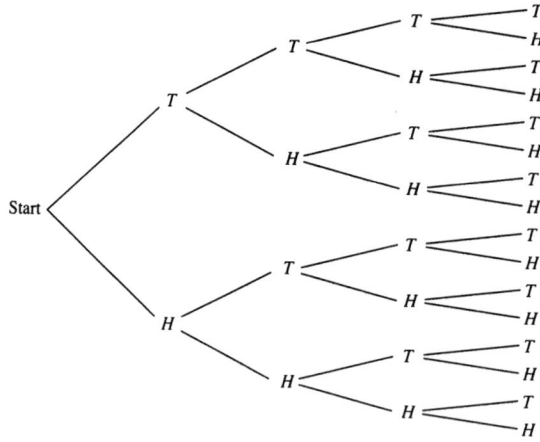

The possible outcomes are

(T, T, T, T)	(H, T, T, T)
(T, T, T, H)	(H, T, T, H)
(T, T, H, T)	(H, T, H, T)
(T, T, H, H)	(H, T, H, H)
(T, H, T, T)	(H, H, T, T)
(T, H, T, H)	(H, H, T, H)
(T, H, H, T)	(H, H, H, T)
(T, H, H, H)	(H, H, H, H)

17.

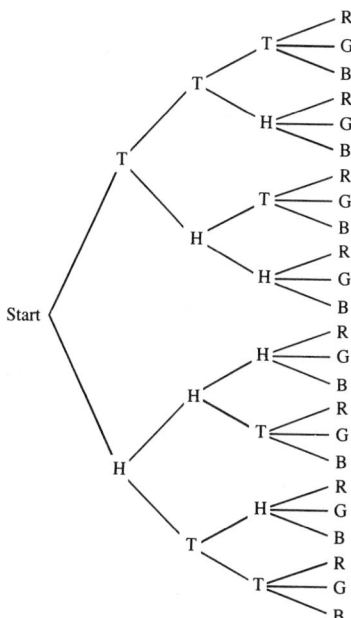

Possible outcomes are
(T, T, T, R)	(T, H, T, R)	(H, H, H, R)	(H, T, H, R)
(T, T, T, G)	(T, H, T, G)	(H, H, H, G)	(H, T, H, G)
(T, T, T, B)	(T, H, T, B)	(H, H, H, B)	(H, T, H, B)
(T, T, H, R)	(T, H, H, R)	(H, H, T, R)	(H, T, T, R)
(T, T, H, G)	(T, H, H, G)	(H, H, T, G)	(H, T, T, G)
(T, T, H, B)	(T, H, H, B)	(H, H, T, B)	(H, T, T, B)

19. $C(n,r) = \dfrac{n!}{(n-r)!r!}$

$$C(n,n-r) = \frac{n!}{(n-(n-r))!(n-r)!} = \frac{n!}{(n-r)!r!} = C(n,r)$$

21. $4! = 24$

$abcd$	$bacd$	$cabd$	$dabc$
$abdc$	$badc$	$cadb$	$dacb$
$acbd$	$bcad$	$cbad$	$dbac$
$acdb$	$bcda$	$cbda$	$dbca$

$$\begin{array}{cccc} adcb & bdac & cdab & dcab \\ adbc & bdca & cdba & dcba \end{array}$$

23. There are 11 letters.
    Two are m
    Two are a
    Two are t
    one is h
    one is e
    one is i
    one is c
    and one is s.

    The number of permutations is $\frac{11!}{2!2!2!} = 11(10)(9)(7)(6)(5)(4)(3)(2) = 4,989,600$

25. $\frac{7!}{3!4!} = \frac{7 \cdot 6 \cdot 5}{3 \cdot 2} = 35$

27. $13 \cdot 12 \cdot 11 = 1,716$ since there are 13 choices for the president. Once the president is selected there are 12 choices for the vice-president. After the president and vice-president have been selected, there are 11 choices for the secretary.

29. $C(10,2) = \frac{10!}{8!2!} = \frac{10 \cdot 9}{2} = 45.$

31. $C(12,3) = \frac{12!}{9!3!} = \frac{12(11)(10)}{6} = 220$

33. $12! = 479,001,600$. He thinks about 30 assignments per minute or 1800 per hour. His total time is $\frac{12!}{1800} = 266,112$ hours or $11,088$ days or approximately 30.4 years.

35. $C(8,6) = \frac{8!}{2!6!} = \frac{8 \cdot 7}{2} = 28$

37. (a) There are $C(18,2) = \frac{18 \cdot 17}{2} = 153$ ways the women may be selected and

    $C(20,3) = 20 \cdot 19 \cdot 18/6 = 1140$ ways the men may be selected. Hence the number of possible committees is $153(1140) = 174,420$.

    (b) If men are the majority, there must be exactly 3,4 or 5 men. Thus the number of committees is $174,420 + C(20,4)C(18,1) + C(20,5)C(18,0) = 174,420 + 87,210 + 15,504 = 277,134.$

39. $C(2,1) \cdot C(7,2) \cdot C(6,2) = 2(21)(15) = 630$

41. $C(52,5)$ is the number of ways 5 cards may be selected for player 1. $C(47,5)$ is then the number of ways 5 cards may be selected for player 2. $C(42,5)$, $C(37,5)$, $C(32,5)$ and $C(27,5)$ are the number of ways 5 cards hands may be selected for players 3, 4, 5 and 6. The number of ways the six five-card hands may be dealt is
    $$C(52,5) \cdot C(47,5) \cdot C(42,5) \cdot C(37,5) \cdot C(32,5) \cdot C(27,5) = \frac{52!}{5!5!5!5!5!5!22!}$$

EXERCISE SET 6.3 CONDITIONAL PROBABILITY

1. Let A be the event a red marble is drawn the first time and B be the event a red marble is drawn the second time. $P(A \cap B) = P(B|A)P(A) = \frac{4}{8} \cdot \frac{5}{12} = \frac{5}{24}$.

3. If E is the event the student takes an English course and M is the event the student takes a mathematics course,

   (a) $P(E|M) = \dfrac{P(E \cap M)}{P(M)} = \dfrac{.18}{.27} = 2/3$

   (b) $P(M|E) = \dfrac{P(E \cap M)}{P(E)} = \dfrac{.18}{.73} = 18/73$

   (c) $P(E \cup M) = P(E) + P(M) - P(E \cap M) = .73 + .27 - .18 = .82$

   $$P(E \cap M | E \cup M) = \frac{P(E \cap M)}{P(E \cup M)} = \frac{.18}{.82} = \frac{9}{41}$$

5. Since the events are independent each with probability $\frac{5}{7}$, the probability is $\frac{5}{7} \cdot \frac{5}{7} = \frac{25}{49}$.

7. $\dfrac{8}{13} \cdot \dfrac{8}{13} = \dfrac{64}{169}$.

9. $P(A \cup J) = P(A) + P(J) + P(A \cap J) = \frac{4}{52} + \frac{4}{52} - 0 = \frac{2}{13}$.

11. $P(H \cup Q) = P(H) + P(Q) - P(H \cap Q) = \frac{13}{52} + \frac{4}{52} - \frac{1}{52} = \frac{4}{13}$.

13. Let A be the event the sum is even and B be the event the sum is less than 7.
   $$P(A \cup B) = P(A) + P(B) - P(A \cap B) = \frac{18}{36} + \frac{15}{36} - \frac{9}{36} = \frac{2}{3}.$$

15. P(none celebrate birthdays on the same day) $=$
   $$\frac{365}{365} \cdot \frac{364}{365} \cdot \frac{363}{365} \cdot \frac{362}{365} \cdot \frac{361}{365} \cdot \frac{360}{365} \cdot \frac{359}{365} \cdot \frac{358}{365} \cdot \frac{357}{365} \cdot \frac{356}{365} = .883051822$$

   P(at least 2 have birthdays on the same day) $= 1 - .883051822 = .116948178$

17. Using the result for Exercise 15, P(none celebrate birthdays on the same day) $=$
   $$.883051822 \cdot \frac{355}{365} \cdot \frac{354}{365} \cdot \frac{353}{365} \cdot \frac{352}{365} \cdot \frac{351}{365} \cdot \frac{350}{365} \cdot \frac{349}{365} = .684992334.$$

   The probability at least 2 have birthdays on the same day is $1 - .684992334 = .31500766$.

19. Using the result of Example 7, P(none celebrate birthdays on the same day) $=$
   $$.4927 \cdot \frac{342 \cdot 341 \cdot 340 \cdot 339 \cdot 338 \cdot 337 \cdot 336}{365 \cdot 365 \cdot 365 \cdot 365 \cdot 365 \cdot 365 \cdot 365} = .293682108$$

   The desired probability is $1 - .293682108 = .70631789$

21. $P(A) = 590/2000 \qquad P(A \cap C) = 380/2000 \qquad P(C) = 730/2000$

(a) $P(C|A) = \dfrac{P(A \cap C)}{P(A)} = \dfrac{380}{590} = \dfrac{38}{59}.$

(b) $P(A'|C) = \dfrac{P(A' \cap C)}{P(C)} = \dfrac{350}{730} = \dfrac{35}{73}.$

(c) $P(C'|A) = \dfrac{P(C' \cap A)}{P(A)} = \dfrac{210}{590} = \dfrac{21}{59}.$

23. P(red on first draw and white on second draw) +

P(white on first draw and red on second draw) =

$\dfrac{3 \cdot 5}{15 \cdot 14} + \dfrac{5 \cdot 3}{15 \cdot 14} = \dfrac{1}{7}.$

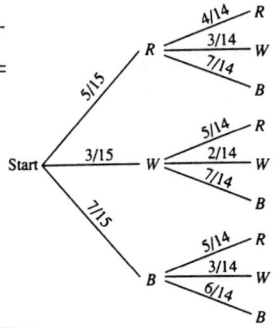

25. If the Supersonics are to win, the outcomes of games 3 through 7 must be one of the following sequences (compare figure) with corresponding probabilities.

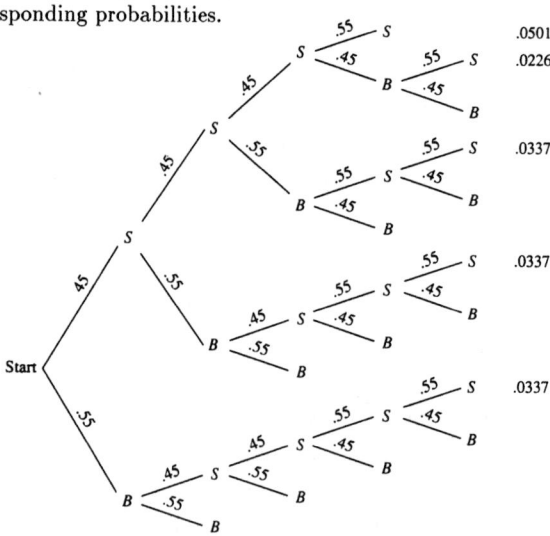

Sequence	Probability
$SSSS$	$(.45)^3(.55) = .0501$
$SSBSS$	$(.45)^2(.55)(.55)^2 = .0337$
$SBSSS$	$(.45)(.55)(.45)(.55)^2 = .0337$
$BSSSS$	$(.55)(.45)^2(.55)^2 = .0337$
$SSSBS$	$(.45)^3(.45)(.55) = .0226$

The probability the Sonics win the championship is the sum of these four probabilities which is .1738

27. The probabilities that the Tigers win games 3, 4, 5, 6, and 7 are .35, .45, .4, .65 and .7 respectively.

Sequence where the Tigers win the Series	Probability
T T T T	$(.35)(.45)(.4)(.65) = .04095$
T E T T T	$(.35)(.55)(.4)(.65)(.7) = .035035$
T T E T T	$(.35)(.45)(.6)(.65)(.7) = .0429975$
T T T E T	$(.35)(.45)(.4)(.35)(.7) = .015435$
E T T T T	$(.65)(.45)(.4)(.65)(.7) = .053235$

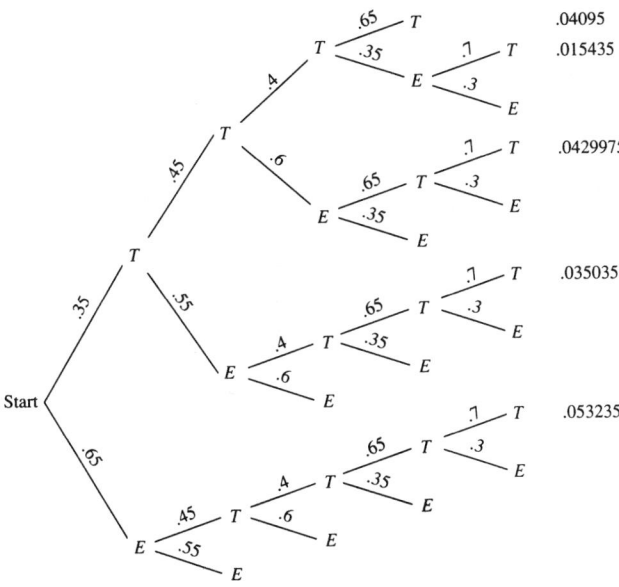

The probability the Tigers win the Series is the sum of these five probabilities which is .1876525

29. The sequences of games 3 through 7 if the Expos win the Series are

Sequence	Probability
E E E E	$(.65)(.55)(.6)(.35) = .075075$
E T E E E	$(.65)(.45)(.6)(.35)(.3) = .0184275$
E E T E E	$(.65)(.55)(.4)(.35)(.3) = .015015$
E E E T E	$(.65)(.55)(.6)(.65)(.3) = .0418275$
T E E E E	$(.35)(.55)(.6)(.35)(.3) = .0121275$

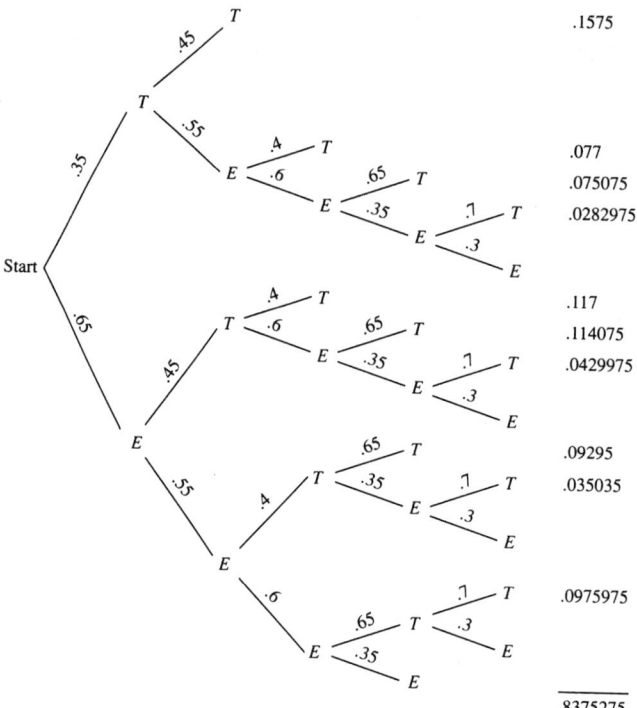

| | | | | | | | | | | | | | .1575 |

The probability the Expos win the Series is .1624725 and the probability the Tigers win is $1 - .1624725 = .8375275$.

31.

Sequences if the Rangers win the Cup	Probability
R R R	$(.38)(.38)(.38) = .054872$
R R N R	$(.38)(.38)(.62)(.62) = .05550736$
R R N N R	$(.38)(.38)(.62)(.38)(.62) = .021092796$
R N R R	$(.38)(.62)(.38)(.62) = .05550736$
R N R N R	$(.38)(.62)(.38)(.38)(.62) = .021092796$
R N N R R	$(.38)(.62)(.62)(.62)(.62) = .056150076$
N R R R	$(.62)(.38)(.38)(.62) = .05550736$
N N R R R	$(.62)(.62)(.38)(.62)(.62) = .056150076$
N R N R R	$(.62)(.38)(.62)(.62)(.62) = .056150076$
N R R N R	$(.62)(.38)(.38)(.38)(.62) = .021092796$

The probability the Rangers win the Cup is the sum of these ten probabilities which is .453122696.

33. $P(\text{Betty wins}) = P(\text{H}) + P(\text{H H H T}) + P(\text{H H H H H H T}) + \cdots =$

$$\frac{1}{2} + (\frac{1}{2})^4 + (\frac{1}{2})^7 + \cdots = \frac{\frac{1}{2}}{1 - \frac{1}{8}} = \frac{4}{7}.$$

35. $P(\text{Jeremy wins}) = \frac{1}{6} + (\frac{5}{6})^2 \cdot \frac{1}{6} + (\frac{5}{6})^4 \cdot \frac{1}{6} + \cdots = \frac{\frac{1}{6}}{1 - 25/36} = \frac{6}{11}.$

37. $P(\text{Julie wins}) = \frac{1}{3} + (\frac{2}{3})^3 \cdot \frac{1}{3} + (\frac{2}{3})^6 \cdot \frac{1}{3} + \cdots = \frac{\frac{1}{3}}{1 - 8/27} = \frac{9}{19}.$

39. $P(\text{Laura wins}) = (\frac{2}{3})^2 \cdot \frac{1}{3} + (\frac{2}{3})^5 \cdot \frac{1}{3} + (\frac{2}{3})^8 \cdot \frac{1}{3} + \cdots = \frac{4/27}{1 - 8/27} = \frac{4}{19}.$

EXERCISE SET 6.4 BAYES' FORMULA

1. Let R, G and W be the events that the right club is selected, the shot is good, and the wrong club is selected, respectfully.

$$P(R|G) = \frac{P(G|R)P(R)}{P(G|R)P(R) + P(G|W)P(W)}.$$

$$\begin{aligned} P(G|R) &= .23 \\ P(G|W) &= .1 \\ P(R) &= 1/6 \\ P(W) &= 5/6 \end{aligned}$$

$$= \frac{.23(1/6)}{.23(1/6) + .1(5/6)} = \frac{.23}{.23 + .5} = \frac{23}{73} \sim .3151$$

3. $P(B|R) = .77, \; P(B|W) = .9$ \qquad where B is the event the shot is bad.

$$P(R|B) = \frac{P(B|R)P(R)}{P(B|R)P(R) + P(B|W)P(W)} = \frac{.77(1/6)}{.77(1/6) + .9(5/6)} = \frac{.77}{.77 + .9} = \frac{77}{167} \sim .4611$$

5. Let U_1 and U_2 be the events that Urn 1 and Urn 2 are selected, respectively, and R be the event the marble selected is red.

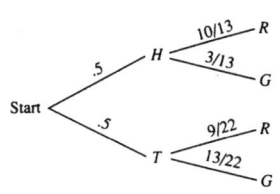

(a) $P(U_1|R) = \dfrac{P(R|U_1)P(U_1)}{P(R|U_1)P(U_1) + P(R|U_2)P(U_2)} =$

$$\frac{.5(10/13)}{.5(10/13) + .5(9/22)} =$$

$$\frac{110}{110 + 58.5} \sim .6528$$

(b) $P(U_2|R) = \dfrac{.5(9/22)}{.5(9/22) + .5(10/13)} \sim .3472$

7. Let B, U_1 and U_2 be the events a blue marble is selected, Urn 1 is selected, and Urn 2 is selected, respectively.

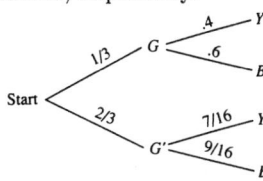

(a) $P(U_1|B) = \dfrac{(1/3)(.6)}{.6(1/3) + (9/16)(2/3)} =$

$$\frac{.6}{.6 + (9/8)} \sim .3478$$

(b) $P(U_2|B) \sim 1 - .3478 = .6522$

9.

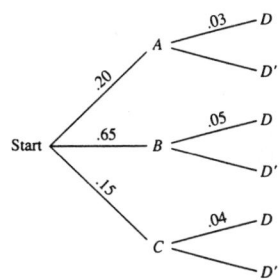

$$P(B|D) = \frac{.65(.05)}{.2(.03) + .65(.05) + (.15)(.04)} = .7303$$

where B is the event machine B produced the gadget selected, and D is the event the gadget is defective.

11. Let A, B, C and D be the events, plants A, B, C and D produced the VCR selected, respectively, and DE is the event it is defective.

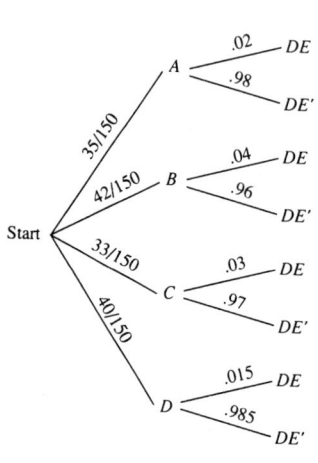

(a) $P(A|DE) = \dfrac{P(DE|A)P(A)}{P(DE)}$

$= \dfrac{.02(35/150)}{.02(\frac{35}{150}) + .04(\frac{42}{150}) + .03(\frac{33}{150}) + .015(\frac{40}{150})}$

$= \dfrac{.02(35)}{(.02)(35) + (.04)42 + .03(33) + .015(40)}$

$= \dfrac{.7}{3.97} \sim .1763$

(b) $P(B|DE) = \dfrac{.04(42)}{3.97} \sim .4232$

(c) $P(C|DE) = \dfrac{.03(33)}{3.97} \sim .2494$

(d) $P(D|DE) = \dfrac{.015(40)}{3.97} \sim .1511$

13. Using the tree diagram of Exercise 11,

(a) $P(A|DE') = \dfrac{P(DE'|A)P(A)}{P(DE')} =$

$\dfrac{.98(35/150)}{.98(35/150) + .96(42/150) + .97(33/150) + .985(40/150)} =$

$\dfrac{.98(35)}{.98(35) + .96(42) + .97(33) + .985(40)} = \dfrac{34.3}{146.03} \sim .2349$

(b) $P(D|DE') = \dfrac{.96(42)}{146.03} \sim .2761$

(c) $P(D|DE') = \dfrac{.97(33)}{146.03} \sim .2192 \qquad P(D|DE') = \dfrac{.985(40)}{146.03} \sim .2698$

15. $P(C) = .6, \quad P(P|C) = .82, \quad P(P|C') = .14$

where C is the event an applicant completes the program and P is the event the applicant passes the test.

$P(C|P) = \dfrac{P(P|C)P(C)}{P(P|C)P(C) + P(P|C')P(C')} = \dfrac{(.82)(.6)}{(.82)(.6) + (.14)(.4)} \sim .8978$

17. $P(D) = .026 \quad P(P|D) = .90, \quad P(P|D') = .05$

where D is the event a person is diabetic and P is the event the person tests positive.

$P(D|P) = \dfrac{P(P \mid D)P(D)}{P(P \mid D)P(D) + P(P \mid D')P(D')} = \dfrac{.90(.026)}{.90(.026) + .05(.974)} = .3245$

19. (a)

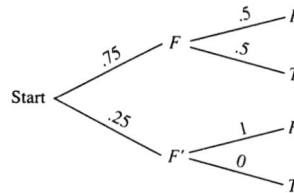

$$P(F|H_1) = \frac{(3/4)(1/2)}{(3/4)(1/2) + (1/4)(1)}$$

$= .6$, where F is the event the coin is fair and H_1 is the event a head occurred on the first toss.

(b)

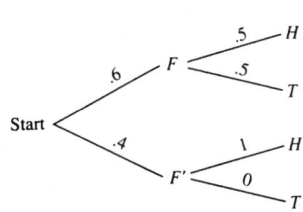

$$P(F|H_2) = \frac{(.6)(.5)}{(.6)(.5) + (.4)(1)}$$

$\sim .4286$
where H_2 is the event a head occurred on both tosses.

(c)　one

EXERCISE SET 6.5   EXPECTED VALUE

1. Let X be the dollar gain. The range of X is $\{8, 3, -1.5\}$.

$E(X) = 8(4/52) + 3(12/52) - 1.5(36/52) = .269$

In the long run one would expect to come out ahead, and in 200 plays expect to win approximately $.269(200) = \$53.80$

3. Let X be the dollar gain.

(a) The range of X is $\{5, -1\}$. $E(X) = (1/3)5 - (2/3)1 = 1$.

(b) The range of X is $\{10, -2, 0\}$ assuming one wins or loses nothing if both coins are black.

$$E(X) = 10 \cdot \frac{3 \cdot 2}{9 \cdot 8} - 2 \cdot \frac{C(6,1) \cdot C(3,1)}{C(9,2)} + 0(\frac{6 \cdot 5}{9 \cdot 8}) = .8 - \frac{2(6)(3)}{36} = .8 - 1 \doteq -.2$$

Play game 1.

5.

x	$f(x) = P(X = x)$
0	$P(\text{TTTTT}) = 1/32$
1	$P(\text{HTTTT, THTTT, TTHTT, TTTHT, TTTTH}) = 5/32$
2	$P(\text{HHTTT, HTHTT, HTTHT, HTTTH, THHTT, THTHT, THTTH,}$ $\text{TTHHT, TTHTH, TTTHH}) = 10/32$
3	$P(\text{HHHTT, HHTHT, HHTTH, HTHHT, HTHTH, HTTHH, THHHT,}$ $\text{THHTH, THTHH, TTHHH}) = 10/32$
4	$P(\text{HHHHT, HHHTH, HHTHH, HTHHH, THHHH}) = 5/32$
5	$P(\text{HHHHH}) = 1/32$

7. The range of X is $\{1, 2, 3, \ldots\}$. $P[X = K] = P[K \text{ non-sixes followed by a } 6]$

$$= (\tfrac{5}{6})^{K-1}(\tfrac{1}{6}), \ K \text{ in } \{1, 2, 3, \ldots\}.$$

9.

x	$f(x) = P(X = x)$
0	1/16
1	4/16
2	6/16
3	4/16
4	1/16

$E(X) = 0(1/16)+1(4/16)+2(6/16)+3(4/16)+4(1/16) = 2$

11.

x	$f(x) = P(X = x)$
2	1/36
3	2/36
4	3/36
5	4/36
6	5/36
7	6/36
8	5/36
9	4/36
10	3/36
11	2/36
12	1/36

$E(X) = (2+6+12+20+30+42+40+36+30+22+12)/36 = 252/36 = 7$

13. If G is the gain, the range of G is $\{360, -324{,}640\}$. $E(G) = 360(.99975)-324{,}640(.00025)$
$= \$278.75$

15. If G is the gain, the range of G is $\{50, -100, -250, -400, -550, -700, -850, -1000\}$.

$E(G) = 50(.953)-100(.0015)-250(.0025)-400(.0035)-550(.005)-700(.0065)-850(.008)$
$-1000(.02) = \$11.375$

17. The range of X is $\{-198.50, -98.50, -48.50, 1.50\}$ where X is the amount the accounting club gains on 1 ticket. $E(X) = -198.50(.002)-98.5(.002)-48.50(.002)+1.50(.994) = 80\cent$.

19.

y	y^2	$g(y)$
1	1	.55
2	4	.1
3	9	.05
4	16	.3

$\text{Var}(Y) = \sum_{i=1}^{n} y_i^2 f(y_i)-(2.1)^2$
$= 1(.55)+4(.1)+9(.05)+16(.3)-4.41$
$= 1.79$

21. (a) $E(X) = 1(.1)+3(.1)+5(.3)+7(.2)+9(.3) = 6$
$E(Y) = 1(.15)+3(.25)+5(.05)+7(.05)+9(.5) = 6$

(b) $\text{Var}(X) = 1(.1)+9(.1)+25(.3)+49(.2)+81(.3)-36 = 6.6$ and $\sigma = 2.569$

(c) $\text{Var}(Y) = 1(.15)+9(.25)+25(.05)+49(.05)+81(.5)-36 = 10.6$ and $\sigma = 3.256$

(d) $\qquad f(x) = P_1(X = x) \qquad\qquad\qquad g(y) = P_2(Y = y)$

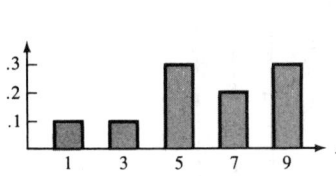 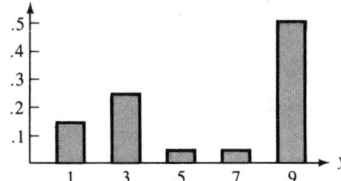

23. By Exercise 11, $E(X) = 7$.

$$\text{Var}(X) = (1/36)(4(1)+9(2)+16(3)+25(4)+36(5)+49(6)+64(5)+81(4)+100(3)$$
$$+121(2)+144(1)-49) = 5.8\overline{3} \text{ and } \sigma = \sqrt{5.8\overline{3}} \sim 2.415$$

25. $\text{Var}(X) = (x_1-E(X))^2 f(x_1)+(x_2-E(X))^2 f(x_2)+(x_3-E(X))^3 f(x_3) =$
$\qquad = (x_1^2-2x_1E(X)+(E(X))^2)f(x_1)+(x_2^2-2x_2E(X)+(E(X))^2)f(x_2)$
$\qquad\quad +(x_3^2-2x_3E(X)+(E(X))^2)f(x_3)$
$\qquad = (x_1^2 f(x_1)+x_2^2 f(x_2)+x_3^2 f(x_3))$
$\qquad\quad -2E(X)(x_1 f(x_1)+x_2 f(x_2)+x_3 f(x_3))+(E(X))^2(f(x_1)+f(x_2)+f(x_3))$
$\qquad = x_1^2 f(x_1) + x_2^2 f(x_2) + x_3^2 f(x_3) - 2E(X)E(X) + (E(X))^2(1)$
$\qquad = x_1^2 f(x_1) + x_2^2 f(x_2) + x_3^2 f(x_3) - (E(X))^2$

EXERCISE SET 6.6 THE BINOMIAL DISTRIBUTION

1. $(a+b)^6 = C(6, 0)a^6+C(6, 1)a^5b+C(6, 2)a^4b^2+C(6, 3)a^3b^3+C(6, 4)a^2b^4$
$\qquad +C(6, 5)ab^5+C(6, 6)b^6 = a^6+6a^5b+15a^4b^2+20a^3b^3+15a^2b^4+6ab^5+b^6$

3. $P(\text{H})=5/7$, $P(\text{T})=2/7 \qquad P(\text{THTTH})=(\frac{2}{7})(\frac{5}{7})(\frac{2}{7})(\frac{2}{7})(\frac{5}{7})=\frac{200}{16807}$

5. $P(\text{S})=1/4 \qquad P(\text{FSFSF})=\frac{3}{4}\cdot\frac{1}{4}\cdot\frac{3}{4}\cdot\frac{1}{4}\cdot\frac{3}{4}=\frac{27}{1024}$

7. $P(X=3)=C(5, 3)(2/7)^3(5/7)^2=10(\frac{8}{343})(\frac{25}{49})=\frac{2000}{16807} \sim .119 \quad$ where X is the number of tails in the five tosses.

9. $P(X=3)=C(8, 3)(5/13)^3(8/13)^5 \sim .281 \quad$ where X is the number of heads in the eight tosses.

11. $P(X=4)=C(7, 4)(1/4)^4(3/4)^3 \sim .0577 \quad$ where X is the number of sixes in the seven tosses.

13. $P(X=2)=C(8, 4)(4/13)^4(9/13)^4 \sim .144 \quad$ where X is the number of twos in the eight tosses.

15. (a) $b(2; 8, .2) = C(8, 2)(.2)^2(.8)^6 \sim .2936$

(b) $b(3; 10, .3) = C(10, 3)(.3)^3(.7)^7 \sim .2668$

(c) $b(9; 13, .4) = C(13, 9)(.4)^9(.6)^4 \sim .0243$

17.

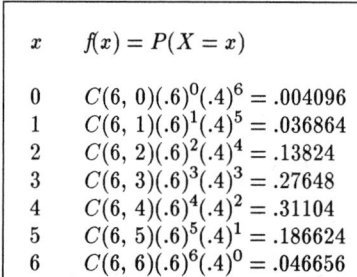

x	$f(x) = P[X = x]$
0	$C(4, 0)(.2)^0(.8)^4 = .4096$
1	$C(4, 1)(.2)^1(.8)^3 = .4096$
2	$C(4, 2)(.2)^2(.8)^2 = .1536$
3	$C(4, 3)(.2)^3(.8)^1 = .0256$
4	$C(4, 4)(.2)^4(.8)^0 = .0016$

19.

x	$f(x) = P(X = x)$
0	$C(6, 0)(.6)^0(.4)^6 = .004096$
1	$C(6, 1)(.6)^1(.4)^5 = .036864$
2	$C(6, 2)(.6)^2(.4)^4 = .13824$
3	$C(6, 3)(.6)^3(.4)^3 = .27648$
4	$C(6, 4)(.6)^4(.4)^2 = .31104$
5	$C(6, 5)(.6)^5(.4)^1 = .186624$
6	$C(6, 6)(.6)^6(.4)^0 = .046656$

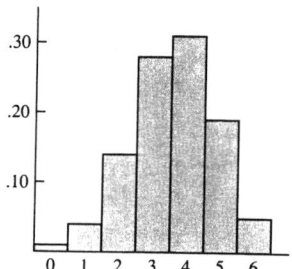

21. Let X be the number of times a one comes up.

$$P(X = 0) + P(X = 1) + P(X = 2) = (1/6)^0(5/6)^{20} + 20(1/6)(5/6)^{19} + 190(1/6)^2(5/6)^{18}$$
$$= .026084053 + .104336213 + .198238805$$
$$\doteq .329$$

$$P(X \geq 3) \doteq 1 - .329 = .671$$

23. (a) $C(10, 7)(.8)^7(.2)^3 = .2013$

(b) $.2013 + C(10, 8)(.8)^8(.2)^2 + C(10, 9)(.8)^9(.2) + (.8)^{10} = .2103 + .3020 + .2684 + .1074$
$$= .8791$$

25. If X is the number of boys,
$$P(X = 0) + P(X = 1) = (.5)^6 + 6(.5)^6 = .109375$$
$$P(X \geq 2) = 1 - .109375 = .890625$$

27. If X is the number of bids won
$$P(X = 0) + P(X = 1) + P(X = 2) + P(X = 3)$$
$$= (.35)^{10} + 10(.65)(.35)^9 + 45(.65)^2(.35)^8 + 120(.65)^3(.35)^7$$
$$= .00003 + .00051 + .00428 + .02120 = .02602$$

$$P(X \geq 4) = 1 - .02602 = .97398$$

29. If X is the number of correct responses

(a) $P(X = 20) = (.2)^{20}$

(b) $P(X = 18) = 190(.2)^{18}(.8)^2$

$P(X = 19) = 20(.2)^{19}(.8)$

$P(X \geq 18) = (.2)^{20} + 20(.2)^{19}(.8) + 190(.2)^{18}(.8)^2 = (.2)^{18}(124.84)$

(c) $P(X = 15) = 15504(.2)^{15}(.8)^5$

$P(X = 16) = 4845(.2)^{16}(.8)^4$

$P(X = 17) = 1140(.2)^{17}(.8)^3$

$P(X \geq 15) = (.2)^{20}\left(1 + \dfrac{20(.8)}{.2} + \dfrac{190(.8)^2}{.04} + \dfrac{1140(.8)^3}{.008} + \dfrac{4845(.8)^4}{.0016} + \dfrac{15504(.8)^5}{.00032}\right)$

$= (.2)^{20}(1 + 80 + 3040 + 72960 + 1240320 + 15876096) = .00000018$

31. $E(X) = 20(.2) = 4$

33. $n = 12$, $p = .4$, $q = .6$. $E(X) = 12(.4) = 4.8$ $\text{Var}(X) = (4.8)(.6) = 2.88$

$\sigma = \sqrt{2.88} \sim 1.697$

35. $n = 18$, $p = .3$, $q = .7$. $E(X) = 18(.3) = 5.4$ $\text{Var}(X) = (5.4)(.7) = 3.78$

$\sigma = \sqrt{3.78} \sim 1.944$

37. If X is the number of defective bulbs found $P(X = 0) + P(X = 1) = (.97)^{15} + 15(.97)^{14}(.03)$
$= .9270$ which is the probability the lot is accepted.

39. (a) Since $C(n, k) = \dfrac{n!}{(n-k)k!}$,

$$\sum_{k=1}^{n} kC(n, k)p^k q^{n-k} = \sum_{k=1}^{n} k\frac{n!}{(n-k)!k!}p^k q^{n-k}$$

(b) $\dfrac{k}{k!} = \dfrac{k}{k(k-1)!} = \dfrac{1}{(k-1)!}$, $n! = n(n-1)!$, and $p^k = p \cdot p^{k-1}$. Hence $\displaystyle\sum_{k=1}^{n} k\frac{n!}{(n-k)!k!}p^k q^{n-k}$

$$= \sum_{k=1}^{n} n \cdot \frac{(n-1)!}{(n-k)!} \cdot \frac{k}{k!}p \cdot p^{k-1}q^{n-k} = np\sum_{k=1}^{n} \frac{(n-1)!}{(n-k)!(k-1)!}p^{k-1}q^{n-k}.$$

(c) $C(n-1, k-1) = \dfrac{(n-1)!}{(n-1-k+1)!(k-1)!} = \dfrac{(n-1)!}{(n-k)!(k-1)!}$. Hence

$$np\sum_{k=1}^{n} \frac{(n-1)!}{(n-k)!(k-1)!}p^{k-1}q^{n-k} = np\sum_{k=1}^{n} C(n-1, k-1)p^{k-1}q^{n-k}$$

$$= np\left(C(n-1, 0)q^{n-1} + C(n-1, 1)p^1 q^{n-2} + \cdots + C(n-1, n-1)p^{n-1}\right)$$

(d) By the binomial theorem the expansion of $(q+p)^{n-1}$ is given by
$C(n-1, 0)q^{n-1} + C(n-1, 1)pq^{n-2} + \cdots + C(n-1, n-1)p^{n-1}$. Thus the expression in
line 4 equals $np(q+p)^{n-1} = np(1)^{n-1} = np$ as $q+p = 1$.

1. $\begin{bmatrix} .3 & .7 \\ 1 & 0 \end{bmatrix} \cdot \begin{bmatrix} .3 & .7 \\ 1 & 0 \end{bmatrix} = \begin{bmatrix} .79 & .21 \\ .3 & .7 \end{bmatrix}$. Since all entries of the latter matrix are positive, the

matrix $\begin{bmatrix} .3 & .7 \\ 1 & 0 \end{bmatrix}$ is regular.

$\begin{bmatrix} x & y \end{bmatrix} \cdot \begin{bmatrix} .3 & .7 \\ 1 & 0 \end{bmatrix} = \begin{bmatrix} x & y \end{bmatrix}$ if $.3x + y = x$, $.7x = y$ and $x + y = 1$

$\begin{cases} -.7x + y = 0 \\ .7x - y = 0 \\ x + y = 1 \end{cases}$ $\quad \begin{array}{l} E_1 + E_2 \to E_3 \\ .7E_3 + E_1 \to E_1 \\ E_3 \to E_2 \end{array}$ $\quad \begin{cases} 1.7y = .7 \\ x + \quad y = 1 \\ 0 = 0 \end{cases}$

Thus $y = 7/17$, $x = 10/17$ and the steady-state vector is $\begin{bmatrix} \frac{10}{17} & \frac{7}{17} \end{bmatrix}$.

3. $\begin{bmatrix} .75 & .25 \\ 1 & 0 \end{bmatrix} \cdot \begin{bmatrix} .75 & .25 \\ 1 & 0 \end{bmatrix} = \begin{bmatrix} .8125 & .1875 \\ .75 & .25 \end{bmatrix}$. Since all entries of the latter matrix are positive,

the matrix $\begin{bmatrix} .75 & .25 \\ 1 & 0 \end{bmatrix}$ is regular.

$\begin{bmatrix} x & y \end{bmatrix} \cdot \begin{bmatrix} .75 & .25 \\ 1 & 0 \end{bmatrix} = \begin{bmatrix} x & y \end{bmatrix}$ if $.75x + y = x$, and $x + y = 1$.

Multiplying the second equation by -1 and adding the result to the first equation, we find $-.25x = x - 1$ or $1.25x = 1$. Thus $x = 4/5$ and $y = 1/5$.

The steady-state vector is $\begin{bmatrix} .8 & .2 \end{bmatrix}$.

5. $\begin{bmatrix} .3 & 0 & .7 \\ 0 & .2 & .8 \\ .4 & .1 & .5 \end{bmatrix} \cdot \begin{bmatrix} .3 & 0 & .7 \\ 0 & .2 & .8 \\ .4 & .1 & .5 \end{bmatrix} = \begin{bmatrix} .37 & .07 & .56 \\ .32 & .12 & .56 \\ .32 & .07 & .61 \end{bmatrix}$.

All entries in the latter matrix are positive. Thus $\begin{bmatrix} .3 & 0 & .7 \\ 0 & .2 & .8 \\ .4 & .1 & .5 \end{bmatrix}$ is regular.

$\begin{bmatrix} x & y & z \end{bmatrix} \begin{bmatrix} .3 & 0 & .7 \\ 0 & .2 & .8 \\ .4 & .1 & .5 \end{bmatrix} = \begin{bmatrix} x & y & z \end{bmatrix}$ if

$.3x + .4z = x$, $.2y + .1z = y$, and $x + y + z = 1$, giving the system

$\begin{cases} -.7x \quad\quad + .4z = 0 \\ \quad\quad - .8y + .1z = 0 \\ x + \quad y + \quad z = 1. \end{cases}$ Using the row-reduction technique we have

$\begin{bmatrix} -.7 & 0 & .4 & | & 0 \\ 0 & -.8 & .1 & | & 0 \\ 1 & 1 & 1 & | & 1 \end{bmatrix}$ $\quad .7R_3 + R_1 \to R_1 \quad$ $\begin{bmatrix} 0 & .7 & 1.1 & | & .7 \\ 0 & -.8 & .1 & | & 0 \\ 1 & 1 & 1 & | & 1 \end{bmatrix}$

$$\begin{array}{l} 8R_1 + 7R_2 \to R_1 \\ .8R_3 + R_2 \to R_3 \end{array} \quad \begin{bmatrix} 0 & 0 & 9.5 & | & 5.6 \\ 0 & -.8 & .1 & | & 0 \\ .8 & 0 & .9 & | & .8 \end{bmatrix} \quad \begin{array}{l} R_1/9.5 \to R_1 \\ 10R_2 \to R_2 \\ 10R_3 \to R_3 \end{array}$$

$$\begin{bmatrix} 0 & 0 & 1 & | & 56/95 \\ 0 & -8 & 1 & | & 0 \\ 8 & 0 & 9 & | & 8 \end{bmatrix} \quad \begin{array}{l} R_1 - R_2 \to R_2 \\ -9R_1 + R_3 \to R_3 \end{array} \quad \begin{bmatrix} 0 & 0 & 1 & | & 56/95 \\ 0 & 8 & 0 & | & 56/95 \\ 8 & 0 & 0 & | & 256/95 \end{bmatrix}$$

$$\begin{array}{l} R_2/8 \to R_2 \\ R_3/8 \to R_3 \end{array} \quad \begin{bmatrix} 0 & 0 & 1 & | & 56/95 \\ 0 & 1 & 0 & | & 7/95 \\ 1 & 0 & 0 & | & 32/95 \end{bmatrix}. \text{ The steady-state vector is } \begin{bmatrix} \frac{32}{95} & \frac{7}{95} & \frac{56}{95} \end{bmatrix}.$$

7. If $T = \begin{bmatrix} .3 & .7 \\ 0 & 1 \end{bmatrix}$, T^n has zero as the entry in the lower left corner for $n = 1, 2, 3, \ldots$.

Thus T is not regular.

9. The sums of the entries in each of the three rows must be one. Hence $x = .1$, $y = .2$ and $z = .4$.

11. Since the sums of the entries in each of the three rows must be one, we have the system

$$\begin{cases} 2x + 3y + .2 = 1 \\ \quad z + .4 + 3x = 1 \\ .4 + 4x + y = 1. \end{cases} \quad \text{Solving}$$

$$\begin{bmatrix} 2 & 3 & 0 & | & .8 \\ 3 & 0 & 1 & | & .6 \\ 4 & 1 & 0 & | & .6 \end{bmatrix} \quad R_1 - 3R_3 \to R_1 \quad \begin{bmatrix} -10 & 0 & 0 & | & -1 \\ 3 & 0 & 1 & | & .6 \\ 4 & 1 & 0 & | & .6 \end{bmatrix} \quad -.1R_1 \to R_1$$

$$\begin{bmatrix} 1 & 0 & 0 & | & .1 \\ 3 & 0 & 1 & | & .6 \\ 4 & 1 & 0 & | & .6 \end{bmatrix} \quad \begin{array}{l} 4R_1 - R_3 \to R_3 \\ 3R_1 - R_2 \to R_2 \end{array} \quad \begin{bmatrix} 1 & 0 & 0 & | & .1 \\ 0 & 0 & -1 & | & -.3 \\ 0 & -1 & 0 & | & -.2 \end{bmatrix}.$$

$x = .1$, $y = .2$ and $z = .3$.

13. (a) $v_0 T = \begin{bmatrix} .2 & .4 & .4 \end{bmatrix} \begin{bmatrix} .3 & .2 & .5 \\ .1 & .7 & .2 \\ .4 & .5 & .1 \end{bmatrix} = \begin{bmatrix} .26 & .52 & .22 \end{bmatrix}$

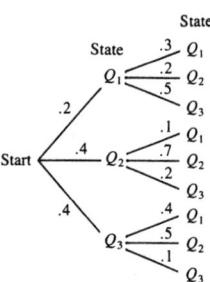

The probability that state Q_1 occurs on the second trial is
$$(.2)(.3)+(.4)(.1)+(.4)(.4) = .26.$$

The probability state Q_2 occurs is $.2(.2)+.4(.7)+.4(.5) = .52.$

The probability state Q_3 occurs is $(.2)(.5)+.4(.2)+.4(.1) = .22.$

(b) $w_0T = \begin{bmatrix} .5 & .3 & .2 \end{bmatrix} \begin{bmatrix} .3 & .2 & .5 \\ .1 & .7 & .2 \\ .4 & .5 & .1 \end{bmatrix} = \begin{bmatrix} .26 & .41 & .33 \end{bmatrix}$

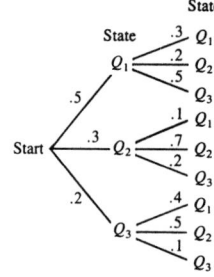

The probability that state Q_1 occurs on the second trial is $.5(.3)+.3(.1)+.2(.4) = .26.$

The probability that state Q_2 occurs is $.5(.2)+.3(.7)+.2(.5) = .41.$

The probability that state Q_3 occurs is $.5(.5)+.3(.2)+.2(.1) = .33.$

(c) $T^2 = \begin{bmatrix} .3 & .2 & .5 \\ .1 & .7 & .2 \\ .4 & .5 & .1 \end{bmatrix} \cdot \begin{bmatrix} .3 & .2 & .5 \\ .1 & .7 & .2 \\ .4 & .5 & .1 \end{bmatrix} = \begin{bmatrix} .31 & .45 & .24 \\ .18 & .61 & .21 \\ .21 & .48 & .31 \end{bmatrix}.$

$w_0T^2 = \begin{bmatrix} .5 & .3 & .2 \end{bmatrix} T^2 = \begin{bmatrix} .251 & .504 & .245 \end{bmatrix}$

The probability state Q_1 occurs on the third trial is .251. The probability state Q_2 occurs is .504 and the probability state Q_3 occurs is .245.

(d) $T^3 = \begin{bmatrix} .3 & .2 & .5 \\ .1 & .7 & .2 \\ .4 & .5 & .1 \end{bmatrix} \cdot \begin{bmatrix} .31 & .45 & .24 \\ .18 & .61 & .21 \\ .21 & .48 & .31 \end{bmatrix} = \begin{bmatrix} .234 & .497 & .269 \\ .199 & .568 & .233 \\ .235 & .533 & .232 \end{bmatrix}.$

$w_0T^3 = \begin{bmatrix} .5 & .3 & .2 \end{bmatrix} T^3 = \begin{bmatrix} .2237 & .5255 & .2508 \end{bmatrix}$

The probabilities that states Q_1, Q_2 and Q_3 occur on the 4th trial are .2237, .5255 and .2508 respectively.

15. (a) $T = \begin{bmatrix} .85 & .15 \\ .025 & .975 \end{bmatrix}$

where state 1 is the state of smoking and state 2 is the state of non-smoking.

$v_0 = \begin{bmatrix} .25 & .75 \end{bmatrix}$

$T^2 = \begin{bmatrix} .85 & .15 \\ .025 & .975 \end{bmatrix} \begin{bmatrix} .85 & .15 \\ .025 & .975 \end{bmatrix} = \begin{bmatrix} .72625 & .27375 \\ .045625 & .954375 \end{bmatrix}$

$T^3 = \begin{bmatrix} .85 & .15 \\ .025 & .975 \end{bmatrix} T^2 = \begin{bmatrix} .62415625 & .37584375 \\ .062640625 & .937359375 \end{bmatrix}$

$v_3 = v_0 T^3 \sim \begin{bmatrix} .203 & .797 \end{bmatrix}$. Hence approximately 20% of the adult population will be smoking after three months.

(b) The steady-state vector is $\begin{bmatrix} x & y \end{bmatrix}$ where

$\begin{bmatrix} x & y \end{bmatrix} \cdot \begin{bmatrix} .85 & .15 \\ .025 & .975 \end{bmatrix} = \begin{bmatrix} x & y \end{bmatrix}$ and $x+y = 1$.

$\begin{cases} .85x + .025y = x \\ .15x + .975y = y \\ \quad\;\; x + \quad\;\; y = 1 \end{cases}$
$\qquad$
$\begin{cases} -.15x + .025y = 0 \\ .15x - .025y = 0 \\ \quad\;\; x + \quad\;\; y = 1 \end{cases}$
$\qquad$
$\begin{array}{l} .15E_3 + E_1 \rightarrow E_1 \\ E_1 + E_2 \rightarrow E_2 \end{array}$

$\begin{cases} .175y = .15 \\ \quad\quad 0 = 0 \\ x + \quad\; y = 1 \end{cases}$
$\qquad$
$y = \dfrac{.15}{.175} = \dfrac{150}{175} = 6/7$

and $x = 1/7 \sim .143$

Approximately 14.3% of the population will be smoking after a long period of time.

17. Let state 1 be the state of using the system, state 2 the state of using one's own car, and state 3, the state of car pooling.

$T^3 = \begin{bmatrix} .8 & .15 & .05 \\ .3 & .6 & .1 \\ .3 & .05 & .65 \end{bmatrix}$. $\quad v_0 = \begin{bmatrix} .2 & .75 & .05 \end{bmatrix}$

$v_1 = \begin{bmatrix} .2 & .75 & .05 \end{bmatrix} T = \begin{bmatrix} .4 & .4825 & .1175 \end{bmatrix}$

$v_2 = \begin{bmatrix} .4 & .4825 & .1175 \end{bmatrix} T = \begin{bmatrix} .5 & .355375 & .144625 \end{bmatrix}$

$v_3 = \begin{bmatrix} .5 & .355375 & .144625 \end{bmatrix} T = \begin{bmatrix} .55 & .29545625 & .15454375 \end{bmatrix}$

(a) After three months the approximate percentages of system users, car drivers and car poolers are 55%, 29.5% and 15.5%, respectively.

(b) We find the steady-state vector,

$\begin{bmatrix} x & y & z \end{bmatrix} \begin{bmatrix} .8 & .15 & .05 \\ .3 & .6 & .1 \\ .3 & .05 & .65 \end{bmatrix} = \begin{bmatrix} x & y & z \end{bmatrix}$, leading to the system

$\begin{cases} .8x + .3y + .3z = x \\ .15x - .6y + .05z = y \\ \quad\; x + \quad y + \quad z = 1 \end{cases}$
$\qquad$ Using row-reduction,

$$\begin{bmatrix} -.2 & .3 & .3 & | & 0 \\ .15 & -.4 & .05 & | & 0 \\ 1 & 1 & 1 & | & 1 \end{bmatrix} \quad \begin{matrix} R_1 + 2R_3 \to R_1 \\ R_2 - .15R_3 \to R_2 \end{matrix} \quad \begin{bmatrix} 0 & .5 & .5 & | & .2 \\ 0 & -.55 & -.1 & | & -.15 \\ 1 & 1 & 1 & | & 1 \end{bmatrix}$$

$$\begin{matrix} -2R_1 + R_3 \to R_3 \\ 1.1R_1 + R_2 \to R_2 \end{matrix} \quad \begin{bmatrix} 0 & .5 & .5 & | & .2 \\ 0 & 0 & .45 & | & .07 \\ 1 & 0 & 0 & | & .6 \end{bmatrix} \quad .9R_1 - R_2 \to R_1 \quad \begin{bmatrix} 0 & .45 & 0 & | & .11 \\ 0 & 0 & .45 & | & .07 \\ 1 & 0 & 0 & | & .6 \end{bmatrix}$$

The steady-state vector is $\left[\begin{matrix} \frac{27}{45} & \frac{11}{45} & \frac{7}{45} \end{matrix} \right] = \left[\begin{matrix} .6 & .2\overline{4} & .1\overline{5} \end{matrix} \right]$. After a long period of time the approximate percentages of system users, car drivers and car poolers will be 60%, 24.4% and 15.6% respectively.

19. If state 1 is the state of arriving on time and state 2 is the state of arriving late,

$$T = \begin{bmatrix} .4 & .6 \\ .9 & .1 \end{bmatrix} \quad \text{and} \quad v_0 = \begin{bmatrix} .8 & .2 \end{bmatrix}$$

If the student is also doing exercises 18 or 20 the most efficient method would be to find T^4 and then $v_0 T^4$. However if only problem 19 is to be done, one can proceed as follows.

$$v_1 = \begin{bmatrix} .8 & .2 \end{bmatrix} \cdot \begin{bmatrix} .4 & .6 \\ .9 & .1 \end{bmatrix} = \begin{bmatrix} .50 & .50 \end{bmatrix} = v_0 T$$

$$v_2 = \begin{bmatrix} .5 & .5 \end{bmatrix} \cdot \begin{bmatrix} .4 & .6 \\ .9 & .1 \end{bmatrix} = \begin{bmatrix} .65 & .35 \end{bmatrix} = v_0 T^2$$

$$v_3 = \begin{bmatrix} .65 & .35 \end{bmatrix} \cdot \begin{bmatrix} .4 & .6 \\ .9 & .1 \end{bmatrix} = \begin{bmatrix} .575 & .425 \end{bmatrix} = v_0 T^3$$

$$v_4 = \begin{bmatrix} .575 & .425 \end{bmatrix} \cdot \begin{bmatrix} .4 & .6 \\ .9 & .1 \end{bmatrix} = \begin{bmatrix} .6125 & .3875 \end{bmatrix} = v_0 T^4$$

The probability the student is on time on the fifth class day is .6125

21. States 1, 2 and 3 are being in categories A_0, A_1, A_+, respectively.

$$T = \begin{bmatrix} .85 & .1 & .05 \\ .9 & .08 & .02 \\ .5 & .3 & .2 \end{bmatrix} \quad \text{and} \quad v_0 = \begin{bmatrix} 1 & 0 & 0 \end{bmatrix}$$

$$v_0 T = \begin{bmatrix} .85 & .1 & .05 \end{bmatrix}$$

$$v_0 T^2 = \begin{bmatrix} .85 & .1 & .05 \end{bmatrix} T = \begin{bmatrix} .8375 & .108 & .0545 \end{bmatrix}$$

$$v_0 T^3 = \begin{bmatrix} .8375 & .108 & .0545 \end{bmatrix} T = \begin{bmatrix} .836325 & .10874 & .054935 \end{bmatrix}$$

(a) After 3 years the probability a motorist belongs to A_0 is .836325

(b) $v_0 T^4 = \begin{bmatrix} .836325 & .10874 & .054935 \end{bmatrix} T$

$= \begin{bmatrix} .83620975 & .1088122 & .05497805 \end{bmatrix}$.

After 4 years the probability a motorist belongs to A_1 is .1088122.

(c) The probability that after 5 years an insured motorist belongs to A_+ is

$\begin{bmatrix} .83620975 & .1088122 & .05497805 \end{bmatrix} \cdot \begin{bmatrix} .05 & .02 & .27 \end{bmatrix}$

$\sim .055$

23. Since the steady-state vector is $\begin{bmatrix} \frac{10}{17} & \frac{7}{17} \end{bmatrix}$, the sequence approaches $\begin{bmatrix} \frac{10}{17} & \frac{7}{17} \\ \frac{10}{17} & \frac{7}{17} \end{bmatrix}$.

25. The steady-state vector is $\begin{bmatrix} .8 & .2 \end{bmatrix}$. The sequence approaches $\begin{bmatrix} .8 & .2 \\ .8 & .2 \end{bmatrix}$.

27. The steady-state vector is $\begin{bmatrix} \frac{32}{95} & \frac{7}{95} & \frac{56}{95} \end{bmatrix}$. The sequence approaches $\begin{bmatrix} \frac{32}{95} & \frac{7}{95} & \frac{56}{95} \\ \frac{32}{95} & \frac{7}{95} & \frac{56}{95} \\ \frac{32}{95} & \frac{7}{95} & \frac{56}{95} \end{bmatrix}$

29. The matrix has $\begin{bmatrix} 0 & 1 \end{bmatrix}$ as a steady-state vector, since

$\begin{bmatrix} 0 & 1 \end{bmatrix} \cdot \begin{bmatrix} .3 & .7 \\ 0 & 1 \end{bmatrix} = \begin{bmatrix} 0 & 1 \end{bmatrix}$.

31. v has size $1 \times n$, W has size $n \times n$, thus vW is defined and has size $1 \times n$ which is the same size as v.

The ith element of vW is $\begin{bmatrix} a_1 & a_2 & \cdots & a_n \end{bmatrix} \cdot \begin{bmatrix} a_i & a_i & \cdots & a_i \end{bmatrix} = a_1 a_i + a_2 a_i + \cdots + a_n a_i = a_i(a_1 + a_2 + \cdots + a_n) = a_i$ which is the ith element of v. Therefore $v = vW$.

EXERCISE SET 6.8 CHAPTER REVIEW

1. (a) $5! = 5 \cdot 4 \cdot 3 \cdot 2 \cdot 1 = 120$

(b) $12!/10! = 12 \cdot (11) = 132$

(c) $\dfrac{17!}{14!5!3!} = \dfrac{17 \cdot 16 \cdot 15}{120(6)} = 5.\overline{6}$

(d) $P(7, 3) = \dfrac{7!}{4!} = 7(6)(5) = 210$

(e) $P(8, 8) = 8! = 40{,}320$

(f) $C(12, 3) = \dfrac{12!}{9!3!} = \dfrac{12(11)(10)}{6} = 220$

(g) $P(15; 3, 5, 7) = \dfrac{15!}{3!5!7!} = 360{,}360$

(h) $C(18; 4, 6, 8) = \dfrac{18!}{4!6!8!} = 9{,}189{,}180$

3. There are one M, two A's, four S's, one C, one H, one U, one E and two T's. The number

of code words is $\dfrac{13!}{1!2!4!1!1!1!1!2!} = 64{,}864{,}800 = P(13; 1, 2, 4, 1, 1, 1, 1, 2)$

5. $C(7, 3)C(6, 3)+C(7, 2)\cdot C(6, 4)+C(7, 1)C(6, 5)+C(7, 0)C(6, 6) =$
$35(20)+21(15)+7(6)+1 = 1,058$

7. $\dfrac{10!}{2!3!5!} = P(10; 2, 3, 5) = 2,520$

9. (a)   The sample space consists of all ordered 4-tuples (x, y, z, w) where x, y, z and w are members of the set $\{H, T\}$.

   (b)   The probabilities are

   (1) $P\{(H, T, T, T), (T, H, T, T), (T, T, H, T), (T, T, T, H)\} = 4/2^4 = 1/4$

   (2) $1-P\{(T, T, T, T)\} = 1-1/16 = 15/16$

   (3) $P\{(T, T, H, H), (T, H, T, H), (T, H, H, T), (H, T, T, H), (H, H, T, T),$
   $(H, T, H, T), (T, T, T, H), (T, H, T, T), (H, T, T, T), (T, T, H, T),$
   $(T, T, T, T)\} = 11/16$

   (4) $P\{(T, T, T, T)\} = 1/16$

11. If $\mathcal{E}$ is the event Lendl wins the championship, $P(\mathcal{E}) = .32$ and $P(\mathcal{E}') = 1-.32 = .68.$
$P(\mathcal{E})/P(\mathcal{E}') = .32/.68 = 32/68 = 8/17.$ The odds Lendl wins the championship are 8 to 17.

13. Let A be the event a black marble was drawn from the 1st urn and B be the event a black marble was drawn from the second urn. $P(A \cap B) = P(A)\cdot P(B|A) = \dfrac{5}{9}\cdot\dfrac{4}{10} = 2/9$

15. $\dfrac{3}{8}\cdot\dfrac{3}{8} = \dfrac{9}{64}$ since the events both have probability 3/8 and are independent.

17. $\dfrac{50}{50}\cdot\dfrac{49}{50}\cdot\dfrac{48}{50}\cdot\dfrac{47}{50}\cdot\dfrac{46}{50} = .8136$

19. $P(\text{rolling a } 6) = 1/6.$ The probability Marian wins is $\dfrac{5}{6}\cdot\dfrac{5}{6}\cdot\dfrac{1}{6}+\dfrac{5}{6}\cdot\dfrac{5}{6}\cdot\dfrac{5}{6}\cdot\dfrac{5}{6}\cdot\dfrac{1}{6}+\cdots.$ This is
the sum of a G.P. with first term $\dfrac{25}{216}$ and common ratio $\dfrac{125}{216}.$ The sum is
$\dfrac{25/216}{1-125/216} = \dfrac{25}{91}.$

21. $P(H/R) = .12,$ $P(R) = 1/6,$ $P(W) = 5/6,$ $P(H/W) = .03$ where H, R and W are the events the target is hit, the right rifle is chosen and the wrong rifle is chosen, respectively.

$$P(R/H) = \dfrac{P(H/R)P(R)}{P(H/R)P(R)+P(H/W)P(W)}$$

$$= \dfrac{(.12)(1/6)}{(.12)(1/6)+(.03)(5/6)}$$

$$= \dfrac{.12}{.27} = \dfrac{4}{9} \sim .44$$

23. Let G be your net gain. The range of G is $\{8, 3, 1, -2\}.$
$E(G) = 8(2/36)+3(4/36)+1(6/36)-2(24/36) = -7/18 = -.3\overline{8}.$

On the average one expects to lose approximately 39¢ per game.

25. If G is the gain per policy the range of G is $\{300, -279{,}700\}$.
$$E(G) = 300(1-.00021) - 279{,}700(.00021) = \$241.20$$

27. $C(4, 2)(.6)^2(.4)^2 = .3456$

29. (a)

x	$f(x) = P(X = x)$
0	$(.7)^3 = .343$
1	$3(.3)(.7)^2 = .441$
2	$3(.3)^2(.7) = .189$
3	$(.3)^3 = .027$

(b) $E(X) = 3(.3) = .9$ $\text{Var}(X) = .9(.7) = .63$ $\sigma = \sqrt{.63} \sim .794$

31. If X is the number of defective bulbs found, $P(X = 0) + P(X = 1) + P(X = 2) + P(X = 3)$
$$= (.95)^{25} + 25(.95)^{24}(.05) + 300(.95)^{23}(.05)^2 + 2300(.95)^{22}(.05)^3$$
$$= (.95)^{22}\big((.95)^3 + 25(.95)^2(.05) + 300(.95)(.0025) + 2300(.05)^3\big)$$
$= .9659$ is the probability the lot is accepted. The probability the lot is rejected is

$1 - .9659 = .0341$

33. $.1 + .3 + \quad x = 1$
$\quad.4 + \quad y + .2 = 1$
$\quad z + .5 + .3 = 1$

Thus $x = .6$, $y = .4$ and $z = .2$

35. (a) $T = \begin{bmatrix} .78 & .22 \\ .02 & .98 \end{bmatrix}$ where P_{11} is the probability a smoker continues to smoke and P_{21} is the probability a non-smoker starts smoking.

$$T^2 = \begin{bmatrix} .78 & .22 \\ .02 & .98 \end{bmatrix}\begin{bmatrix} .78 & .22 \\ .02 & .98 \end{bmatrix} = \begin{bmatrix} .6128 & .3872 \\ .0352 & .9648 \end{bmatrix}$$

$$T^3 = \begin{bmatrix} .78 & .22 \\ .02 & .98 \end{bmatrix}\begin{bmatrix} .6128 & .3872 \\ .0352 & .9648 \end{bmatrix} = \begin{bmatrix} .485728 & .514872 \\ .046752 & .953248 \end{bmatrix}$$

$$T^4 = \begin{bmatrix} .78 & .22 \\ .02 & .98 \end{bmatrix} \cdot T^3 = \begin{bmatrix} .38915328 & .61084672 \\ .05553152 & .94446848 \end{bmatrix}$$

$$v_0 = \begin{bmatrix} .2 & .8 \end{bmatrix}$$

$$v_0 T^4 = \begin{bmatrix} .2 & .8 \end{bmatrix} \cdot T^4 = \begin{bmatrix} .122255872 & .877744128 \end{bmatrix}.$$

The percentage smoking after four months will be approximately 12.2.

(b) If $\begin{bmatrix} x & y \end{bmatrix}$ is the steady-state vector,

$$\begin{bmatrix} x & y \end{bmatrix} \cdot \begin{bmatrix} .78 & .22 \\ .02 & .98 \end{bmatrix} = \begin{bmatrix} x & y \end{bmatrix}.$$ We have the system

$$\begin{cases} .78x + .02y = x \\ .22x + .98y = y \\ x + y = 1 \end{cases} \qquad \begin{cases} -.22x + .02y = 0 \\ .22x - .02y = 0 \\ x + y = 1 \end{cases} \qquad \begin{matrix} E_1 + .22E_3 \rightarrow E_1 \\ E_1 + E_2 \rightarrow E_3 \\ E_3 \rightarrow E_2 \end{matrix}$$

$$\begin{cases} .24y = .22 \\ x + y = 1 \\ 0 = 0 \end{cases}.$$

Hence $y = 22/24$, $x = 2/24 = .08\overline{3}$, and the percentage smoking after a long period of time will be approximately $8\frac{1}{3}\%$.

CHAPTER 7

ELEMENTARY THEORY OF GAMES AND GRAPHS

EXERCISE SET 7.1.   MATRIX GAMES

1. (a) R has 3 choices and C has 4 choices.

 (b) $a_{21} = -3$.   So, R pays \$3 to C.

 (c) $a_{33} = 9$.   So, C pays \$9 to R.

3. Let $R_i = i+2$ and $C_j = j+2$.   Then, $a_{ii} = 0$ for $i = 1, 2, 3$, since in this case they selected the same number. $R_1 + C_2 = 3+4 = 7$.   Since 7 is odd, R wins and C pays R \$7.   So, $a_{12} = 7$. $R_1 + C_3 = 3+5 = 8$.   Since 8 is even, C wins and R pays C \$8.   So, $a_{13} = -8$. The matrix is
$$\begin{bmatrix} 0 & 7 & -8 \\ 7 & 0 & 9 \\ -8 & 9 & 0 \end{bmatrix}$$

5. Rows represent Rosemary, Columns represent Cindy.
$R_1 = 1, \quad R_2 = 3, \quad R_3 = 5$
$C_1 = 1, \quad C_2 = 3, \quad C_3 = 5$
$a_{11} = -1$ since in this case they selected the same number and Rosemary pays Cindy \$1. Similarly, $a_{22} = -3$ and $a_{33} = -5$.   Also, $a_{12} = 1$, $a_{13} = 1$; $a_{21} = 3$, $a_{23} = 3$, $a_{31} = 5$, $a_{32} = 5$.   The matrix is
$$\begin{bmatrix} -1 & 1 & 1 \\ 3 & -3 & 3 \\ 5 & 5 & -5 \end{bmatrix}$$

7. Circle the smallest entry in each row.

$$\begin{bmatrix} 3 & 2 & ① & 4 & 5 \\ -2 & ㊋ & 0 & 3 & -2 \\ 4 & 5 & ㊄ & 1 & 2 \end{bmatrix}$$

R will choose the largest of these encircled numbers ① and will always play row 1.

Box the largest entry in each column.

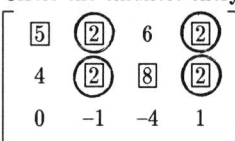

$$\begin{bmatrix} 3 & 2 & \boxed{1} & \boxed{4} & \boxed{5} \\ -2 & -3 & 0 & 3 & -2 \\ \boxed{4} & \boxed{5} & -5 & 1 & 2 \end{bmatrix}$$

C will choose the smallest of those boxed numbers and will always play column 3.

9. Circle the smallest entry in each row and box the largest entry in each column.

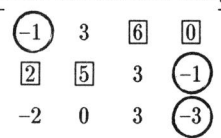

$$\begin{bmatrix} \boxed{5} & \boxed{\!\textcircled{2}\!} & 6 & \boxed{\!\textcircled{2}\!} \\ 4 & \boxed{\!\textcircled{2}\!} & \boxed{8} & \boxed{\!\textcircled{2}\!} \\ 0 & -1 & -4 & 1 \end{bmatrix}$$

The value of the game is 2. R should play either row 1 or row 2. C should play either column 2 or column 4.

11. Circle the smallest entry in each row and box the largest entry in each column.

$$\begin{bmatrix} \textcircled{-1} & 3 & \boxed{6} & \boxed{0} \\ \boxed{2} & \boxed{5} & 3 & \textcircled{-1} \\ -2 & 0 & 3 & \textcircled{-3} \end{bmatrix}$$

No entry is smallest in its row and largest in its column. So, the game is not strictly determined.

13.

$$\begin{bmatrix} 3 & -2 & 4 & 1 & 6 \\ 1 & -4 & 3 & 0 & 2 \\ 4 & -3 & 5 & -4 & -5 \end{bmatrix}$$ Row 2 is recessive. Delete it.

$$\begin{bmatrix} 3 & -2 & 4 & 1 & 6 \\ 4 & -3 & 5 & -4 & -5 \end{bmatrix}$$ Columns 1 and 3 are recessive. Delete them.

$$\begin{bmatrix} -2 & 1 & 6 \\ -3 & -4 & -5 \end{bmatrix}$$ Row 2 is recessive. Delete it.

$$\begin{bmatrix} -2 & 1 & 6 \end{bmatrix}$$

15. Circle the smallest entry in each row and box the largest entry in each column.

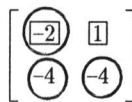

$$\begin{bmatrix} \boxed{\!\textcircled{-2}\!} & \boxed{1} \\ \textcircled{-4} & \textcircled{-4} \end{bmatrix}$$

(a) -2 is the saddle joint.

(b) R should always play row 1 and C should always play column 1.

(c)   The value of the game is -2.

17.
$$\begin{bmatrix} 1 & 0 & -3 & -1 \\ 3 & 1 & -2 & 0 \\ -1 & -4 & 0 & -1 \\ 0 & -3 & 4 & 1 \end{bmatrix}$$
$\leftarrow$ recessive row

$\leftarrow$ recessive row

$\uparrow$
recessive column

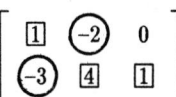

This game is not strictly determined because no entry is row minimum and column maximum.

19.
$$\begin{bmatrix} 1 & 0 & -1 & 2 \\ 0 & 5 & -2 & -4 \\ 2 & 2 & 1 & 3 \\ 3 & -1 & 0 & 18 \end{bmatrix}$$
$\leftarrow$ recessive row

$\uparrow$
recessive column

Delete recessive row and column, to get

$$\begin{bmatrix} \boxed{5} & -2 & \left(-4\right) \\ 2 & \boxed{1} & 3 \\ \left(-1\right) & 0 & \boxed{18} \end{bmatrix}$$

(a)   1 is the saddle joint.

(b) R should always play row 3 and C should always play column 3.

(c)   The value of the game is 1.

21.
$$\begin{bmatrix} 1 & 4 & 0 & -1 & 3 \\ 4 & 5 & 2 & 1 & 4 \\ 0 & 0 & 6 & -2 & -4 \\ -9 & 17 & -2 & 0 & 15 \\ -10 & 2 & -3 & -1 & 2 \end{bmatrix}$$

Delete recessive rows and columns, to obtain

$$\begin{bmatrix} \boxed{4} & 2 & \boxed{\textstyle\bigcirc\!\!1} & 4 \\ 0 & \boxed{6} & -2 & \enclose{circle}{-4} \\ \enclose{circle}{-9} & -2 & 0 & \boxed{15} \end{bmatrix}$$

(a) 1 is the saddle point.

(b) R should always play row 1 and C should always play column 3 (or row 2, column 4, of the original matrix)

(c) The value of the game is 1.

23. Let W be the row player and T be the column player. Rows 1 and 2 represent locations 1 and 2 for W and columns 1 and 2 represent locations 3 and 4 for T. If W builds in 1 and T in 3, W gets 60% of the membership which is 10% over 50%. So, $a_{11} = 10$. Similarly, $a_{12} = -5$, $a_{21} = 12$ and $a_{22} = 20$. The game matrix is

$$\begin{bmatrix} 10 & -5 \\ 12 & 20 \end{bmatrix}$$

Circle the smallest entry in each row and largest entry in each column, to get

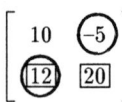

So, 12 is the saddle point. W should build in location 2 and T in location 3. W will get 62% of the membership.

25. Let rows represent A and columns represent B. If a store is built in city 1, 2, or 3 by A, it is represented by row 1, 2, or 3 respectively. Similarly, columns 1, 2, or 3 represent the store being built in city 1, 2, or 3 by B. If both stores are built in the same city, A gets 60% of the total business which is 10% over 50%. So, $a_{11} = a_{22} = a_{33} = 10$.

If A builds in 1 and B in 2, then A gets 75% of 1, 25% of 2, and 45% of 3. But, 25% of the population shops in 1, 35% in 2, and 40% in 3. So, A gets
$(.75)(.25) + (.25)(.35) + (.45)(.40) = .455$ (45.5%).
So, $a_{12} = -4.5$ because 45.5% is 4.5% below 50%.

If A builds in 1 and B in 3, then A gets
$(.75)(.25) + (.45)(.35) + (.25)(.40) = .445$ (44.5%). So, $a_{13} = -5.5$.

If A builds in 2 and B in 1, then A gets
$(.75)(.35) + (.45)(.40) + (.25)(.25) = .505$ (50.5%). So, $a_{21} = .5$.

If A builds in 2 and B in 3, then A gets
$(.75)(.35) + (.45)(.25) + (.25)(.40) = .475$ (47.5%). So, $a_{23} = -2.5$.

If A builds in 3 and B in 1, then A gets
$(.75)(.40) + (.45)(.35) + (.25)(.25) = .52$ (52%). So, $a_{31} = 2$.

Finally, if A builds in 3 and B in 2, then A gets
$(.75)(.40) + (.45)(.25) + (.25)(.35) = .5$ (50%). So, $a_{32} = 0$.

The game matrix is

$$\begin{bmatrix} 10 & -4.5 & -5.5 \\ .5 & 10 & -2.5 \\ 2 & 0 & 10 \end{bmatrix}$$

Circle the smallest entry in each row and box the largest entry in each column. We get

$$\begin{bmatrix} \boxed{10} & -4.5 & \boxed{-5.5} \\ .5 & \boxed{10} & -2.5 \\ 2 & 0 & \boxed{10} \end{bmatrix}$$

The game is not strictly determined.

27.

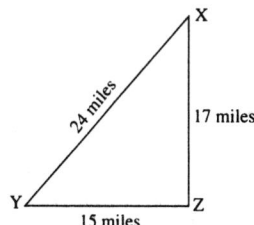

Store R will be the row player and store C will be the column player. Cities X, Y, Z will be represented by rows 1, 2, 3 and columns 1, 2, 3, respectively. Clearly $a_{11} = a_{22} = a_{33} = 0$.

If R builds in X and C in Y, then C gets the business from Z. So, $a_{12} = -1$.

If R builds in X and C in Z, then C gets the business from Y. So, $a_{13} = -1$.

If R builds in Y and C in X, then R gets the business from Z. So, $a_{21} = 1$.

If R builds in Y and C in Z, then C gets the business from X. So, $a_{23} = -1$.

If R builds in Z and C in X, then R gets the business from Y. So, $a_{31} = 1$.

If R builds in Z and C in Y, then R gets the business from X. So, $a_{32} = 1$.

The game matrix is

$$\begin{bmatrix} 0 & -1 & -1 \\ 1 & 0 & -1 \\ 1 & 1 & 0 \end{bmatrix}$$

Circle the smallest entry in each row and box the largest entry in each column.

$$\begin{bmatrix} 0 & -1 & -1 \\ \boxed{1} & 0 & -1 \\ \boxed{1} & \boxed{1} & \boxed{0} \end{bmatrix}$$

The saddle point is 0. Both stores should build in Z and they split the business evenly.

29. R will be the row player and C will be the column player. Rows 1 and 2 will represent region A and B, respectively. Similarly, columns 1 and 2 will represent region A and B, respectively.

If R and C concentrate in A, then R gets 65% of the players. (15% over 50%). So, $a_{11} = 15$.

If R concentrates in A and C in B, then R gets 70% of the players. So, $a_{12} = 20$.

If R concentrates in B and C in A, then R gets 60% of the players. So, $a_{21} = 10$.

If R concentrates in B and C in B, then R gets 40% of the players. So, $a_{22} = -10$.

The game matrix is

$$\begin{bmatrix} 15 & 20 \\ 10 & -10 \end{bmatrix}$$

Circle the smallest entry in each row and box the largest entry in each column.

$$\begin{bmatrix} \boxed{15} & \boxed{20} \\ 10 & -10 \end{bmatrix}$$

15 is the saddle point. Both coaches should concentrate in A and R will get 65% of the players.

EXERCISE SET 7.2   MIXED STRATEGY GAMES

1. **(a)** The probability of the payoff being $35 to R is $\frac{2}{5} \cdot \frac{3}{7} = \frac{6}{35}$ because this payment will occur if R plays row 1 and C plays column 1.

(b) The payment is $70 to C if R plays row 1 and C plays column 2 with probability $\frac{2}{5} \cdot \frac{4}{7} = \frac{8}{35}$.

(c) The payment is $105 to C if R plays row 2 and C plays column 1 with probability $\frac{3}{5} \cdot \frac{3}{7} = \frac{9}{35}$.

(d) The payment is $140 to R if R plays row 2 and C plays column 2 with probability $\frac{3}{5} \cdot \frac{4}{7} = \frac{12}{35}$.

(e) The expected value of the game is

$$E = 35(\tfrac{6}{35}) + (-70)(\tfrac{8}{35}) + (-105)(\tfrac{9}{35}) + 140(\tfrac{12}{35}) = 11.$$

The expected value of the game (for player R) is $11.

(f) $PAQ = \begin{bmatrix} \frac{2}{5} & \frac{3}{5} \end{bmatrix} \begin{bmatrix} 35 & -70 \\ -105 & 140 \end{bmatrix} \begin{bmatrix} \frac{3}{7} \\ \frac{4}{7} \end{bmatrix} = \begin{bmatrix} -49 & 56 \end{bmatrix} \begin{bmatrix} \frac{3}{7} \\ \frac{4}{7} \end{bmatrix} = [11] = 11.$

3. $E = \begin{bmatrix} \frac{1}{4} & \frac{3}{4} \end{bmatrix} \begin{bmatrix} 40 & -60 & 120 \\ 80 & -40 & 160 \end{bmatrix} \begin{bmatrix} \frac{1}{5} \\ \frac{3}{5} \\ \frac{1}{5} \end{bmatrix} = \begin{bmatrix} 70 & -45 & 150 \end{bmatrix} \begin{bmatrix} \frac{1}{5} \\ \frac{3}{5} \\ \frac{1}{5} \end{bmatrix} = [7] = 7$

5. $\begin{bmatrix} 3 & 8 \\ 10 & 5 \end{bmatrix}$ is nonstrictly determined since minimum $\{8,10\}$ > maximum$\{-3,5\}$.

7. $\begin{bmatrix} 6 & -3 \\ 5 & 7 \end{bmatrix}$ is nonstrictly determined since minimum $\{6,7\}$ > maximum$\{-3,5\}$.

9. $A = \begin{bmatrix} 5 & 2 \\ -3 & 7 \end{bmatrix}$

Let the probability that R chooses row 1 be p. Then, he will choose row 2 with probability $1 - p$. If C chooses column 1, R's expected earnings will be

$$E_R = 5p + (-3)(1 - p) = 8p - 3 \qquad (1)$$

If C chooses column 2, R's expected earnings will be

$$E_R = p(2) + (1 - p)(7) = -5p + 7 \qquad (2)$$

Graph both equations. The point of intersection is $(\frac{10}{13}, \frac{41}{13})$

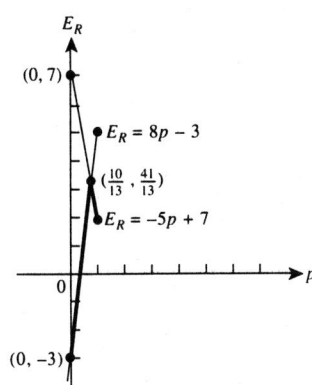

— 289 —

The optimal strategy for R is $p^* = \begin{bmatrix} \frac{10}{13} & \frac{3}{13} \end{bmatrix}$

Suppose C chooses column 1 with probability q. Then she will choose column 2 with probability $1 - q$.

If R chooses row 1, R's expected earning will be

$$E_R = 5q + 2(1 - q) = 3q + 2 \qquad (1)$$

If R chooses row 2, R's expected earnings will be

$$E_R = -3q + 7(1 - q) = -10q + 7 \qquad (2)$$

Graphing these equations, the point of intersection is $(\frac{5}{13}, \frac{41}{13})$.

So, an optimum strategy for C is $Q^* = \begin{bmatrix} \frac{5}{13} \\ \frac{8}{13} \end{bmatrix}$.

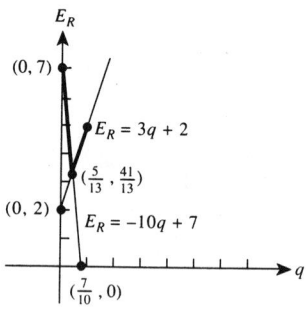

The value of the game is

$$\begin{bmatrix} \frac{10}{13} & \frac{3}{13} \end{bmatrix} \begin{bmatrix} 5 & 2 \\ -3 & 7 \end{bmatrix} \begin{bmatrix} \frac{5}{13} \\ \frac{8}{13} \end{bmatrix} = \begin{bmatrix} \frac{41}{13} & \frac{41}{13} \end{bmatrix} \begin{bmatrix} \frac{5}{13} \\ \frac{8}{13} \end{bmatrix} = \begin{bmatrix} \frac{41}{13} \end{bmatrix}$$

So, the value of the game is $\$\frac{41}{13}$.

11.

$$\begin{bmatrix} ②&3&⑤&⑦ \\ ③&⑥&(-1)&0 \\ 1&5&-2&(-3) \end{bmatrix}$$

This game is nonstrictly determined.   Row 3 is recessive. Delete it.

$$\begin{bmatrix} 2&3&5&7 \\ 3&6&-1&0 \end{bmatrix}$$

Columns 2 and 4 are recessive.   Delete them.

$$\begin{bmatrix} 2&5 \\ 3&-1 \end{bmatrix} \quad a = 2,\ b = 5,\ c = 3,\ d = -1;$$

$$D = [2 + (-1)] - (5 + 3) = 1 - 8 = -7;$$

$$p_1{}^* = \frac{-1-3}{-7} = \frac{4}{7}; \quad p_2{}^* = 1 - \frac{4}{7} = \frac{3}{7};$$

$$q_1{}^* = \frac{-1-5}{-7} = \frac{6}{7}; \quad q_2{}^* = 1 - \frac{6}{7} = \frac{1}{7};$$

$$V = \frac{(2)(-1) - (5)(3)}{-7} = \frac{-2 - 15}{-7} = \frac{17}{7}.$$

In terms of the original matrix, the optimum strategy for R is $\begin{bmatrix} \frac{4}{7} & \frac{3}{7} & 0 \end{bmatrix}$

and for C it is $\begin{bmatrix} \frac{6}{7} \\ 0 \\ \frac{1}{7} \\ 0 \end{bmatrix}$.

The value of the game is found as follows:

$$\begin{bmatrix} \frac{4}{7} & \frac{3}{7} & 0 \end{bmatrix} \begin{bmatrix} 2&3&5&7 \\ 3&6&-1&0 \\ 1&5&-2&-3 \end{bmatrix} \begin{bmatrix} \frac{6}{7} \\ 0 \\ \frac{1}{7} \\ 0 \end{bmatrix} = \begin{bmatrix} \frac{17}{7} & \frac{30}{7} & \frac{17}{7} & \frac{28}{7} \end{bmatrix} \begin{bmatrix} \frac{6}{7} \\ 0 \\ \frac{1}{7} \\ 0 \end{bmatrix}$$

$$= \begin{bmatrix} \frac{17}{7} \end{bmatrix}$$

13. (a) $$\begin{bmatrix} 3&(-2)&1&②\\ 2&(-4)&③&②\\ ④&-6&1&(-8) \end{bmatrix}$$

The game is strictly determined.   Its value is -2.   The optimal strategies are for R to play row 1 and for C to play column 2.

15. (a)

$$\begin{bmatrix} 0 & 1 & 2 & \boxed{-1} \\ \boxed{1} & 2 & \boxed{7} & 3 \\ 5 & 3 & 2 & \boxed{-2} \\ \boxed{9} & \boxed{11} & \boxed{5} & \boxed{10} \end{bmatrix}$$

The game is nonstrictly determined. Rows 1 and 3 are recessive. Delete them.

(b)

$$\begin{bmatrix} 1 & 2 & 7 & 3 \\ 9 & 11 & 5 & 10 \end{bmatrix}$$

Columns 2 and 4 are recessive. Delete them.

$$\begin{bmatrix} 1 & 7 \\ 9 & 5 \end{bmatrix} \quad a = 1,\ b = 7,\ c = 9,\ d = 5;$$

$$D = (1+5) - (7+9) = 6 - 16 = -10;$$

$$p_1{}^* = \frac{5-9}{-10} = \frac{-4}{-10} = \frac{2}{5}; \quad p_2{}^* = 1 - \frac{2}{5} = \frac{3}{5};$$

$$q_1{}^* = \frac{5-7}{-10} = \frac{-2}{-10} = \frac{1}{5}; \quad q_2{}^* = 1 - \frac{1}{5} = \frac{4}{5};$$

$$V = \frac{(1)(5) - (7)(9)}{-10} = \frac{5 - 63}{-10} = \frac{-58}{-10} = \frac{29}{5}.$$

In terms of the original game, the optimal strategies are

$$p^* = \begin{bmatrix} 0 & \frac{2}{5} & 0 & \frac{3}{5} \end{bmatrix} \text{ for } R, \quad q^* = \begin{bmatrix} \frac{1}{5} \\ 0 \\ \frac{4}{5} \\ 0 \end{bmatrix} \text{ for } C.\ \text{The value of the game is } \frac{29}{5}.$$

Check:

$$\begin{bmatrix} 0 & \frac{2}{5} & 0 & \frac{3}{5} \end{bmatrix} \begin{bmatrix} 0 & 1 & 2 & -1 \\ 1 & 2 & 7 & 3 \\ 5 & 3 & 2 & -2 \\ 9 & 11 & 5 & 10 \end{bmatrix} \begin{bmatrix} \frac{1}{5} \\ 0 \\ \frac{4}{5} \\ 0 \end{bmatrix}$$

$$\begin{bmatrix} \frac{29}{5} & \frac{37}{5} & \frac{29}{5} & \frac{36}{5} \end{bmatrix} \begin{bmatrix} \frac{1}{5} \\ 0 \\ \frac{4}{5} \\ 0 \end{bmatrix} = \begin{bmatrix} \frac{29}{5} \end{bmatrix}$$

17.

$$\begin{bmatrix} 2 & \boxed{-1} & 0 & \boxed{5} & \boxed{7} \\ 1 & 2 & 1 & \boxed{0} & 3 \\ \boxed{4} & \boxed{3} & \boxed{4} & \boxed{2} & 4 \\ 2 & \boxed{0} & 1 & 1 & 3 \end{bmatrix}$$

The game is nonstrictly determined. Rows 2 and 4 are recessive. Delete them.

$$\begin{bmatrix} 2 & -1 & 0 & 5 & 7 \\ 4 & 3 & 4 & 2 & 4 \end{bmatrix}$$ Columns 1, 3, and 5 are recessive. Delete them. We get

$$A = \begin{bmatrix} -1 & 5 \\ 3 & 2 \end{bmatrix} \qquad a = -1, \ b = 5, \ c = 3, \ d = 2.$$

$$D = (-1 + 2) - (5 + 3) = -7.$$

$$p_1^* = \frac{2-3}{-7} = \frac{1}{7}; \qquad p_2^* = 1 - \frac{1}{7} = \frac{6}{7};$$

$$q_1^* = \frac{2-5}{-7} = \frac{3}{7}; \qquad q_2^* = 1 - \frac{3}{7} = \frac{4}{7};$$

$$V = \frac{(-1)(2) - (5)(3)}{-7} = \frac{-17}{-7} = \frac{17}{7}.$$

The value of the game is $\frac{17}{7}$. The optimal strategies for R and C (in terms of the original game) are

$$p^* = \begin{bmatrix} \frac{1}{7} & 0 & \frac{6}{7} & 0 \end{bmatrix} \text{ and } q^* = \begin{bmatrix} 0 \\ \frac{3}{7} \\ 0 \\ \frac{4}{7} \\ 0 \end{bmatrix}, \text{ respectively.}$$

Check:

$$\begin{bmatrix} \frac{1}{7} & 0 & \frac{6}{7} & 0 \end{bmatrix} \begin{bmatrix} 2 & -1 & 0 & 5 & 7 \\ 1 & 2 & 1 & 0 & 3 \\ 4 & 3 & 4 & 2 & 4 \\ 2 & 0 & 1 & 1 & 3 \end{bmatrix} \begin{bmatrix} 0 \\ \frac{3}{7} \\ 0 \\ \frac{4}{7} \\ 0 \end{bmatrix} =$$

$$\begin{bmatrix} \frac{26}{7} & \frac{17}{7} & \frac{24}{7} & \frac{17}{7} & \frac{31}{7} \end{bmatrix} \begin{bmatrix} 0 \\ \frac{3}{7} \\ 0 \\ \frac{4}{7} \\ 0 \end{bmatrix} = \begin{bmatrix} \frac{17}{7} \end{bmatrix}.$$

19. Let Ron be player R and Carol be player C. Row 1 and column 1 represent 1 finger showing and row 2 and column 2 represent 2 fingers showing.

$a_{11} = 2$ since $1 + 1 = 2$ is even and Ron gets paid 2 dollars. Similarly $a_{22} = 4$.

Also, $a_{12} = -3$ and $a_{21} = -3$, since $1 + 2 = 2 + 1 = 3$ is odd.

$$A = \begin{bmatrix} 2 & -3 \\ -3 & 4 \end{bmatrix} \qquad a = 2, \ b = -3, \ c = -3, \ d = 4.$$

$$D = (2 + 4) - (-3 + -3) = 12;$$

$$p_1{}^* = \frac{4-(-3)}{12} = \frac{7}{12}; \qquad p_2{}^* = 1 - \frac{7}{12} = \frac{5}{12};$$

$$q_1{}^* = \frac{4-(-3)}{12} = \frac{7}{12}; \qquad q_2{}^* = 1 - \frac{7}{12} = \frac{5}{12};$$

$$V = \frac{(2)(4)-(-3)(-3)}{12} = \frac{8-9}{12} = \frac{-1}{12}.$$

The value of the game is $\frac{-1}{12}$. The optimal strategies are

$$p^* = \begin{bmatrix} \frac{7}{12} & \frac{5}{12} \end{bmatrix} \text{ for Ron and } \begin{bmatrix} \frac{7}{12} \\ \frac{5}{12} \end{bmatrix} \text{ for Carol.}$$

21. (a) The probabilities that Rosemary draws a blue, red, white marbles are $\frac{3}{9}, \frac{4}{9},$ and $\frac{2}{9}$, respectively. So, her strategy is $p = \begin{bmatrix} \frac{3}{9} & \frac{4}{9} & \frac{2}{9} \end{bmatrix}$. The probabilities that Charlie draws a blue, red, white, green marbles are $\frac{1}{11}, \frac{3}{11}, \frac{4}{11},$ and $\frac{3}{11}$ respectively. So, his strategy is

$$q = \begin{bmatrix} \frac{1}{11} \\ \frac{3}{11} \\ \frac{4}{11} \\ \frac{3}{11} \end{bmatrix}$$

Rosemary's expected value of the game is obtained as follows:

$$\begin{bmatrix} \frac{3}{9} & \frac{4}{9} & \frac{2}{9} \end{bmatrix} \begin{bmatrix} 2 & 1 & -3 & 0 \\ 4 & 2 & 5 & 1 \\ 2 & -6 & 1 & 4 \end{bmatrix} \begin{bmatrix} \frac{1}{11} \\ \frac{3}{11} \\ \frac{4}{11} \\ \frac{3}{11} \end{bmatrix}$$

$$= \begin{bmatrix} \frac{26}{9} & \frac{-1}{9} & \frac{13}{9} & \frac{12}{9} \end{bmatrix} \begin{bmatrix} \frac{1}{11} \\ \frac{3}{11} \\ \frac{4}{11} \\ \frac{3}{11} \end{bmatrix} = \begin{bmatrix} \frac{176}{99} \end{bmatrix} = \begin{bmatrix} \frac{16}{9} \end{bmatrix}.$$

Rosemary's expected value of the game is $\$\frac{16}{9}$.

(b) The game is nonstrictly determined. Deleting recessive rows and columns, we obtain the matrix

$$A = \begin{bmatrix} 2 & 1 \\ -6 & 4 \end{bmatrix} \quad a = 2, \ b = 1, \ c = -6, \ d = 4.$$

$$D = [2+4] - [-6+1] = 6+5 = 11;$$

$$p_1{}^* = \frac{4-(-6)}{11} = \frac{10}{11}; \qquad p_2{}^* = 1 - \frac{10}{11} = \frac{1}{11}.$$

In terms of the original matrix, R optimum strategy is $\begin{bmatrix} 0 & \frac{10}{11} & \frac{1}{11} \end{bmatrix}$.

$$q_1{}^* = \frac{4-1}{11} = \frac{3}{11}; \qquad q_2{}^* = 1 - \frac{3}{11} = \frac{8}{11}.$$

In terms of the original matrix, C optimal strategy is $q^* = \begin{bmatrix} 0 \\ \frac{3}{11} \\ 0 \\ \frac{8}{11} \end{bmatrix}$.

So, if Charlie changes to his optional strategy and Rosemary continues to play by drawing marbles, we calculate

$$\begin{bmatrix} \frac{3}{9} & \frac{4}{9} & \frac{2}{9} \end{bmatrix} \begin{bmatrix} 2 & 1 & -3 & 0 \\ 4 & 2 & 5 & 1 \\ 2 & -6 & 1 & 4 \end{bmatrix} \begin{bmatrix} 0 \\ \frac{3}{11} \\ 0 \\ \frac{8}{11} \end{bmatrix}$$

$$= \begin{bmatrix} \frac{26}{9} & \frac{-1}{9} & \frac{13}{9} & \frac{12}{9} \end{bmatrix} \begin{bmatrix} 0 \\ \frac{3}{11} \\ 0 \\ \frac{8}{11} \end{bmatrix} = \begin{bmatrix} \frac{93}{99} \end{bmatrix}.$$

So, Rosemary's new expected value of the game is $\$\frac{93}{99}$.

(c)   If Rosemary changes to her optimal strategy and Charlie continues to play by drawing marbles, we calculate

$$\begin{bmatrix} 0 & \frac{10}{11} & \frac{1}{11} \end{bmatrix} \begin{bmatrix} 2 & 1 & -3 & 0 \\ 4 & 2 & 5 & 1 \\ 2 & -6 & 1 & 4 \end{bmatrix} \begin{bmatrix} \frac{1}{11} \\ \frac{3}{11} \\ \frac{4}{11} \\ \frac{3}{11} \end{bmatrix}$$

$$= \begin{bmatrix} \frac{42}{11} & \frac{14}{11} & \frac{51}{11} & \frac{14}{11} \end{bmatrix} \begin{bmatrix} \frac{1}{11} \\ \frac{3}{11} \\ \frac{4}{11} \\ \frac{3}{11} \end{bmatrix} = \begin{bmatrix} \frac{330}{121} \end{bmatrix}.$$

So, Rosemary's new expected value of the game is $\$\frac{330}{121} = \$\frac{30}{11}$.

(d)   If they both switch to their optimal strategies, we calculate

$$\begin{bmatrix} 0 & \frac{10}{11} & \frac{1}{11} \end{bmatrix} \begin{bmatrix} 2 & 1 & -3 & 0 \\ 4 & 2 & 5 & 1 \\ 2 & -6 & 1 & 4 \end{bmatrix} \begin{bmatrix} 0 \\ \frac{3}{11} \\ 0 \\ \frac{8}{11} \end{bmatrix}$$

$$\begin{bmatrix} \dfrac{42}{11} & \dfrac{14}{11} & \dfrac{51}{11} & \dfrac{14}{11} \end{bmatrix} \begin{bmatrix} 0 \\ \dfrac{3}{11} \\ 0 \\ \dfrac{8}{11} \end{bmatrix} = \begin{bmatrix} \dfrac{154}{121} \end{bmatrix} = \begin{bmatrix} \dfrac{14}{11} \end{bmatrix}.$$

So, the value of the game is $\$\dfrac{14}{11}$.

23. (a)
$$\begin{bmatrix} 3 & 6 & -10 & 0 \\ 4 & 8 & -6 & 10 \\ -8 & 2 & 1 & -4 \\ 3 & 5 & 7 & 9 \end{bmatrix}$$
Rows 1 and 3 are recessive. Delete them.

$$\begin{bmatrix} 4 & 8 & -6 & 10 \\ 3 & 5 & 7 & 9 \end{bmatrix}$$
Columns 2 and 4 are recessive. Delete them.

$$A = \begin{bmatrix} 4 & -6 \\ 3 & 7 \end{bmatrix} \qquad a = 4, \; b = -6, \; c = 3, \; d = 7.$$

$D = (4 + 7) - (3 + (-6)) = 11 + 3 = 14;$

$p_1^* = \dfrac{7 - 3}{14} = \dfrac{4}{14} = \dfrac{2}{7};$ $\qquad p_2^* = 1 - \dfrac{2}{7} = \dfrac{5}{7}.$ In terms of the original matrix,

$p^* = \begin{bmatrix} 0 & \dfrac{2}{7} & 0 & \dfrac{5}{7} \end{bmatrix}.$ Also,

$q_1^* = \dfrac{7 - (-6)}{14} = \dfrac{13}{14};$ $\qquad q_2^* = 1 - \dfrac{13}{14} = \dfrac{1}{14}.$ In terms of the original matrix

$q^* = \begin{bmatrix} \dfrac{13}{14} \\ 0 \\ \dfrac{1}{14} \\ 0 \end{bmatrix}.$ The value of the game is $\dfrac{4(7) - 3(-6)}{14} = \dfrac{23}{7}.$

Check:

$$\begin{bmatrix} 0 & \dfrac{2}{7} & 0 & \dfrac{5}{7} \end{bmatrix} \begin{bmatrix} 3 & 6 & -10 & 0 \\ 4 & 8 & -6 & 10 \\ -8 & 2 & 1 & -4 \\ 3 & 5 & 7 & 9 \end{bmatrix} \begin{bmatrix} \dfrac{13}{14} \\ 0 \\ \dfrac{1}{14} \\ 0 \end{bmatrix}$$

$$\begin{bmatrix} \dfrac{23}{7} & \dfrac{41}{7} & \dfrac{23}{7} & \dfrac{65}{7} \end{bmatrix} \begin{bmatrix} \dfrac{13}{14} \\ 0 \\ \dfrac{1}{14} \\ 0 \end{bmatrix} = \begin{bmatrix} \dfrac{322}{98} \end{bmatrix} = \begin{bmatrix} \dfrac{23}{7} \end{bmatrix}.$$

(b) If R always uses the newspaper, the expected value of the game for R is found as follows:

$$\begin{bmatrix} 0 & 0 & 0 & 1 \end{bmatrix} \begin{bmatrix} 3 & 6 & -10 & 0 \\ 4 & 8 & -6 & 10 \\ -8 & 2 & 1 & -4 \\ 3 & 5 & 7 & 9 \end{bmatrix} \begin{bmatrix} \frac{13}{14} \\ 0 \\ \frac{1}{14} \\ 0 \end{bmatrix}$$

$$= \begin{bmatrix} 3 & 5 & 7 & 9 \end{bmatrix} \begin{bmatrix} \frac{13}{14} \\ 0 \\ \frac{1}{14} \\ 0 \end{bmatrix} = \begin{bmatrix} \frac{46}{14} \end{bmatrix} = \begin{bmatrix} \frac{23}{7} \end{bmatrix}.$$

So, the value of the game remains the same $\left(\frac{23}{7}\right)$ if R consistently uses the paper.

(c) If C always chooses television and R uses the optimum strategy, the value of the game for R will be

$$\begin{bmatrix} 0 & \frac{2}{7} & 0 & \frac{5}{7} \end{bmatrix} \begin{bmatrix} 3 & 6 & -10 & 0 \\ 4 & 8 & -6 & 10 \\ -8 & 2 & 1 & -4 \\ 3 & 5 & 7 & 9 \end{bmatrix} \begin{bmatrix} 0 \\ 1 \\ 0 \\ 0 \end{bmatrix}$$

$$= \begin{bmatrix} 0 & \frac{2}{7} & 0 & \frac{5}{7} \end{bmatrix} \begin{bmatrix} 6 \\ 8 \\ 2 \\ 5 \end{bmatrix} = \begin{bmatrix} \frac{41}{7} \end{bmatrix}.$$

The value of the game is $\frac{41}{7}$.

25.

$$\begin{array}{cc} & Z \quad W \\ \begin{array}{c} X \\ Y \end{array} & \begin{bmatrix} 15 & -18 \\ 10 & 13 \end{bmatrix} \end{array} \qquad a = 15, \ b = -18, \ c = 10, \ d = 13.$$

$D = (15 + 13) - (10 + -18) = 36;$

$$p_1{}^* = \frac{13 - 10}{36} = \frac{3}{36} = \frac{1}{12}; \qquad p_2{}^* = 1 - \frac{1}{12} = \frac{11}{12};$$

$$p^* = \begin{bmatrix} \frac{1}{12} & \frac{11}{12} \end{bmatrix}$$

$$q_1{}^* = \frac{13 - (-18)}{36} = \frac{31}{36}; \qquad q_2{}^* = 1 - \frac{31}{36} = \frac{5}{36};$$

$$q^* = \begin{bmatrix} \frac{31}{36} \\ \frac{5}{36} \end{bmatrix}.$$

The value of the game is found as follows:

$$\begin{bmatrix} \frac{1}{12} & \frac{11}{12} \end{bmatrix} \begin{bmatrix} 15 & -18 \\ 10 & 13 \end{bmatrix} \begin{bmatrix} \frac{31}{36} \\ \frac{5}{36} \end{bmatrix}$$

$$\begin{bmatrix} \frac{125}{12} & \frac{125}{12} \end{bmatrix} \begin{bmatrix} \frac{31}{36} \\ \frac{5}{36} \end{bmatrix} \begin{bmatrix} \frac{125}{12} \end{bmatrix}$$

So, the value of the game is $\frac{125}{12}$.

27. $D = (a + d) - (b + c)$

$$p^* = \begin{bmatrix} \frac{d-b}{D}, & 1 - \frac{d-c}{D} \end{bmatrix} = \begin{bmatrix} \frac{d-c}{D}, & 1 - \frac{a-b}{D} \end{bmatrix}$$

$$q^* = \begin{bmatrix} \frac{d-b}{D} \\ 1 - \frac{d-b}{D} \end{bmatrix} = \begin{bmatrix} \frac{d-b}{D} \\ \frac{a-c}{D} \end{bmatrix}$$

According to theorem 4, $V = \frac{ad - bc}{D}$.

Also, $p^*A = \begin{bmatrix} \frac{d-c}{D} & \frac{a-b}{D} \end{bmatrix} \begin{bmatrix} a & b \\ c & d \end{bmatrix} = \begin{bmatrix} \frac{ad - ac + ac - bc}{D}, & \frac{db - bc + ad - bd}{D} \end{bmatrix}$

$$= \begin{bmatrix} \frac{ad - bc}{D} & \frac{ad - bc}{D} \end{bmatrix} = [V \quad V].$$

and, $Aq^* = \begin{bmatrix} a & b \\ c & d \end{bmatrix} \begin{bmatrix} \frac{d-b}{D} \\ \frac{a-c}{D} \end{bmatrix} = \begin{bmatrix} \frac{ad - ab + ab - abc}{D} \\ \frac{cd - bc + ad - dc}{D} \end{bmatrix} = \begin{bmatrix} \frac{ad - bc}{D} \\ \frac{ad - bc}{D} \end{bmatrix} \begin{bmatrix} V \\ V \end{bmatrix}$

29. $\dfrac{d - c}{(a + d) - (b + c)} = \dfrac{d - c}{(a - b) + (d - c)}.$

Recall that $\max \{b, c\} < \min \{a, d\}$ (i)

    or $\max \{a, d\} < \min \{b, c\}$ (ii)

In case (i), $a + d > b + c$, so that $(a + d) - (b + c) > 0$.

Also $d - c > 0$, so $\dfrac{d - c}{(a + b) - (b + c)} > 0$.

Since $a - b > 0$, $d - c < (a - b) + (d - c)$ and $\dfrac{d - c}{(a - b) + (d - c)} < 1$.

Similarly for case (ii).

31. $$\begin{bmatrix} a' & b' \\ c' & d' \end{bmatrix} = \begin{bmatrix} a + k & b + k \\ c + k & d + k \end{bmatrix}$$

$$p_1'^* = \frac{d'-c'}{D'} = \frac{(d+k)-(c+k)}{[(a+k)+(d+k)]-[(c+k)+(b+k)]} = \frac{d-c}{(a+d)-(b+c)} = p_1^*.$$

$$q_1'^* = \frac{d'-b'}{D'} = \frac{(d+k)-(b+k)}{[(a+k)+(d+k)]-[(c+k)+(b+k)]} = \frac{d-b}{(a+d)-(b+c)} = q_1^*.$$

$$V' = \frac{a'd'-b'c'}{D'} = \frac{(a+k)(d+k)-(b+k)(c+k)}{[(a+k)+(d+k)]-[(c+k)+(b+k)]}$$

$$= \frac{ad+k(a+d)+k^2-bc-k(b+c)-k^2}{(a+d)-(b+c)} = \frac{ad-bc}{(a+d)-(b+c)} + k\frac{(a+d)-(b+c)}{(a+d)-(b+c)}$$

$$= V+k.$$

33. Consider the matrix of Exercise 25, $A = \begin{bmatrix} 15 & -18 \\ 10 & 13 \end{bmatrix}$. We found that $p^* = \begin{bmatrix} \frac{1}{12} & \frac{11}{12} \end{bmatrix}$,

$q^* = \begin{bmatrix} \frac{31}{36} \\ \frac{5}{36} \end{bmatrix}$ and $V = \frac{125}{12}$. Not that $p^*A\begin{bmatrix} 1 \\ 0 \end{bmatrix} = \frac{125}{12}$ and that $\begin{bmatrix} 1 \\ 0 \end{bmatrix} \neq \begin{bmatrix} \frac{31}{36} \\ \frac{5}{36} \end{bmatrix}$.

EXERCISE SET 7.3 SOLUTION of 2×2 GAME MATRIX USING LINEAR PROGRAMMING

1. Matrix A of Exercise 9, Section 7.2

$$A = \begin{bmatrix} 5 & 2 \\ -3 & 7 \end{bmatrix}$$

(a) min $\{5, 7\} >$ max $\{-3, 2\}$. So, the game is nonstrictly determined.

(b) Let $k = 4$, $B = \begin{bmatrix} 9 & 6 \\ 1 & 11 \end{bmatrix}$

$a = 9$, $b = 6$, $c = 1$, $d = 11$.

(c) The linear programming problems are

I. Minimize $z = x_1 + x_2$ subject to

$$9x_1 + x_2 \geq 1$$
$$6x_1 + 11x_2 \geq 1$$
$$x_1 \geq 0, \; x_2 \geq 1$$

II. Maximize $Z = y_1 + y_2$

$$9y_1 + 6y_2 \leq 1$$
$$y_1 + 11y_2 \leq 1$$
$$y_1 \leq 0, \; y_2 \geq 0$$

(d)   We solve each linear programming problem geometrically.

First problem

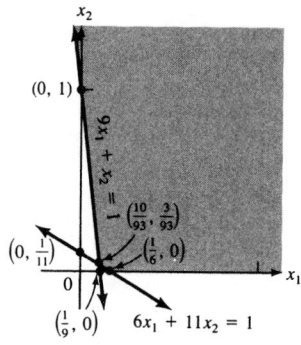

Z attains a minimum $\frac{13}{19}$
where $x_1 = \frac{10}{93}$, $x_2 = \frac{3}{93}$

Figure 1

The region of Figure 1 is unbounded.

At $(0, 1)$, $z = x_1 + x_2 = 1$.

At $(\frac{10}{93}, \frac{3}{93})$, $z = \frac{13}{93}$.

At $(\frac{1}{6}, 0)$, $z = \frac{1}{6} + 0 = \frac{1}{6}$.

The minimum value of z is $\frac{13}{93}$.   The graph of any line $x_1 + x_2 = c$ with $c < \frac{13}{93}$ does not intersect the region.   So, the minimum value of z is $\frac{13}{93}$.
It occurs at $x_1 = \frac{10}{93}$, $x_2 = \frac{3}{93}$.

Second problem

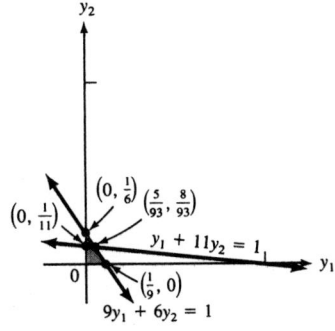

Z attains a maximum $\frac{13}{19}$
where $y_1 = \frac{5}{93}$, $y_2 = \frac{8}{93}$

Figure 2

At $(0, 0)$, $Z = 0 + 0 = 0$, $\qquad$ At $(0, \frac{1}{11})$, $Z = 0 + \frac{1}{11} = \frac{1}{11}$,

at $(\frac{1}{9}, 0)$, $Z = \frac{1}{9} + 0 = \frac{1}{9}$, $\qquad$ At $(\frac{5}{93}, \frac{8}{93})$, $Z = \frac{5}{93} + \frac{8}{93} = \frac{13}{93}$.

The maximum value of Z is $\frac{13}{93}$; it occurs at $y_1 = \frac{5}{93}$, $y_2 = \frac{8}{93}$.

(e) $\quad$ Since $\frac{1}{v} = Z$ and $Z = \frac{13}{93}$, we have $v = \frac{93}{13}$. $\quad$ Further, $p_1{}^* = vx_1 = (\frac{93}{13})(\frac{10}{93}) = \frac{10}{13}$

and $p_2{}^* = vx_2 = \frac{93}{13} \cdot \frac{3}{93} = \frac{3}{13}$. $\quad$ So, $p^* = \begin{bmatrix} \frac{10}{13} & \frac{3}{13} \end{bmatrix}$.

Also, $q_1{}^* = vy_1 = \frac{93}{13}(\frac{5}{93}) = \frac{5}{13}$, $q_2{}^* = \frac{93}{13}(\frac{8}{93}) = \frac{8}{13}$. $\quad$ Therefore, $Q = \begin{bmatrix} \frac{5}{13} \\ \frac{8}{13} \end{bmatrix}$

The value of the original game is $\frac{93}{13} - 4 = \frac{41}{13}$.

3. Matrix A of Exercise 11, section 7.2.

$$\begin{bmatrix} 2 & 3 & 5 & 7 \\ 3 & 6 & -1 & 0 \\ 1 & 5 & -2 & -3 \end{bmatrix}$$

Deleting recessive rows and columns, we obtain

$$A = \begin{bmatrix} 2 & 5 \\ 3 & -1 \end{bmatrix}$$

Let $k = 2$,

$$B = \begin{bmatrix} 4 & 7 \\ 5 & 1 \end{bmatrix}$$

Problem 1:   minimize $z = x_1 + x_2$ subject to

$$4x_1 + 5x_2 \geq 1$$
$$7x_1 + x_2 \geq 1$$
$$x_1 \geq 0,\ x_2 \geq 0$$

Problem 2:   maximize $Z = y_1 + y_2$ subject to

$$4y_1 + 7y_2 \leq 1$$
$$5y_1 + y_2 \leq 1$$
$$y_1 \geq 0,\ y_2 \geq 0$$

We solve both problems geometrically.

First problem

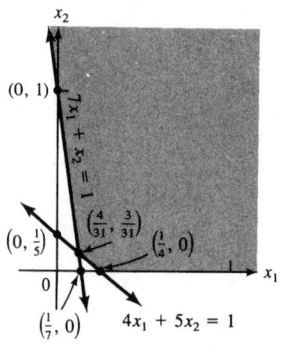

Z attains a minimum $\frac{7}{31}$
where $x_1 = \frac{4}{31}$, $x_2 = \frac{3}{31}$

Figure 1

At $(0, 1)$, $z = 1$, at $(\frac{4}{31}, \frac{3}{31})$, $z = \frac{7}{31}$ and at $(\frac{1}{4}, 0)$, $x = \frac{1}{4}$. The minimum is $\frac{7}{31}$.
It occurs at $x_1 = \frac{4}{31}$, $x_2 = \frac{3}{31}$.

Second problem

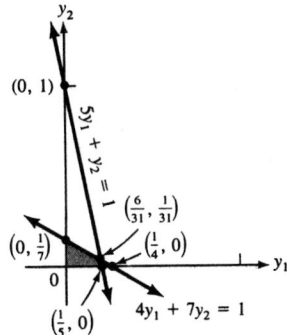

Z attains a maximum $\frac{7}{31}$
where $y_1 = \frac{6}{31}$, $y_2 = \frac{1}{31}$

Figure 2

At $(0, 0)$, $Z = 0$, at $(0, \frac{1}{7})$, $Z = \frac{1}{7}$, at $(\frac{6}{31}, \frac{1}{31})$, $Z = \frac{7}{31}$, at $(\frac{1}{5}, 0)$, $Z = \frac{1}{5}$. The maximum

value of Z is $\frac{7}{31}$. It occurs at $y_1 = \frac{6}{31}$, $y_2 = \frac{1}{31}$.

Since $z = \frac{1}{v}$ and $z = \frac{7}{31}$, $v = \frac{31}{7}$.

$$p_1{}^* = x_1 v = \frac{4}{31}(\frac{31}{7}) = \frac{4}{7}, \ p_2{}^* = x_2 v = \frac{3}{31}(\frac{31}{7}) = \frac{3}{7}. \ \text{So, } p^* = \begin{bmatrix} \frac{4}{7} & \frac{3}{7} \end{bmatrix}.$$

$$q_1{}^* = y_1 v = \frac{6}{31}(\frac{31}{7}) = \frac{6}{7}, \ q_2{}^* = y_2 v = \frac{1}{31}(\frac{31}{7}) = \frac{1}{7}.$$

Therefore, $q^* = \begin{bmatrix} \frac{6}{7} \\ \frac{1}{7} \end{bmatrix}$. The value of the original game is $\frac{31}{7} - 2 = \frac{17}{7}$.

In terms of the original matrix, $p^* = \begin{bmatrix} \frac{4}{7} & \frac{3}{7} & 0 \end{bmatrix}$ and

$$q^* = \begin{bmatrix} \frac{6}{7} \\ 0 \\ \frac{1}{7} \\ 0 \end{bmatrix}.$$

5. Use matrix A of Exercise 15, Section 7.2.

$$\begin{bmatrix} 0 & 1 & 2 & -1 \\ 1 & 2 & 7 & 3 \\ 5 & 3 & 2 & -2 \\ 9 & 11 & 5 & 10 \end{bmatrix}$$ Deleting recessive rows and columns we obtain

$$A = \begin{bmatrix} 1 & 7 \\ 9 & 5 \end{bmatrix}$$

Problem 1. Minimize $z = x_1 + x_2$ subject to

$$x_1 + 9x_2 \geq 1$$
$$7x_1 + 5x_2 \geq 1$$
$$x_1 \geq 0,\ x_2 \geq 0$$

Problem 2. Maximize $Z = y_1 + y_2$ subject to

$$y_1 + 7y_2 \leq 1$$
$$9y_1 + 5y_2 \leq 1$$
$$y_1 \geq 0,\ y_2 \geq 0$$

We solve the problems geometrically.

First problem

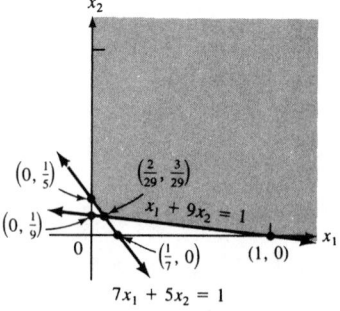

Z attains a minimum $\frac{5}{29}$
where $x_1 = \frac{2}{29}, x_2 = \frac{3}{29}$

Figure 1

Second problem

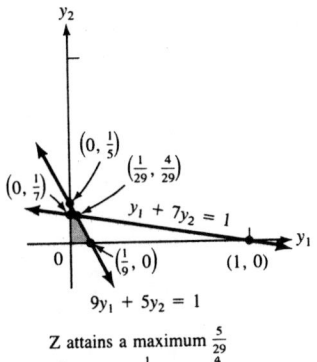

$y_1 + 7y_2 = 1$

$9y_1 + 5y_2 = 1$

Z attains a maximum $\frac{5}{29}$

where $y_1 = \frac{1}{29}$, $y_2 = \frac{4}{29}$

Figure 2

The region of Figure 1 is unbounded, At $(0, \frac{1}{5})$, $z = 0 + \frac{1}{5} = \frac{1}{5}$, at $(\frac{2}{29}, \frac{3}{29})$,

$z = \frac{2}{29} + \frac{3}{29} = \frac{5}{29}$ and at $(1, 0)$, $z = 1 + 0 = 1$. The minimum is $\frac{5}{29}$. Any line $x_1 + x_2 = c$

with $c < \frac{5}{29}$ does not intersect the region. The minimum occurs at $x_1 = \frac{2}{29}$, $x_2 = \frac{3}{29}$.

Also, $v = \frac{1}{z} = \frac{29}{5}$. $p_1{}^* = x_1 v = \frac{2}{29}(\frac{29}{5}) = \frac{2}{5}$, $p_2{}^* = x_2 v = (\frac{3}{29})(\frac{29}{5}) = \frac{3}{5}$.

$p^* = \begin{bmatrix} \frac{2}{5} & \frac{3}{5} \end{bmatrix}$.

The region of Figure 2 is bounded. At $(0,0)$, $Z = 0$, at $(0, \frac{1}{7})$, $Z = \frac{1}{7}$, at $(\frac{1}{29}, \frac{4}{29})$, $Z = \frac{5}{29}$,

and at $(\frac{1}{9}, 0)$, $Z = \frac{1}{9}$. The maximum is $\frac{5}{29}$. It occurs at $y_1 = \frac{1}{29}$, $y_2 = \frac{4}{29}$.

So, $q_1{}^* = y_1 v = (\frac{1}{29})(\frac{29}{5}) = \frac{1}{5}$, $q_2{}^* = (\frac{4}{29})(\frac{29}{5}) = \frac{4}{5}$ and $q^* = \begin{bmatrix} \frac{1}{5} \\ \frac{4}{5} \end{bmatrix}$.

The value of the original game is $\frac{29}{5}$. In terms of the original matrix,

$p^* = \begin{bmatrix} 0 & \frac{2}{5} & 0 & \frac{3}{5} \end{bmatrix}$ and

$$q^* = \begin{bmatrix} \frac{1}{5} \\ 0 \\ \frac{4}{5} \\ 0 \end{bmatrix}.$$

7. In Exercise 19, Section 7.2, we obtained the matrix

$$A = \begin{bmatrix} 2 & -3 \\ -3 & 4 \end{bmatrix} \qquad \text{Let } k = 4 \text{ to get} \qquad B = \begin{bmatrix} 6 & 1 \\ 1 & 8 \end{bmatrix}$$

The corresponding linear programming problems are:

Problem 1. Minimize $z = x_1 + x_2$ subject to

$$6x_1 + x_2 \geq 1$$
$$x_1 + 8x_2 \geq 1$$
$$x_1 \geq 0, \ x_2 \geq 0$$

Problem 2. Maximize $Z = y_1 + y_2$ subject to

$$6y_1 + y_2 \leq 1$$
$$y_1 + 8y_2 \leq 1$$
$$y_1 \geq 0, \ y_2 \geq 0$$

Problem 2 is the dual of problem 1. So, we let x_1 and x_2 be slack variables for problem 2 and get

Maximize $Z = y_1 + y_2$ subject to

$$6y_1 + \ y_2 + x_1 \qquad = 1$$
$$y_1 + 8y_2 \qquad + x_2 = 1$$
$$z - y_1 - \ y_2 \qquad = 0$$

The initial tableau is

Tableau 1

Basis	y_1	y_2	x_1	x_2	Values of basic variables
x_1	⑥	1	1	0	1
x_2	1	8	0	1	1
Z	-1	-1	0	0	0

Pivoting, we get

Tableau 2

Basis	y_1	y_2	x_1	x_2	Values of basic variables
y_1	1	$\frac{1}{6}$	$\frac{1}{6}$	0	$\frac{1}{6}$
x_2	0	$\boxed{\frac{47}{6}}$	$\frac{-1}{6}$	1	$\frac{5}{6}$
Z	0	$\frac{-5}{6}$	$\frac{1}{6}$	0	$\frac{1}{6}$

$$\frac{\frac{5}{6}}{\frac{47}{6}} < \frac{\frac{1}{6}}{\frac{1}{6}}, \text{ so } \frac{47}{6} \text{ is the pivot.}$$

Pivoting, we obtain the 3rd tableau.

Basis	y_1	y_2	x_1	x_2	Values of basic variables
			Tableau 3		
y_1	1	0	$\frac{8}{47}$	$\frac{-1}{47}$	$\frac{7}{47}$
y_2	0	1	$\frac{-1}{47}$	$\frac{6}{47}$	$\frac{5}{47}$
Z	0	0	$\frac{7}{47}$	$\frac{5}{47}$	$\frac{12}{47}$

There are no negative indicators in the bottom row. So, the maximum value of Z (and the minimum value of z) is $\frac{12}{47}$. It occurs at $y_1 = \frac{7}{47}$, $y_2 = \frac{5}{47}$, $x_1 = \frac{7}{47}$, $x_2 = \frac{5}{47}$.

Therefore, $v = \frac{1}{z} = \frac{47}{12}$,

$p_1{}^* = x_1 v = \frac{7}{47}(\frac{47}{12}) = \frac{7}{12}$, $p_2{}^* = x_2 v = (\frac{5}{47})(\frac{47}{12}) = \frac{5}{12}$ and $p^* = \begin{bmatrix} \frac{7}{12} & \frac{5}{12} \end{bmatrix}$.

$q_1{}^* = y_1 v = \frac{7}{47}(\frac{47}{12}) = \frac{7}{12}$, $q_2{}^* = x_2 v = (\frac{5}{47})(\frac{47}{12}) = \frac{5}{12}$ and $q^* = \begin{bmatrix} \frac{7}{12} \\ \frac{5}{12} \end{bmatrix}$.

The value of the game is $v - k = v - 4 = \frac{-1}{12}$.

9. Do Exercise 21, Section 7.2.

Parts a, b, c are as in Exercise 21, Exercise set 7.2. We use linear programming for part (d). As in Exercise 21, we delete recessive rows and columns to get

$$A = \begin{bmatrix} 2 & 1 \\ -6 & 4 \end{bmatrix} \quad \text{Let } k = 7 \text{ and obtain} \quad \begin{bmatrix} 9 & 8 \\ 1 & 11 \end{bmatrix}$$

The corresponding linear programming problems are:

Problem 1. Minimize $z = x_1 + x_2$ subject to

$$9x_1 + x_2 \geq 1$$
$$8x_1 + 11x_2 \geq 1$$
$$x_1 \geq 0, \ x_2 \geq 0$$

Problem 2 (dual): Maximize $Z = y_1 + y_2$ subject to

$$9y_1 + 8y_2 \leq 1$$
$$y_1 + 11y_2 \leq 1$$
$$y_1 \geq 0, \ y_2 \geq 0$$

We solve problem 2, letting x_1, x_2 be slack variables.

$$\begin{cases} 9y_1 + 8y_2 + x_1 = 1 \\ y_1 + 11y_2 + x_2 = 1 \\ Z - y_1 - y_2 = 0 \end{cases}$$

The initial tableau is

Tableau 1

Basis	y_1	y_2	x_1	x_2	Values of basic variables
x_1	⑨	8	1	0	1
x_2	1	11	0	1	1
Z	-1	-1	0	0	0

We choose 9 as the pivot. Pivoting, we get

Tableau 2

Basis	y_1	y_2	x_1	x_2	Values of basic variables
y_1	1	$\frac{8}{9}$	$\frac{1}{9}$	0	$\frac{1}{9}$
x_2	0	$\left(\frac{91}{9}\right)$	$\frac{-1}{9}$	1	$\frac{8}{9}$
Z	0	$\frac{-1}{9}$	$\frac{1}{9}$	0	$\frac{1}{9}$

$\dfrac{\frac{8}{9}}{\frac{91}{9}} < \dfrac{\frac{1}{9}}{\frac{8}{9}}$. So, $\frac{91}{9}$ is the pivot. Pivoting, we get

	Tableau 3				
Basis	y_1	y_2	x_1	x_2	Values of basic variables
y_1	1	0	$\frac{11}{91}$	$\frac{-8}{91}$	$\frac{3}{91}$
y_2	0	1	$\frac{-1}{91}$	$\frac{9}{91}$	$\frac{8}{91}$
Z	0	0	$\frac{10}{91}$	$\frac{1}{91}$	$\frac{11}{91}$

There are no negative indicators in the bottom row. So, the maximum value of Z (and the minimum value of z) is $\frac{11}{91}$. It occurs at $y_1 = \frac{3}{91}$, $y_2 = \frac{8}{91}$, $x_1 = \frac{10}{91}$, $x_2 = \frac{1}{91}$.

Also, $v = \frac{1}{z} = \frac{91}{11}$.

$p_1{}^* = x_1 v = \frac{10}{91} \cdot \frac{91}{11} = \frac{10}{11}$, $p_2{}^* = \frac{1}{91} \cdot \frac{91}{11} = \frac{1}{11}$ and $p^* = \begin{bmatrix} \frac{10}{11} & \frac{1}{11} \end{bmatrix}$.

$q_1{}^* = y_1 v = \frac{3}{91} \cdot \frac{91}{11} = \frac{3}{11}$, $q_2{}^* = y_2 v = \frac{8}{91} \cdot \frac{91}{11} = \frac{8}{11}$ and $q^* = \begin{bmatrix} \frac{3}{11} \\ \frac{8}{11} \end{bmatrix}$.

The value of the game is $\frac{91}{11} - 7 = \frac{14}{11}$. With respect to the original matrix,

$$p^* = \begin{bmatrix} 0 & \frac{10}{11} & \frac{1}{11} \end{bmatrix} \text{ and } q^* = \begin{bmatrix} 0 \\ \frac{3}{11} \\ 0 \\ \frac{8}{11} \end{bmatrix}$$

11. Exercise 23, Section 7.2. We need only do part (a) using linear programming. Deleting recessive rows and columns, we get

$$\begin{bmatrix} 4 & -6 \\ 3 & 7 \end{bmatrix}. \quad \text{Let } k = 7, \ B = \begin{bmatrix} 11 & 1 \\ 10 & 14 \end{bmatrix}$$

The corresponding linear programming problems are:

Problem 1. Minimize $z = x_1 + x_2$ subject to

$$11x_1 + 10x_2 \geq 1$$
$$x_1 + 14x_2 \geq 1$$
$$x_1 \geq 0, \ x_2 \geq 0$$

Problem 2: Maximize $Z = y_1 + y_2$ subject to

$$11y_1 + y_2 \leq 1$$
$$10y_1 + 14y_2 \leq 1$$
$$y_1 \geq 0, \ y_2 \geq 0$$

We solve the second problem (the dual of the first). We use x_1, x_2 as slack variables and obtain

$$\begin{cases} 11y_1 + y_2 + x_1 & = 1 \\ 10y_1 + 14y_2 & + x_2 = 1 \\ Z - y_1 - y_2 & = 0 \end{cases}$$

We get the first tableau

Tableau 1

Basis	y_1	y_2	x_1	x_2	Values of basic variables
x_1	11	1	1	0	1
x_2	10	14	0	1	1
Z	-1	-1	0	0	0

$\uparrow$

We choose 14 as the pivot element. Pivoting, we get

Tableau 2

Basis	y_1	y_2	x_1	x_2	Values of basic variables
x_1	$\left(\dfrac{72}{7}\right)$	0	1	$\dfrac{-1}{14}$	$\dfrac{13}{14}$
y_2	$\dfrac{5}{7}$	1	0	$\dfrac{1}{14}$	$\dfrac{1}{14}$
Z	$\dfrac{-2}{7}$	0	0	$\dfrac{1}{14}$	$\dfrac{1}{14}$

↑

$\dfrac{\frac{13}{14}}{\frac{72}{7}} < \dfrac{\frac{1}{14}}{\frac{5}{7}}$. So, $\dfrac{72}{7}$ is the pivot element. Pivoting we get

Tableau 3

Basis	y_1	y_2	x_1	x_2	Values of basic variables
y_1	1	0	$\dfrac{7}{72}$	$\dfrac{-1}{144}$	$\dfrac{13}{144}$
y_2	0	1	$\dfrac{-5}{12}$	$\dfrac{77}{1008}$	$\dfrac{1}{144}$
Z	0	0	$\dfrac{1}{36}$	$\dfrac{5}{72}$	$\dfrac{7}{72}$

There are no negative indicators in the bottom row. So, the maximum value of Z (and the minimum value of z) is $\dfrac{7}{72}$. It occurs at $y_1 = \dfrac{13}{144}$, $y_2 = \dfrac{1}{144}$, $x_1 = \dfrac{1}{36}$, $x_2 = \dfrac{5}{72}$. Also, $v = \dfrac{1}{z} = \dfrac{72}{7}$.

$p_1{}^* = x_1 v = \dfrac{1}{36}\left(\dfrac{72}{7}\right) = \dfrac{2}{7}$, $p_2{}^* = x_2 v = \dfrac{5}{72}\left(\dfrac{72}{7}\right) = \dfrac{5}{7}$ and $p^* = \begin{bmatrix} \dfrac{2}{7} & \dfrac{5}{7} \end{bmatrix}$.

$q_1{}^* = y_1 v = \dfrac{13}{144} \cdot \dfrac{72}{7} = \dfrac{13}{14}$, $q_2{}^* = y_2 v = \dfrac{1}{144}\left(\dfrac{72}{7}\right) = \dfrac{1}{14}$ and $q^* = \begin{bmatrix} \dfrac{13}{14} \\ \dfrac{1}{14} \end{bmatrix}$.

The value of the game is $\dfrac{72}{7} - 7 = \dfrac{23}{7}$. With respect to the original matrix,

$$p^* = \begin{bmatrix} 0 & \dfrac{2}{7} & 0 & \dfrac{5}{7} \end{bmatrix} \text{ and } q^* = \begin{bmatrix} \dfrac{13}{14} \\ 0 \\ \dfrac{1}{14} \\ 0 \end{bmatrix}.$$

13. Exercise 25, Exercise set 7.2.

$$\begin{array}{c} \\ X \\ Y \end{array} \begin{array}{cc} Z & W \\ \left[\begin{array}{cc} 15 & -18 \\ 10 & 13 \end{array} \right] \end{array}$$

Let $k = 19$.

$$B = \left[\begin{array}{cc} 34 & 1 \\ 29 & 32 \end{array} \right]$$

The corresponding linear programming problems are

Problem 1.   Minimize $z = x_1 + x_2$ subject to
$$34x_1 + 29x_2 \geq 1$$
$$x_1 + 32x_2 \geq 1$$
$$x_1 \geq 0, \; x_2 \geq 0$$

Problem 2:   Maximize $Z = y_1 + y_2$ subject to

$$34y_1 + y_2 \leq 1$$
$$29y_1 + 32y_2 \leq 1$$
$$y_1 \geq 0, \; y_2 \geq 0$$

We solve the second problem (the dual of the first) using x_1, x_2 as slack variables.

$$\begin{cases} 34y_1 + \; y_2 + x_1 \qquad = 1 \\ 29y_1 + 32y_2 \qquad + x_2 = 1 \\ Z - y_1 - \; y_2 \qquad \qquad = 0 \end{cases}$$

The first tableau is

Tableau 1

Basis	y_1	y_2	x_1	x_2	Values of basic variables
x_1	(34)	1	1	0	1
x_2	29	32	0	1	1
Z	−1	−1	0	0	0

We choose 34 as the pivot.   Pivoting, we get

Tableau 2

Basis	y_1	y_2	x_1	x_2	Values of basic variables
y_1	1	$\dfrac{1}{34}$	$\dfrac{1}{34}$	0	$\dfrac{1}{34}$
x_2	0	$\left(\dfrac{1059}{34}\right)$	$\dfrac{-29}{34}$	1	$\dfrac{5}{34}$
Z	0	$\dfrac{-33}{34}$	$\dfrac{1}{34}$	0	$\dfrac{1}{34}$

$$\frac{\frac{5}{34}}{\frac{1059}{34}} < \frac{\frac{1}{34}}{\frac{1}{34}}.$$ So, $\frac{1059}{34}$ is the pivot. Pivoting, we get

Tableau 3

Basis	y_1	y_2	x_1	x_2	Values of basic variables
y_1	1	0	$\frac{32}{1059}$	$\frac{-1}{1059}$	$\frac{31}{1059}$
y_2	0	1	$\frac{-29}{1059}$	$\frac{34}{1059}$	$\frac{5}{1059}$
Z	0	0	$\frac{3}{1059}$	$\frac{33}{1059}$	$\frac{36}{1059}$

There are no negative indicators in the bottom row. So, the maximum value of Z (and the minimum value of z) is $\frac{36}{1059} = \frac{12}{353}$. It occurs at $x_1 = \frac{3}{1059} = \frac{1}{353}$, $x_2 = \frac{33}{1059} = \frac{11}{353}$, $y_1 = \frac{31}{1059}$, $y_2 = \frac{5}{1059}$. So,

$$v = \frac{1}{z} = \frac{353}{12}, \quad p_1{}^* = x_1 v = \frac{1}{353}\left(\frac{353}{12}\right) = \frac{1}{12}, \quad p_2{}^* = x_2 v = \frac{11}{353}\left(\frac{353}{12}\right) = \frac{11}{12} \text{ and } p^* = \begin{bmatrix} \frac{1}{12} & \frac{11}{12} \end{bmatrix}.$$

Also, $q_1{}^* = q_1 v = \frac{31}{1059}\left(\frac{353}{12}\right) = \frac{31}{36}$, $q_2{}^* = \frac{5}{1059}\left(\frac{353}{12}\right) = \frac{5}{36}$ and $q^* = \begin{bmatrix} \frac{31}{36} \\ \frac{5}{36} \end{bmatrix}$.

The value of the game is $\frac{353}{12} - 19 = \frac{125}{12}$.

15. If a is positive, then $a + k$ is positive. If a is negative, note that $e \leq a$. So, e is negative also and $|e| = -e$. Therefore, $k = |e| + 1 = -e + 1$ and $a + k = a - e + 1$. But, $e \leq a$ implies $0 \leq a - e$ and $1 \leq a - e + 1$. So $a + k$ is positive. Similarly, for $b + k$, $c + k$, and $d + k$.

17. If $PBQ \leq v$ for any choice of strategy P by R, then the inequality holds for $P = P_1$ and $P = P_2$. Conversely, suppose $P_1 BQ \leq v$ and $P_2 BQ \leq v$ and let $P = [a_1 \ a_2]$ be any strategy for R.

Then, $\quad P = a_1 P_1 + a_2 P_2$

So, $\quad PB = (a_1 P_1 + a_2 P_2)B$
$\quad\quad\quad = a_1 PB + a_2 P_2 B$

and $\quad PBQ = (a_1 P_1 B + a_2 P_2 B)Q$
$\quad\quad\quad\quad = a_1 P_1 BQ + a_2 P_2 BQ$
$\quad\quad\quad\quad \leq a_1 v + a_2 v = (a_1 + a_2)v = 1(v) = v.$

19. $[p_1 \ p_2 \ ... p_m]K = A$ is a 1 by n vector. Each entry of that vector is $p_1 k + p_2 k + ... + p_m k = k(p_1 + p_2 + ... + p_m) = k(1) = k$.

Now, $A\begin{bmatrix} q_1 \\ q_2 \\ \vdots \\ q_n \end{bmatrix}$ is 1×1 and its only entry is

$kq_1 + kq_2 + ... + kq_n = k(q_1 + q_2 + ... + q_n) = k(1) = k.$

1.

$$A = \begin{bmatrix} 0 & -1 & 3 \\ -1 & 1 & -2 \end{bmatrix}. \quad \text{Let } k = 3. \quad B = \begin{bmatrix} 3 & 2 & 6 \\ 2 & 4 & 1 \end{bmatrix}$$

The corresponding linear programming problems are:

Problem 1.   Minimize $z = x_1 + x_2$ subject to

$$3x_1 + 2x_2 \geq 1$$
$$2x_1 + 4x_2 \geq 1$$
$$6x_1 + x_2 \geq 1$$
$$x_1 \geq 0,\ x_2 \geq 0$$

Problem 2:   Maximize $Z = y_1 + y_2 + y_3$ subject to

$$3y_1 + 2y_2 + 6y_3 \leq 1$$
$$2y_1 + 4y_2 + \quad y_3 \leq 1$$
$$y_1 \geq 0,\ y_2 \geq 0,\ y_3 \geq 0$$

We solve the second problem, using x_1, x_2 as slack variables.

$$\begin{cases} 3y_1 + 2y_2 + 6y_3 + x_1 \quad\ = 1 \\ 2y_1 + 4y_2 + \ y_3 \quad\ + x_2 = 1 \\ Z - y_1 - \ y_2 - y_3 \quad\quad = 0 \end{cases}$$

The first tableau is

Tableau 1

Basis	y_1	y_2	y_3	x_1	x_2	Values of basic variables
x_1	3	2	6	1	0	1
x_2	2	4	1	0	1	1
Z	-1	-1	-1	0	0	0

We choose column 2 as the pivot column and row 2 as the pivot row.   Pivoting, we get

Tableau 2

Basis	y_1	y_2	y_3	x_1	x_2	Values of basic variables
x_1	2	0	$\frac{11}{2}$	1	$\frac{-1}{2}$	$\frac{1}{2}$
y_2	$\frac{1}{2}$	1	$\frac{1}{4}$	0	$\frac{1}{4}$	$\frac{1}{4}$
Z	$\frac{-1}{2}$	0	$\frac{-3}{4}$	0	$\frac{1}{4}$	$\frac{1}{4}$

$\frac{11}{2}$ is the pivot.   Pivoting we get

Tableau 3

Basis	y_1	y_2	y_3	x_1	x_2	Values of basic variables
y_3	$\boxed{\frac{4}{11}}$	0	1	$\frac{2}{11}$	$\frac{-1}{11}$	$\frac{1}{11}$
y_2	$\frac{9}{22}$	1	0	$\frac{-1}{22}$	$\frac{3}{11}$	$\frac{5}{22}$
Z	$\frac{-5}{22}$	0	0	$\frac{3}{22}$	$\frac{2}{11}$	$\frac{7}{22}$

$\frac{4}{11}$ is the pivot. Pivoting we get

Tableau 4

Basis	y_1	y_2	y_3	x_1	x_2	Values of basic variables
y_1	1	0	$\frac{11}{4}$	$\frac{1}{2}$	$\frac{-1}{4}$	$\frac{1}{4}$
y_2	0	1	$\frac{-9}{8}$	$\frac{-1}{4}$	$\frac{3}{8}$	$\frac{1}{8}$
Z	0	0	$\frac{5}{8}$	$\frac{1}{4}$	$\frac{1}{8}$	$\frac{3}{8}$

There are no negative indicators in the bottom row. So, the maximum value of Z (and the minimum value of z) is $\frac{3}{8}$. It occurs at $x_1 = \frac{1}{4}$, $x_2 = \frac{1}{8}$, $y_1 = \frac{1}{4}$, $y_2 = \frac{1}{8}$, $y_3 = 0$.

So, $v = \frac{1}{z} = \frac{1}{3/8} = \frac{8}{3}$.

$$p^* = \begin{bmatrix} \frac{8}{3}(\frac{1}{4}) & \frac{8}{3}(\frac{1}{8}) \end{bmatrix} = \begin{bmatrix} \frac{2}{3} & \frac{1}{3} \end{bmatrix}.$$

$$q^* = \begin{bmatrix} \frac{8}{3}(\frac{1}{4}) \\ \frac{8}{3}(\frac{1}{8}) \\ \frac{8}{3}(0) \end{bmatrix} = \begin{bmatrix} \frac{2}{3} \\ \frac{1}{3} \\ 0 \end{bmatrix}.$$

The value of the game is $\frac{8}{3} - 3 = \frac{-1}{3}$.

3.

$$A = \begin{bmatrix} -1 & 2 & -1 \\ -1 & -2 & 0 \\ 0 & -1 & -1 \end{bmatrix} \quad \text{Let } k = 3. \quad B = \begin{bmatrix} 2 & 5 & 2 \\ 2 & 1 & 3 \\ 3 & 2 & 2 \end{bmatrix}$$

Problem 1.   Minimize $z = x_1 + x_2 + x_3$ subject to

$$2x_1 + 2x_2 + 3x_3 \geq 1$$
$$5x_1 + x_2 + 2x_3 \geq 1$$
$$2x_1 + 3x_2 + 2x_3 \geq 1$$
$$x_1 \geq 0, \; x_2 \geq 0, x_3 \geq 0$$

Problem 2: Maximize $Z = y_1 + y_2 + y_3$ subject to

$$2y_1 + 5y_2 + 2y_3 \leq 1$$
$$2y_1 + y_2 + 3y_3 \leq 1$$
$$3y_1 + 2y_2 + 2y_3 \leq 1$$
$$y_1 \geq 0,\ y_2 \geq 0,\ y_3 \geq 0$$

We solve the second problem, using $x_1,\ x_2,\ x_3$ as slack variables. We have

$$2y_1 + 5y_2 + 2y_3 + x_1 \qquad\qquad = 1$$
$$2y_1 + y_2 + 3y_3 \qquad + x_2 \qquad = 1$$
$$3y_1 + 2y_2 + 2y_3 \qquad\qquad + x_3 = 1$$
$$Z - y_1 - y_2 - y_3 \qquad\qquad\qquad = 0$$

The first tableau is

Tableau 1

Basis	y_1	y_2	y_3	x_1	x_2	x_3	Values of basic variables
x_1	2	(5)	2	1	0	0	1
x_2	2	1	3	0	1	0	1
x_3	3	2	2	0	0	1	1
Z	-1	-1	-1	0	0	0	0

We choose 5 as the pivot. Pivoting, we get

Tableau 2

Basis	y_1	y_2	y_3	x_1	x_2	x_3	Values of basic variables
y_2	$\frac{2}{5}$	1	$\frac{2}{5}$	$\frac{1}{5}$	0	0	$\frac{1}{5}$
x_2	$\frac{8}{5}$	0	$\left(\frac{13}{5}\right)$	$\frac{-1}{5}$	1	0	$\frac{4}{5}$
x_3	$\frac{11}{5}$	0	$\frac{6}{5}$	$\frac{-2}{5}$	0	1	$\frac{3}{5}$
Z	$\frac{-3}{5}$	0	$\frac{-3}{5}$	$\frac{1}{5}$	0	0	$\frac{1}{5}$

$\frac{13}{5}$ is the pivot. Pivoting we get

Tableau 3

Basis	y_1	y_2	y_3	x_1	x_2	x_3	Values of basic variables
y_2	$\frac{2}{13}$	1	0	$\frac{3}{13}$	$\frac{-2}{13}$	0	$\frac{1}{13}$
y_3	$\frac{8}{13}$	0	1	$\frac{-1}{13}$	$\frac{5}{13}$	0	$\frac{4}{13}$
x_3	$\left(\frac{19}{13}\right)$	0	0	$\frac{-4}{13}$	$\frac{-6}{13}$	1	$\frac{3}{13}$
Z	$\frac{-3}{13}$	0	0	$\frac{2}{13}$	$\frac{3}{13}$	0	$\frac{5}{13}$

$\frac{19}{13}$ is the pivot. Pivoting we get

Tableau 4

Basis	y_1	y_2	y_3	x_1	x_2	x_3	Values of basic variables
y_2	0	1	0	$\frac{5}{19}$	$\frac{-2}{19}$	$\frac{-2}{19}$	$\frac{1}{19}$
y_3	0	0	1	$\frac{1}{19}$	$\frac{11}{19}$	$\frac{-8}{19}$	$\frac{4}{19}$
y_1	1	0	0	$\frac{-4}{19}$	$\frac{-6}{19}$	$\frac{13}{19}$	$\frac{3}{19}$
Z	0	0	0	$\frac{2}{19}$	$\frac{3}{19}$	$\frac{3}{19}$	$\frac{8}{19}$

There are no negative indicators in the bottom row. Thus, the maximum value of Z has been reached. It is $\frac{8}{19}$. It occurs when $x_1 = \frac{2}{19}$, $x_2 = \frac{3}{19}$, $x_3 = \frac{3}{19}$, $y_1 = \frac{3}{19}$. $y_2 = \frac{1}{19}$, $y_3 = \frac{4}{19}$. Thus, $v = \frac{1}{z} = \frac{19}{8}$ and

$$p^* = \left[\; \frac{2}{19}(\frac{19}{8}) \quad \frac{3}{19}(\frac{19}{8}) \quad \frac{3}{19}(\frac{19}{8}) \;\right] = \left[\; \frac{1}{4} \quad \frac{3}{8} \quad \frac{3}{8} \;\right] \quad \text{and}$$

$$q^* = \begin{bmatrix} \frac{3}{19}(\frac{19}{8}) \\ \frac{1}{19}(\frac{19}{8}) \\ \frac{4}{19}(\frac{19}{8}) \end{bmatrix} = \begin{bmatrix} \frac{3}{8} \\ \frac{1}{8} \\ \frac{1}{2} \end{bmatrix}.$$

The value of the game is $v - k = \frac{19}{8} - 3 = \frac{-5}{8}$.

5.

$$A = \begin{bmatrix} -1 & 3 & 0 \\ 0 & 1 & -2 \\ 1 & 4 & -3 \end{bmatrix}$$

Column 2 is recessive. We delete it and get

$$\begin{bmatrix} -1 & 0 \\ 0 & -2 \\ 1 & -3 \end{bmatrix} \qquad \text{Let } k = 4, \text{ so that} \quad B = \begin{bmatrix} 3 & 4 \\ 4 & 2 \\ 5 & 1 \end{bmatrix}$$

So, we have

Problem 1. Minimize $z = x_1 + x_2 + x_3$ subject to

$$3x_1 + 4x_2 + 5x_3 \geq 1$$
$$4x_1 + 2x_2 + x_3 \geq 1$$
$$x_1 \geq 0,\ x_2 \geq 0,\ x_3 \geq 0$$

Problem 2: Maximize $Z = y_1 + y_2$ subject to

$$3y_1 + 4y_2 \leq 1$$
$$4y_1 + 2y_2 \leq 1$$
$$5y_1 + y_2 \leq 1$$
$$y_1 \geq 0,\ y_2 \geq 0$$

We solve problem 2, using x_1, x_2, x_3 as slack variables.

$$3y_1 + 4y_2 + x_1 = 1$$
$$4y_1 + 2y_2 + x_2 = 1$$
$$5y_1 + y_2 + x_3 = 1$$
$$Z - y_1 - y_2 = 0$$

The first tableau is

Tableau 1

Basis	y_1	y_2	x_1	x_2	x_3	Values of basic variables
x_1	3	4	1	0	0	1
x_2	4	2	0	1	0	1
x_3	⑤	1	0	0	1	1
Z	−1	−1	0	0	0	0

We choose 5 as our first pivot. Pivoting, we get

Tableau 2

Basis	y_1	y_2	x_1	x_2	x_3	Values of basic variables
x_1	0	$\frac{17}{5}$	1	0	$\frac{-3}{5}$	$\frac{2}{5}$
x_2	0	$\frac{6}{5}$	0	1	$\frac{-4}{5}$	$\frac{1}{5}$
y_2	1	$\frac{1}{5}$	0	0	$\frac{1}{5}$	$\frac{1}{5}$
Z	0	$\frac{-4}{5}$	0	0	$\frac{1}{5}$	$\frac{1}{5}$

$\frac{17}{5}$ is our next pivot. Pivoting we get

Tableau 3

Basis	y_1	y_2	x_1	x_2	x_3	Values of basic variables
y_2	0	1	$\frac{5}{17}$	0	$\frac{-3}{17}$	$\frac{2}{17}$
x_2	0	0	$\frac{-6}{17}$	1	$\frac{-10}{17}$	$\frac{1}{17}$
y_1	1	0	$\frac{-1}{17}$	0	$\frac{4}{17}$	$\frac{3}{17}$
Z	0	0	$\frac{4}{17}$	0	$\frac{1}{17}$	$\frac{5}{17}$

There are no negative indicators in the bottom row. The maximum value of Z (and the minimum value of z) has been attained. It is $\frac{5}{17}$ and occurs when $x_1 = \frac{4}{17}$, $x_2 = 0$, $x_3 = \frac{1}{17}$, $y_1 = \frac{3}{17}$, $y_2 = \frac{2}{17}$. $v = \frac{1}{z} = \frac{17}{5}$.

$$p^* = \left[\frac{4}{17}(\frac{17}{5}) \quad 0(\frac{17}{5}) \quad \frac{1}{17}(\frac{17}{5}) \right] = \left[\frac{4}{5} \quad 0 \quad \frac{1}{5} \right].$$

$$q^* = \begin{bmatrix} (\frac{3}{17})(\frac{17}{5}) \\ (\frac{2}{17})(\frac{17}{5}) \end{bmatrix} = \begin{bmatrix} \frac{3}{5} \\ \frac{2}{5} \end{bmatrix}.$$

The value of the game is $v - k = \frac{17}{5} - 4 = \frac{-3}{5}$. In terms of the original matrix, the optimal strategies are

$$p^* = \begin{bmatrix} \frac{4}{5} & 0 & \frac{1}{5} \end{bmatrix} \text{ and } q^* = \begin{bmatrix} \frac{3}{5} \\ 0 \\ \frac{2}{5} \end{bmatrix}$$

7.
$$A = \begin{bmatrix} 2 & 3 & 0 & 7 \\ -1 & 5 & -2 & 5 \\ 0 & 2 & 3 & -2 \end{bmatrix}$$

Delete column 2 because it is recessive

$$\begin{bmatrix} 2 & 0 & 7 \\ -1 & -2 & 5 \\ 0 & 3 & -2 \end{bmatrix} \quad \text{Now row 2 is recessive.} \quad \text{Delete it} \quad \begin{bmatrix} 2 & 0 & 7 \\ 0 & 3 & -2 \end{bmatrix}$$

Let $k = 3$. We get $\begin{bmatrix} 5 & 3 & 10 \\ 3 & 6 & 1 \end{bmatrix}$

We get the following:

Problem 1. Minimize $z = x_1 + x_2$ subject to

$$5x_1 + 3x_2 \geq 1$$
$$3x_1 + 6x_2 \geq 1$$
$$10x_1 + x_2 \geq 0$$
$$x_1 \geq 0, \; x_2 \geq 0$$

Problem 2: Maximize $Z = y_1 + y_2 + y_3$ subject to

$$5y_1 + 3y_2 + 10y_3 \leq 1$$
$$3y_1 + 6y_2 + y_3 \leq 1$$
$$y_1 \geq 0, \; y_2 \geq 0, \; y_3 \geq 0$$

We solve the second problem, using x_1, x_2 as slack variables.

$$\begin{cases} 5y_1 + 3y_2 + 10y_3 + x_1 && = 1 \\ 3y_1 + 6y_2 + y_3 && + x_2 = 1 \\ Z - y_1 - y_2 - y_3 && = 0 \end{cases}$$

The first tableau is

Tableau 1

Basis	y_1	y_2	y_3	x_1	x_2	Values of basic variables
x_1	5	3	⑩	1	0	1
x_2	3	6	1	0	1	1
Z	-1	-1	-1	0	0	0

← (at x_1 row)

↑ (under y_3)

We choose 10 as our first pivot. Pivoting, we get

Tableau 2

Basis	y_1	y_2	y_3	x_1	x_2	Values of basic variables
y_3	$\frac{1}{2}$	$\frac{3}{10}$	1	$\frac{1}{10}$	0	$\frac{1}{10}$
x_2	$\frac{5}{2}$	$\boxed{\frac{57}{10}}$	0	$\frac{-1}{10}$	1	$\frac{9}{10}$
Z	$\frac{-1}{2}$	$\frac{-7}{10}$	0	$\frac{1}{10}$	0	$\frac{1}{10}$

← (at x_2 row)

↑ (under y_2)

$\frac{-7}{10}$ is the most negative indicator. Also $\frac{9}{10}/\frac{57}{10}$ is less that $\frac{1}{10}/\frac{3}{10}$. So, $\frac{57}{10}$ is the pivot.
Pivoting, we get

Tableau 3

Basis	y_1	y_2	y_3	x_1	x_2	Values of basic variables
y_3	$\boxed{\frac{7}{19}}$	0	1	$\frac{2}{19}$	$\frac{-1}{19}$	$\frac{1}{19}$
y_2	$\frac{25}{57}$	1	0	$\frac{-1}{57}$	$\frac{10}{57}$	$\frac{3}{19}$
Z	$\frac{-11}{57}$	0	0	$\frac{5}{57}$	$\frac{7}{57}$	$\frac{12}{57}$

← (at y_3 row)

↑ (under y_1)

$\frac{-11}{57}$ is the only negative indicator. Since $\frac{2}{57}/\frac{7}{19} < \frac{9}{57}/\frac{25}{57}$, $\frac{7}{19}$ is the pivot element.
Pivoting, we get

Tableau 4

Basis	y_1	y_2	y_3	x_1	x_2	Values of basic variables
y_1	1	0	$\frac{19}{7}$	$\frac{2}{7}$	$\frac{-1}{7}$	$\frac{1}{7}$
y_2	0	1	$\frac{-25}{21}$	$\frac{-1}{7}$	$\frac{5}{21}$	$\frac{2}{21}$
Z	0	0	$\frac{11}{21}$	$\frac{1}{7}$	$\frac{2}{21}$	$\frac{5}{21}$

There are no negative indicators. So, the maximum value of Z, and the minimum value
of z has been attained. It is $\frac{5}{21}$. It occurs when $y_1 = \frac{1}{7}$, $y_2 = \frac{2}{21}$, $y_3 = 0$, $x_1 = \frac{1}{7}$, $x_2 = \frac{2}{21}$.

So, $v = \frac{1}{z} = \frac{21}{5}$. Also, $p^* = \left[\begin{array}{cc} \frac{1}{7}(\frac{21}{5}) & \frac{2}{21}(\frac{21}{5}) \end{array}\right] = \left[\begin{array}{cc} \frac{3}{5} & \frac{2}{5} \end{array}\right]$ and $q^* = \left[\begin{array}{c} \frac{1}{7}(\frac{21}{5}) \\ \frac{2}{21}(\frac{21}{5}) \\ 0(\frac{21}{5}) \end{array}\right] = \left[\begin{array}{c} \frac{3}{5} \\ \frac{2}{5} \\ 0 \end{array}\right]$.

The value of the game is $v - k = \frac{21}{5} - 3 = \frac{6}{5}$. With respect to the original matrix,

$$p^* = \left[\begin{array}{ccc} \frac{3}{5} & 0 & \frac{2}{5} \end{array}\right] \text{ and } q^* = \left[\begin{array}{c} \frac{3}{5} \\ 0 \\ \frac{2}{5} \\ 0 \end{array}\right]$$

9. (a) Let Roberta be the row player, and Charlie the column player. Promises 1, 2, 3 by Roberta are represented by rows 1, 2, 3, respectively. Similarly, promises A, B, C by Charlie are represented by columns 1, 2, 3, respectively. The matrix is

$$A = \left[\begin{array}{ccc} 3 & 8 & -7 \\ -2 & 3 & 18 \\ 13 & -12 & 3 \end{array}\right]$$

(b) Let $k = 13$. We get

$$B = \left[\begin{array}{ccc} 16 & 21 & 6 \\ 11 & 16 & 31 \\ 26 & 1 & 16 \end{array}\right]. \text{ We get}$$

Problem 1. Minimize $z = x_1 + x_2 + x_3$ subject to
$$16x_1 + 11x_2 + 26x_3 \geq 1$$
$$21x_1 + 16x_2 + x_3 \geq 1$$
$$6x_1 + 31x_2 + 16x_3 \geq 1$$
$$x_1 \geq 0,\ x_2 \geq 0,\ x_3 \geq 0$$

Problem 2: Maximize $Z = y_1 + y_2 + y_3$ subject to

$$16y_1 + 21y_2 + 6y_3 \leq 1$$
$$11y_1 + 16y_2 + 31y_3 \leq 1$$
$$26y_1 + y_2 + 16y_3 \leq 1$$
$$y_1 \geq 0,\ y_2 \geq 0,\ y_3 \geq 0$$

We solve the second problem, using x_1, x_2, x_3 as slack variables.
$$16y_1 + 21y_2 + 6y_3 + x_1 = 1$$
$$11y_1 + 16y_2 + 31y_3 + x_2 = 1$$
$$26y_1 + y_2 + 16y_3 + x_3 = 1$$
$$Z - y_1 - y_2 - y_3 = 0$$

The first tableau is

Tableau 1

Basis	y_1	y_2	y_3	x_1	x_2	x_3	Values of basic variables
x_1	16	⟨21⟩	6	1	0	0	1
x_2	11	16	31	0	1	0	1
x_3	26	1	16	0	0	1	1
Z	-1	-1	-1	0	0	0	0

We choose 21 as our first pivot. Pivoting, we get

Tableau 2

Basis	y_1	y_2	y_3	x_1	x_2	x_3	Values of basic variables
y_2	$\frac{16}{21}$	1	$\frac{2}{7}$	$\frac{1}{21}$	0	0	$\frac{1}{21}$
x_2	$\frac{-25}{21}$	0	$\left(\frac{185}{7}\right)$	$\frac{-16}{21}$	1	0	$\frac{5}{21}$
x_3	$\frac{530}{21}$	0	$\frac{110}{7}$	$\frac{-1}{21}$	0	1	$\frac{20}{21}$
Z	$\frac{-5}{21}$	0	$\frac{-5}{7}$	$\frac{1}{21}$	0	0	$\frac{1}{21}$

$\frac{-5}{7}$ is the most negative indicator in the bottom row. $\frac{185}{7}$ must be the pivot.

Pivoting, we get

Tableau 3

Basis	y_1	y_2	y_3	x_1	x_2	x_3	Values of basic variables
y_2	$\frac{86}{111}$	1	0	$\frac{31}{555}$	$\frac{-2}{185}$	0	$\frac{5}{111}$
y_3	$\frac{-5}{111}$	0	1	$\frac{-16}{555}$	$\frac{7}{185}$	0	$\frac{1}{111}$
x_3	$\left(\frac{960}{37}\right)$	0	0	$\frac{45}{111}$	$\frac{-22}{37}$	1	$\frac{30}{37}$
Z	$\frac{-10}{37}$	0	0	$\frac{1}{37}$	$\frac{1}{37}$	0	$\frac{2}{37}$

The first column must be the pivot column. Since $\frac{30}{37}/\frac{960}{37} < \frac{5}{111}/\frac{86}{111}$, $\frac{960}{37}$ is the pivot. Pivoting, we get the next tableau.

Tableau 4

Basis	y_1	y_2	y_3	x_1	x_2	x_3	Values of basic variables
y_2	0	1	0	$\frac{7}{160}$	$\frac{1}{144}$	$\frac{-43}{1440}$	$\frac{1}{48}$
y_3	0	0	1	$\frac{-9}{320}$	$\frac{53}{1440}$	$\frac{1}{576}$	$\frac{1}{96}$
y_1	1	0	0	$\frac{1}{64}$	$\frac{-11}{480}$	$\frac{37}{960}$	$\frac{1}{32}$
Z	0	0	0	$\frac{1}{32}$	$\frac{1}{48}$	$\frac{1}{96}$	$\frac{1}{16}$

There are no negative indicators in the bottom row. Thus, the maximum value of Z (and minimum value of z) has been attained. It is $\frac{1}{16}$. It occurs when $y_1 = \frac{1}{32}$, $y_2 = \frac{1}{48}$, $y_3 = \frac{1}{96}$, $x_1 = \frac{1}{32}$, $x_2 = \frac{1}{48}$, $x_3 = \frac{1}{96}$. Further $v = \frac{1}{z} = 16$. Also,

$$p^* = \left[\ \frac{1}{32}(16)\quad \frac{1}{48}(16)\quad \frac{1}{96}(16)\ \right] = \left[\ \frac{1}{2}\quad \frac{1}{3}\quad \frac{1}{6}\ \right] \text{ and }$$

$$q^* = \begin{bmatrix} \frac{1}{32}(16) \\ \frac{1}{48}(16) \\ \frac{1}{96}(16) \end{bmatrix} = \begin{bmatrix} \frac{1}{2} \\ \frac{1}{3} \\ \frac{1}{6} \end{bmatrix}.$$

The value of the game is $v - k = 16 - 13 = 3$.

(c) If Roberta continues to use her optional strategy but Charlie consistantly chooses column 1, the value of the game will be

$$\begin{bmatrix} \frac{1}{2} & \frac{1}{3} & \frac{1}{6} \end{bmatrix} \begin{bmatrix} 3 & 8 & -7 \\ -2 & 3 & 18 \\ 13 & -12 & 3 \end{bmatrix} \begin{bmatrix} 1 \\ 0 \\ 0 \end{bmatrix} = \begin{bmatrix} 3 & 3 & 3 \end{bmatrix} \begin{bmatrix} 1 \\ 0 \\ 0 \end{bmatrix} = 3$$

Similarly, if Charlie consistantly chooses column 2, or column 3.

11. Let the real estate development company be player R and the state be player C. The shopping center, restaurants, and sports arena are represented by rows 1, 2, and 3, respectively. Column 1 represents build the highway, column 2 represents do not build the highway.

$$A = \begin{bmatrix} .63 & .03 \\ .48 & .33 \\ .18 & .48 \end{bmatrix}$$

Problem 1. Minimize $z = x_1 + x_2 + x_3$ subject to

$$\begin{cases} .63x_1 + .48x_2 + .18x_3 \geq 1 \\ .03x_1 + .33x_2 + .48x_3 \geq 1 \\ \qquad\qquad x_1 \geq 0,\ x_2 \geq 0,\ x_3 \geq 0 \end{cases}$$

Problem 2: Maximize $Z = y_1 + y_2$ subject to

$$\begin{aligned} .63y_1 + .03y_2 &\leq 1 \\ .48y_1 + .33y_2 &\leq 1 \\ .18y_1 + .48y_2 &\leq 1 \\ y_1 \geq 0,\ y_2 &\geq 0 \end{aligned}$$

We solve the second problem, using x_1, x_2, x_3 as slack variables.

$$\begin{cases} .63y_1 + .03y_2 + x_1 && = 1 \\ .48y_1 + .33y_2 & + x_2 & = 1 \\ .18y_1 + .48y_2 & + x_3 & = 1 \\ Z - y_1 - y_2 && = 0 \end{cases}$$

The first tableau is

Tableau 1

Basis	y_1	y_2	x_1	x_2	x_3	Values of basic variables
x_1	.63	.03	1	0	0	1
x_2	.48	.33	0	1	0	1
x_3	.18	(.48)	0	0	1	1
Z	-1	-1	0	0	0	0

We choose .48 to be our first pivot. Pivoting, we get

Tableau 2

Basis	y_1	y_2	x_1	x_2	x_3	Values of basic variables
x_1	$\frac{99}{160}$	0	1	0	$\frac{-3}{48}$	$\frac{15}{16}$
x_2	$\left(\frac{57}{160}\right)$	0	0	1	$\frac{-11}{16}$	$\frac{5}{16}$
y_2	$\frac{3}{8}$	1	0	0	$\frac{25}{12}$	$\frac{25}{12}$
Z	$\frac{-5}{8}$	0	0	0	$\frac{25}{12}$	$\frac{25}{12}$

We see that $\dfrac{57}{160}$ must be the pivot. Pivoting, we get

Tableau 3

Basis	y_1	y_2	x_1	x_2	x_3	Values of basic variables
x_1	0	0	1	$\frac{-33}{19}$	$\frac{43}{38}$	$\frac{15}{38}$
y_1	1	0	0	$\frac{160}{57}$	$\frac{-110}{57}$	$\frac{50}{57}$
y_2	0	1	0	$\frac{-20}{19}$	$\frac{160}{57}$	$\frac{100}{57}$
Z	0	0	0	$\frac{100}{57}$	$\frac{50}{57}$	$\frac{150}{57}$

There are no negative indicators. The maximum value of Z has been attained. It is $\frac{150}{57}$. It occurs when $x_1 = 0$, $x_2 = \frac{100}{57}$, $x_3 = \frac{50}{57}$, $y_1 = \frac{50}{57}$, $y_2 = \frac{100}{57}$. We have $v = \frac{1}{z} = \frac{57}{150}$

$= \frac{19}{50}$. So, $p^* = \begin{bmatrix} 0(\frac{57}{150}) & \frac{100}{57}(\frac{57}{150}) & \frac{50}{57}(\frac{57}{150}) \end{bmatrix} = \begin{bmatrix} 0 & \frac{2}{3} & \frac{1}{3} \end{bmatrix}$ and

$q^* = \begin{bmatrix} \frac{50}{57}(\frac{57}{150}) \\ \frac{100}{57}(\frac{57}{150}) \end{bmatrix} = \begin{bmatrix} \frac{1}{3} \\ \frac{2}{3} \end{bmatrix}.$

So, no part of the land should be used for the shopping center, $\frac{2}{3}$ should be used for the restaurants and $\frac{1}{3}$ should be used for the sports arena. The value of the game is $\frac{19}{50}$. So, the development company will get a 38% profit.

13. Let radio, television, and newspaper be represented by rows and columns 1, 2, and 3, respectively. The matrix is

$$A = \begin{bmatrix} 2 & -10 & 14 \\ -10 & 26 & 2 \\ 14 & -22 & -10 \end{bmatrix} \quad \text{Let } k = 23. \text{ Then } B = \begin{bmatrix} 25 & 13 & 37 \\ 13 & 49 & 25 \\ 37 & 1 & 13 \end{bmatrix}$$

The problems are

Problem 1.   Minimize $z = x_1 + x_2 + x_3$ subject to

$$25x_1 + 13x_2 + 37x_3 \geq 1$$
$$13x_1 + 49x_2 + \quad x_3 \geq 1$$
$$37x_1 + 25x_2 + 13x_3 \geq 1$$
$$x_1 \geq 0, \ x_2 \geq 0, \ x_3 \geq 0$$

Problem 2:   Maximize $Z = y_1 + y_2 + y_3$ subject to

$$25y_1 + 13y_2 + 37y_3 \leq 1$$
$$13y_1 + 49y_2 + 25y_3 \leq 1$$
$$37y_1 + \quad y_2 + 13y_3 \leq 1$$
$$y_1 \geq 0, \ y_2 \geq 0, \ y_3 \geq 0$$

We solve the second problem, using x_1, x_2, x_3 as slack variables.
$$25y_1 + \quad 13y_2 + 37y_3 + x_1 \qquad\qquad = 1$$
$$13y_1 + \quad 49y_2 + 25y_3 \qquad + x_2 \qquad = 1$$
$$37y_1 + \quad y_2 + 13y_3 \qquad\qquad + x_3 = 1$$
$$Z - y_1 - \quad y_2 \quad - y_3 \qquad\qquad\qquad = 0$$

The first tableau is

Tableau 1

Basis	y_1	y_2	y_3	x_1	x_2	x_3	Values of basic variables
x_1	25	13	37	1	0	0	1
x_2	13	49	25	0	1	0	1
x_3	(37)	1	13	0	0	1	1
Z	-1	-1	-1	0	0	0	0

We choose 37 in row 3 column 1 to be the first pivot. Pivoting, we get

Tableau 2

Basis	y_1	y_2	y_3	x_1	x_2	x_3	Values of basic variables
x_1	0	$\frac{456}{37}$	$\frac{1044}{37}$	1	0	$\frac{-25}{37}$	$\frac{12}{37}$
x_2	0	$\left(\frac{1800}{37}\right)$	$\frac{756}{37}$	0	1	$\frac{-13}{37}$	$\frac{24}{37}$
y_1	1	$\frac{1}{37}$	$\frac{13}{37}$	0	0	$\frac{1}{37}$	$\frac{1}{37}$
Z	0	$\frac{-36}{37}$	$\frac{-24}{37}$	0	0	$\frac{1}{37}$	$\frac{1}{37}$

$\frac{-36}{37}$ is the most negative indicator in the bottom row. Also $\frac{24}{37}/\frac{1800}{37}$ is less than both $\frac{1}{37}/\frac{1}{37}$ and $\frac{12}{37}/\frac{456}{37}$. Thus, $\frac{1800}{37}$ is the pivot. Pivoting, we get

Tableau 3

Basis	y_1	y_2	y_3	x_1	x_2	x_3	Values of basic variables
x_1	0	0	$\left(\frac{576}{25}\right)$	1	$\frac{-19}{75}$	$\frac{-44}{75}$	$\frac{4}{25}$
y_2	0	1	$\frac{21}{50}$	0	$\frac{37}{1800}$	$\frac{-13}{1800}$	$\frac{1}{75}$
y_1	1	0	$\frac{17}{50}$	0	$\frac{-1}{1800}$	$\frac{49}{1800}$	$\frac{2}{75}$
Z	0	0	$\frac{-6}{25}$	0	$\frac{1}{50}$	$\frac{1}{50}$	$\frac{1}{25}$

There is only one negative indicator in the bottom row. So, the third column is the pivot column. Since $\frac{4}{25}/\frac{576}{25}$ is less than both $\frac{1}{75}/\frac{21}{50}$ and $\frac{1}{2}/\frac{17}{50}$, the first row is the pivot row. Pivoting, we get

Tableau 4

Basis	y_1	y_2	y_3	x_1	x_2	x_3	Values of basic variables
y_3	0	0	1	$\frac{25}{576}$	$\frac{-19}{1728}$	$\frac{-11}{432}$	$\frac{1}{144}$
y_2	0	1	0	$\frac{-7}{384}$	$\frac{29}{1152}$	$\frac{1}{288}$	$\frac{1}{96}$
y_1	1	0	0	$\frac{-17}{1152}$	$\frac{11}{3456}$	$\frac{31}{864}$	$\frac{7}{288}$
Z	0	0	0	$\frac{1}{96}$	$\frac{5}{288}$	$\frac{1}{72}$	$\frac{1}{24}$

There are no negative indicators in the bottom row. So, the maximum value of Z has been attained. It is $\frac{1}{24}$. It occurs when $x_1 = \frac{1}{96}$, $x_2 = \frac{5}{288}$, $x_3 = \frac{1}{72}$, $y_1 = \frac{7}{288}$, $y_2 = \frac{1}{96}$, $y_3 = \frac{1}{144}$. Thus, $v = \frac{1}{z} = 24$.

$$p^* = \left[\frac{1}{96}(24) \quad \frac{5}{288}(24) \quad \frac{1}{72}(24) \right] = \left[\frac{1}{4} \quad \frac{5}{12} \quad \frac{1}{3} \right] \text{ and}$$

$$q^* = \begin{bmatrix} \frac{7}{288}(24) \\ \frac{1}{96}(24) \\ \frac{1}{144}(24) \end{bmatrix} = \begin{bmatrix} \frac{7}{12} \\ \frac{1}{4} \\ \frac{1}{6} \end{bmatrix}.$$

The value of the game is $v - k = 24 - 23 = 1$.

15. Assume the investor is the row player and the legislature is the column player.

Proposition

		I	II	III
	A	7	1	13
Stock	B	1	19	7
	C	13	-5	1

Let $k = 6$. Then $B = \begin{bmatrix} 13 & 7 & 19 \\ 7 & 25 & 13 \\ 19 & 1 & 7 \end{bmatrix}$

Problem 1. Minimize $z = x_1 + x_2 + x_3$ subject to
$$13x_1 + 7x_2 + 19x_3 \geq 1$$
$$7x_1 + 25x_2 + \quad x_3 \geq 1$$
$$19x_1 + 13x_2 + 7x_3 \geq 1$$
$$x_1 \geq 0, \ x_2 \geq 0, \ x_3 \geq 0$$

Problem 2: Maximize $Z = y_1 + y_2 + y_3$ subject to
$$13y_1 + 7y_2 + 19y_3 \leq 1$$
$$7y_1 + 25y_2 + 13y_3 \leq 1$$
$$19y_1 + \quad y_2 + 7y_3 \leq 1$$
$$y_1 \geq 0, \ y_2 \geq 0, \ y_3 \geq 0$$

We solve the second problem, using x_1, x_2, x_3 as slack variables.

$$\begin{aligned}
13y_1 + 7y_2 + 19y_3 + x_1 &= 1 \\
7y_1 + 25y_2 + 13y_3 \qquad + x_2 &= 1 \\
19y_1 + y_2 + 7y_3 \qquad\qquad + x_3 &= 1 \\
Z - y_1 - y_2 - y_3 \qquad\qquad\qquad &= 0
\end{aligned}$$

The first tableau is

Tableau 1

Basis	y_1	y_2	y_3	x_1	x_2	x_3	Values of basic variables
x_1	13	7	19	1	0	0	1
x_2	7	(25)	13	0	1	0	1
x_3	19	1	7	0	0	1	1
Z	−1	−1	−1	0	0	0	0

We choose 25 as the first pivot. Pivoting, we get

Tableau 2

Basis	y_1	y_2	y_3	x_1	x_2	x_3	Values of basic variables
x_1	$\frac{276}{25}$	0	$\frac{384}{25}$	1	$\frac{-7}{25}$	0	$\frac{18}{25}$
y_2	$\frac{7}{25}$	1	$\frac{13}{25}$	0	$\frac{1}{25}$	0	$\frac{1}{25}$
x_3	$\left(\frac{468}{25}\right)$	0	$\frac{162}{25}$	0	$\frac{-1}{25}$	1	$\frac{24}{25}$
Z	$\frac{-18}{25}$	0	$\frac{-12}{25}$	0	$\frac{1}{25}$	0	$\frac{1}{25}$

$-\frac{18}{25}$ is the most negative indicator in the bottom row. $\frac{468}{25}$ must be the pivot. Pivoting, we get

Tableau 3

Basis	y_1	y_2	y_3	x_1	x_2	x_3	Values of basic variables
x_1	0	0	$\left(\frac{150}{13}\right)$	1	$\frac{-10}{39}$	$\frac{-23}{39}$	$\frac{2}{13}$
y_2	0	1	$\frac{11}{26}$	0	$\frac{19}{468}$	$\frac{-7}{468}$	$\frac{1}{39}$
y_1	1	0	$\frac{9}{26}$	0	$\frac{-1}{468}$	$\frac{25}{468}$	$\frac{2}{39}$
Z	0	0	$\frac{-3}{13}$	0	$\frac{1}{26}$	$\frac{1}{26}$	$\frac{1}{13}$

Pivoting, we get

Tableau 4

Basis	y_1	y_2	y_3	x_1	x_2	x_3	Values of basic variables
y_3	0	0	1	$\frac{13}{150}$	$\frac{-1}{45}$	$\frac{-23}{450}$	$\frac{1}{75}$
y_2	0	1	0	$\frac{-11}{300}$	$\frac{1}{20}$	$\frac{1}{150}$	$\frac{1}{50}$
y_1	1	0	0	$\frac{-3}{100}$	$\frac{1}{180}$	$\frac{16}{225}$	$\frac{7}{150}$
Z	0	0	0	$\frac{1}{50}$	$\frac{1}{30}$	$\frac{2}{75}$	$\frac{2}{25}$

There are no negative indicators in the bottom row. The maximum value of Z (and the minimum value of z) has been attained. It is $\frac{2}{25}$. It occurs when $x_1 = \frac{1}{50}$, $x_2 = \frac{1}{30}$, $x_3 = \frac{2}{75}$, $y_1 = \frac{7}{150}$, $y_2 = \frac{1}{50}$, $x_3 = \frac{1}{75}$. Thus, $v = \frac{1}{z} = \frac{25}{2}$.

$$p^* = \left[\begin{array}{ccc} \frac{25}{2}(\frac{1}{50}) & \frac{25}{2}(\frac{1}{30}) & \frac{25}{2}(\frac{2}{75}) \end{array} \right] = \left[\begin{array}{ccc} \frac{1}{4} & \frac{5}{12} & \frac{1}{3} \end{array} \right] \quad \text{and}$$

$$q^* = \left[\begin{array}{c} \frac{25}{2}(\frac{7}{150}) \\ \frac{25}{2}(\frac{1}{50}) \\ \frac{25}{2}(\frac{1}{75}) \end{array} \right] = \left[\begin{array}{c} \frac{7}{12} \\ \frac{1}{4} \\ \frac{1}{6} \end{array} \right].$$

The value of the game is $v - k = \frac{25}{2} - 6 = \frac{13}{2} = 6.5$. The investor should invest \$3,000 in stock A (since $\frac{1}{4}(12,000) = 3,000$), \$5,000 in stock B (since $\frac{5}{12}(12,000) = 5,000$), and \$4,000 in stock C (since $\frac{1}{3}(12,000) = 4,000$).

EXERCISE SET 7.5 GRAPHS

1.

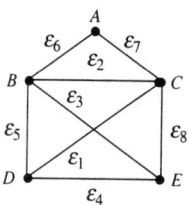

The graph is traceable. The path D, ϵ_1, C, ϵ_2, B, ϵ_3, E, ϵ_4, D, ϵ_5, B, ϵ_6, A, ϵ_7, C, ϵ_8, E contains every vertex and every edge, and no edge is repeated.

3. The graph is not traceable. There are four odd vertices.

5. The graph is traceable. The path F, D, B, A, C, B, F, G, C, E, G contains every vertex and every edge and no edge is repeated.

7. 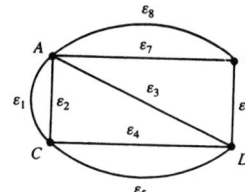 The graph is traceable. The path A, ϵ_1, C, ϵ_2, A, ϵ_3, D, ϵ_4, C, ϵ_5, D, ϵ_6, B, ϵ_7, A, ϵ_8, B contains every vertex and every edge and no edge is repeated.

9. 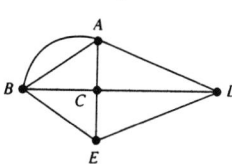 It is not possible. The only traceable path must start at Island D and end at Land E (or vice versa) as D and E are odd vertices.

11. 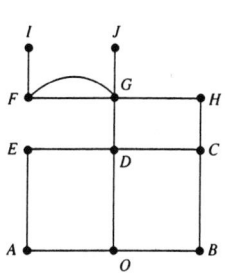 Vertices A, B, C, D, E, F, G, and H represent suites. Vertices I and J are possible exits and vertex O is the entrance.

The vertices O, C, F, G, I and J are odd. Thus, the graph is not traceable and the task is not possible.

13. $244 + 210 + 110 + 212 = 776$ is the value of the path.

15. The graph is a tree. It is connected and contains no cycles.

17. The graph is not a tree. The path D, G, F, E, D, is a cycle.

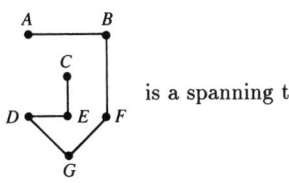 is a spanning tree.

19. The graph is not a tree. The path A, B, C, A is a cycle.

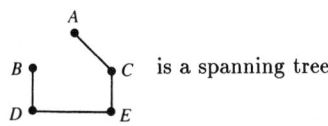 is a spanning tree.

21.

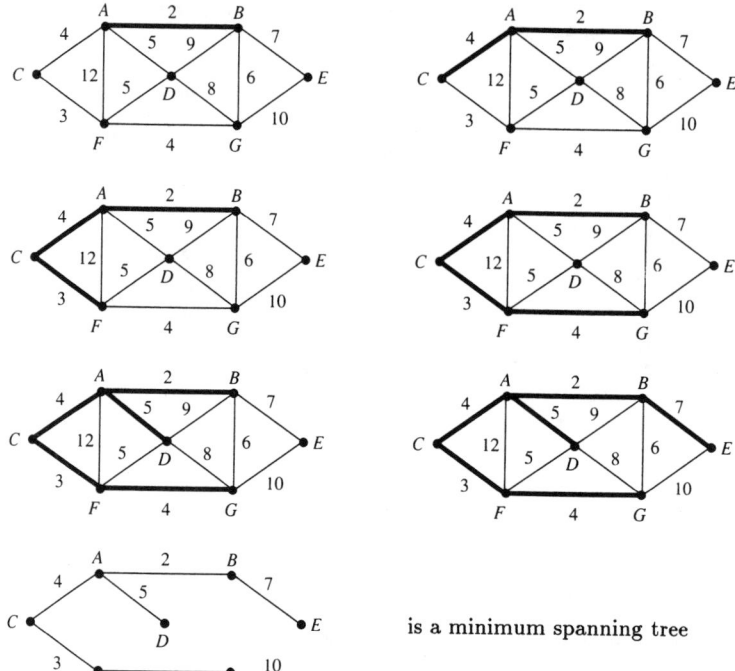

is a minimum spanning tree

23.

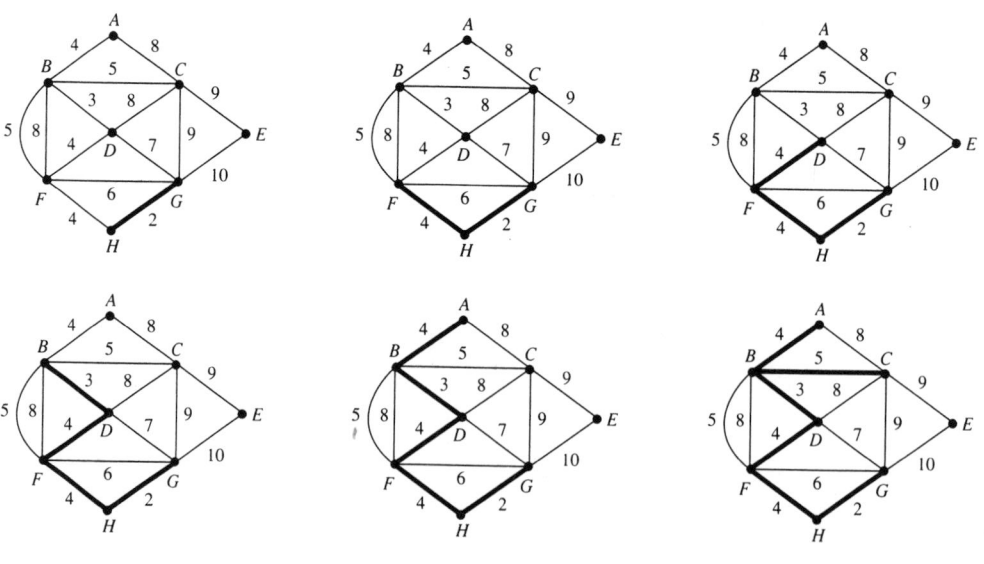

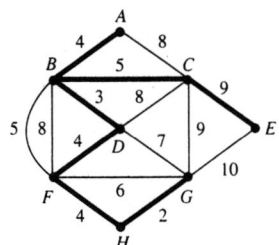 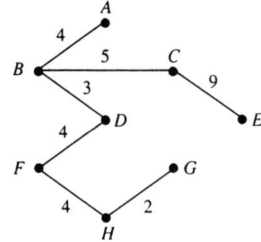

is the minimum
spanning tree.

25. A line of minimum length connects Dallas with Fort Worth, Forth Worth with Cisco, Cisco with Brownwood, Brownwood with San Antonio, San Antonio with Austin, San Antonio with Victoria, and Victoria with Houston.

27.

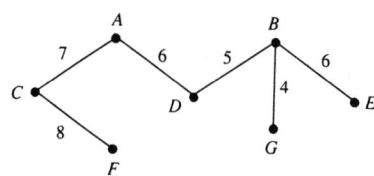

Install the network as indicated.

29.

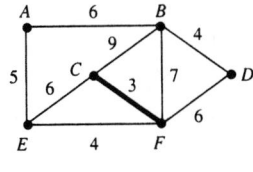 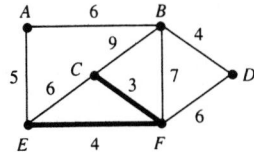 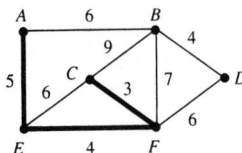

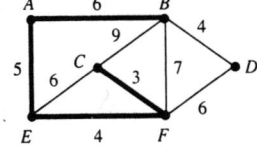 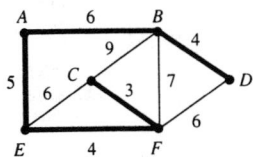 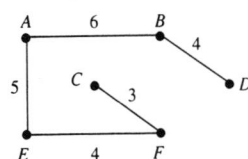

is another minimum
spanning tree.

EXERCISE SET 7.6   DIGRAPHS

1.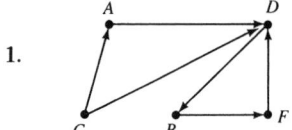

3.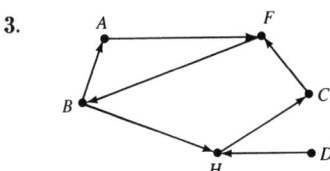

5. $V = \{P, V_1, V_2, V_3, A, B, C, D_1, D_2, D_3, X, Y, Z\}$
      $E = \{(P, V_1), (P, V_2), (P, V_3), (V_1, A), (V_1, B), (V_1, B) (V_1, C),$
      $(V_2, D_1), (V_2, D_2), (V_2, D_3), (V_3, X), (V_3, Y), (V_3, Z)\}$

7.   (a) $(R, T), (T, V)$ is a directed path of length 2 from R *to* V.
             $(R, U), (U, V)$ is a directed path of length 2 from R to V.
             $(R, T), (T, U), (U, V)$ is directed path of length 3 from R to V.
      (b)   Yes
      (c)   No
      (d) $d(R, V) = 2$, $d(T, U) = 1$

9.   (a)   Two directed paths of length 3 are $(S, U), (U, W), (W, T)$ and $(S, R), (R, V),$
             (V, T).   A directed path of length 4 is $(S, R), (R, U), (U, W), (W, T)$.

      (b)   Yes
      (c)   Yes
      (d) $d(S, T) = 3$
             $d(T, S) = \infty$

11.

	R	S	T	U	V
R	0	0	1	1	0
S	0	0	0	0	0
T	0	0	0	1	1
U	0	1	0	0	1
V	0	0	0	0	0

13.

	R	S	T	U	V	W
R	0	0	0	1	1	0
S	1	0	0	1	0	0
T	1	0	0	0	0	0
U	0	0	0	0	0	1
V	0	0	1	0	0	0
W	0	0	1	0	0	0

15.

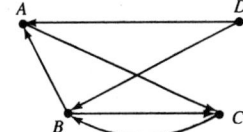

17.

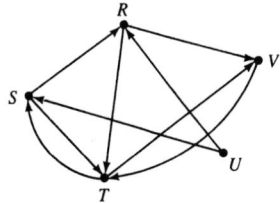

19.

$$M = \begin{bmatrix} 0 & 0 & 1 & 0 \\ 1 & 0 & 1 & 0 \\ 0 & 1 & 0 & 0 \\ 1 & 1 & 0 & 0 \end{bmatrix}$$

$$M^2 = \begin{bmatrix} 0 & 0 & 1 & 0 \\ 1 & 0 & 1 & 0 \\ 0 & 1 & 0 & 0 \\ 1 & 1 & 0 & 0 \end{bmatrix} \cdot \begin{bmatrix} 0 & 0 & 1 & 0 \\ 1 & 0 & 1 & 0 \\ 0 & 1 & 0 & 0 \\ 1 & 1 & 0 & 0 \end{bmatrix} = \begin{bmatrix} 0 & 1 & 0 & 0 \\ 0 & 1 & 1 & 0 \\ 1 & 0 & 1 & 0 \\ 1 & 0 & 2 & 0 \end{bmatrix}$$

There is one path of length 2 from A to B, B to B, B to C, C to A, C to C, and D to A. There are two paths of length 2 from D to C. There are no other paths of length 2.

$$M^3 = \begin{bmatrix} 0 & 0 & 1 & 0 \\ 1 & 0 & 1 & 0 \\ 0 & 1 & 0 & 0 \\ 1 & 1 & 0 & 0 \end{bmatrix} \cdot \begin{bmatrix} 0 & 1 & 0 & 0 \\ 0 & 1 & 1 & 0 \\ 1 & 0 & 1 & 0 \\ 1 & 0 & 2 & 0 \end{bmatrix} = \begin{bmatrix} 1 & 0 & 1 & 0 \\ 1 & 1 & 1 & 0 \\ 0 & 1 & 1 & 0 \\ 0 & 2 & 1 & 0 \end{bmatrix}$$

There is one path of length 3 from A to A, A to C, B to B, B to C, B to A, C to B, C to C and D to C. There are two paths of length 3 from D to B. There are no other paths of length 3.

21.

$$M^2 = \begin{bmatrix} 0 & 0 & 1 & 0 & 1 \\ 1 & 0 & 1 & 0 & 0 \\ 0 & 1 & 0 & 0 & 1 \\ 1 & 1 & 0 & 0 & 0 \\ 0 & 0 & 1 & 0 & 0 \end{bmatrix} \cdot \begin{bmatrix} 0 & 0 & 1 & 0 & 1 \\ 1 & 0 & 1 & 0 & 0 \\ 0 & 1 & 0 & 0 & 1 \\ 1 & 1 & 0 & 0 & 0 \\ 0 & 0 & 1 & 0 & 0 \end{bmatrix}$$

$$= \begin{bmatrix} 0 & 1 & 1 & 0 & 1 \\ 0 & 1 & 1 & 0 & 2 \\ 1 & 0 & 2 & 0 & 0 \\ 1 & 0 & 2 & 0 & 1 \\ 0 & 1 & 0 & 0 & 1 \end{bmatrix}$$

There are exactly two directed paths of length two from S to V, T to T, and U to T. There is exactly one directed path of length two from R to S, R to T, R to V, S to S, S to T, T to R, U to V, U to R, V to S and V to V. There are no other directed paths of length two.

$$M^3 = \begin{bmatrix} 0 & 0 & 1 & 0 & 1 \\ 1 & 0 & 1 & 0 & 0 \\ 0 & 1 & 0 & 0 & 1 \\ 1 & 1 & 0 & 0 & 0 \\ 0 & 0 & 1 & 0 & 0 \end{bmatrix} \cdot \begin{bmatrix} 0 & 1 & 1 & 0 & 1 \\ 0 & 1 & 1 & 0 & 2 \\ 1 & 0 & 2 & 0 & 0 \\ 1 & 0 & 2 & 0 & 1 \\ 0 & 1 & 0 & 0 & 1 \end{bmatrix}$$

$$= \begin{bmatrix} 1 & 1 & 2 & 0 & 1 \\ 1 & 1 & 3 & 0 & 1 \\ 0 & 2 & 1 & 0 & 3 \\ 0 & 2 & 2 & 0 & 3 \\ 1 & 0 & 2 & 0 & 0 \end{bmatrix}$$

There is exactly one directed path of length three from R to R, R to S, R to V, S to R, S to S, S to V, T to T, and V to R. There are exactly two directed paths of length three from R to T, T to S, U to S, and V to T and U to T. There are exactly three directed paths of length three form S to T, T to V, and U to V. There are no other directed paths of length three.

23.

$$M = \begin{array}{c c} & \begin{array}{c c c c c c} A & B & C & D & E & F \end{array} \\ \begin{array}{c} A \\ B \\ C \\ D \\ E \\ F \end{array} & \left[\begin{array}{c c c c c c} 0 & 0 & 1 & 1 & 0 & 0 \\ 1 & 0 & 0 & 0 & 0 & 1 \\ 0 & 1 & 0 & 0 & 1 & 1 \\ 0 & 0 & 1 & 0 & 1 & 1 \\ 1 & 0 & 0 & 0 & 0 & 1 \\ 1 & 0 & 0 & 0 & 0 & 0 \end{array}\right] \end{array}$$

$$M^2 = \left[\begin{array}{c c c c c c} 0 & 1 & 1 & 0 & 2 & 2 \\ 1 & 0 & 1 & 1 & 0 & 0 \\ 3 & 0 & 0 & 0 & 0 & 2 \\ 2 & 1 & 0 & 0 & 1 & 2 \\ 1 & 0 & 1 & 1 & 0 & 0 \\ 0 & 0 & 1 & 1 & 0 & 0 \end{array}\right]$$

Committee Member	Amount of one-or two-stage influence
Allen	$2 + 6 = 8$
Bob	$2 + 3 = 5$
Carl	$3 + 5 = 8$
Doug	$3 + 6 = 9$
Earl	$2 + 3 = 5$
Fran	$1 + 2 = 3$

Doug is the most influential committee member.

25. (a)

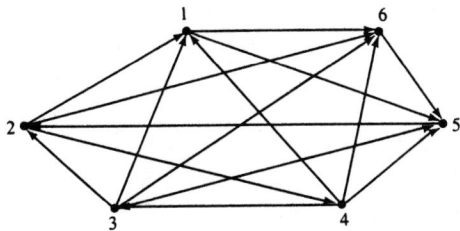

(b)

Player	Number of Wins	
Phillip	2	
Gregory	3	
Lee Michael	4	The tournament results in a tie between
Timothy	4	Lee Michael and Timothy.
Katie	1	
Beth	1	

(c)

$$T^2 = \begin{bmatrix} 0 & 1 & 0 & 0 & 1 & 0 \\ 1 & 0 & 1 & 0 & 3 & 2 \\ 1 & 1 & 0 & 1 & 2 & 2 \\ 1 & 2 & 0 & 0 & 3 & 2 \\ 1 & 0 & 0 & 1 & 0 & 1 \\ 0 & 1 & 0 & 0 & 0 & 0 \end{bmatrix},$$

$$T + T^2 = \begin{bmatrix} 0 & 1 & 0 & 0 & 2 & 1 \\ 2 & 0 & 1 & 1 & 3 & 3 \\ 2 & 2 & 0 & 1 & 3 & 1 \\ 2 & 2 & 1 & 0 & 4 & 3 \\ 1 & 1 & 0 & 1 & 0 & 1 \\ 0 & 1 & 0 & 0 & 1 & 0 \end{bmatrix}$$

Player	One and two-stage dominances	
Phillip	4	
Gregory	10	
Lee Michael	11	Timothy is the winner of the tournament.
Timothy	12	
Katie	4	
Beth	2	

EXERCISE SET 7.7 MORE APPLICATIONS

1. $S(A) = d(A, B) + d(A, C) + d(A, D) + d(A, E) + d(A, F) + d(A, a) + d(A, b) + d(A, c) +$
   $d(A, d) + d(A, e) = 1 + 1 + 2 + 2 + 2 + 3 + 3 + 3 + 3 + 3 = 23$
   $S(B) = 1 + 1 + 2 + 2 + 2 + 2 = 10$
   $S(C) = 1 + 2 + 2 = 5$
   $S(D) = 2, \quad S(E) = 3, \quad S(F) = 2$
   $S(a) = S(b) = S(c) = S(d) = S(e) = 0$

3. $S(v_1) = 1 + 1 + 2 + 2 + 2 + 3 + 3 + 3 + 3 = 20$
   $S(v_2) = 1 + 1 + 2 + 2 + 2 = 8$
   $S(v_3) = 1 + 1 + 2 + 2 = 6$
   $S(v_4) = 2, \quad S(v_5) = 1, \quad S(v_6) = 2, \quad S(v_7) = S(v_8) = S(v_9) = S(v_{10}) = 0$

5.

$$\begin{array}{c} \\ A \\ B \\ C \\ D \end{array} \begin{array}{cccc} A & B & C & D \\ \end{array}$$

$$\begin{array}{c} A \\ B \\ C \\ D \end{array} \begin{bmatrix} 0 & 1 & 2 & 1 \\ \infty & 0 & \infty & \infty \\ \infty & \infty & 0 & \infty \\ \infty & \infty & 1 & 0 \end{bmatrix}$$

7.

$$
\begin{array}{c}
\\ A \\ B \\ C \\ D \\ E
\end{array}
\begin{array}{ccccc}
A & B & C & D & E \\
\end{array}
$$

$$
\begin{bmatrix}
0 & 1 & 1 & 2 & 2 \\
\infty & 0 & \infty & 1 & \infty \\
\infty & \infty & 0 & \infty & 1 \\
\infty & \infty & \infty & 0 & \infty \\
\infty & \infty & \infty & \infty & 0
\end{bmatrix}
$$

9.

$$
A = \begin{bmatrix}
0 & 1 & 0 & 1 \\
0 & 0 & 0 & 0 \\
0 & 0 & 0 & 0 \\
0 & 0 & 1 & 0
\end{bmatrix}
\qquad
D = \begin{bmatrix}
0 & 1 & & 1 \\
 & 0 & & \\
 & & 0 & \\
 & & 1 & 0
\end{bmatrix}
$$

$$
A^2 = \begin{bmatrix}
0 & 0 & 1 & 0 \\
0 & 0 & 0 & 0 \\
0 & 0 & 0 & 0 \\
0 & 0 & 0 & 0
\end{bmatrix}
\qquad
D = \begin{bmatrix}
0 & 1 & 2 & 1 \\
 & 0 & & \\
 & & 0 & \\
 & & 1 & 0
\end{bmatrix}
$$

$$
A^3 = \begin{bmatrix}
0 & 0 & 0 & 0 \\
0 & 0 & 0 & 0 \\
0 & 0 & 0 & 0 \\
0 & 0 & 0 & 0
\end{bmatrix}
\qquad
D = \begin{bmatrix}
0 & 1 & 2 & 1 \\
\infty & 0 & \infty & \infty \\
\infty & \infty & 0 & \infty \\
\infty & \infty & 1 & 0
\end{bmatrix}
$$

is the distance matrix.

11.

$$
A = \begin{bmatrix}
0 & 1 & 1 & 0 & 0 \\
0 & 0 & 0 & 1 & 0 \\
0 & 0 & 0 & 0 & 1 \\
0 & 0 & 0 & 0 & 0 \\
0 & 0 & 0 & 0 & 0
\end{bmatrix}
\qquad
D = \begin{bmatrix}
0 & 1 & 1 & & \\
 & 0 & & 1 & \\
 & & 0 & & 1 \\
 & & & 0 & \\
 & & & & 0
\end{bmatrix}
$$

$$
A^2 = \begin{bmatrix}
0 & 0 & 0 & 1 & 1 \\
0 & 0 & 0 & 0 & 0 \\
0 & 0 & 0 & 0 & 0 \\
0 & 0 & 0 & 0 & 0 \\
0 & 0 & 0 & 0 & 0
\end{bmatrix}
\qquad
D = \begin{bmatrix}
0 & 1 & 1 & 2 & 2 \\
 & 0 & & 1 & \\
 & & 0 & & 1 \\
 & & & 0 & \\
 & & & & 0
\end{bmatrix}
$$

$$A^3 = \begin{bmatrix} 0 & 0 & 0 & 0 & 0 \\ 0 & 0 & 0 & 0 & 0 \\ 0 & 0 & 0 & 0 & 0 \\ 0 & 0 & 0 & 0 & 0 \\ 0 & 0 & 0 & 0 & 0 \end{bmatrix} \qquad D = \begin{bmatrix} 0 & 1 & 1 & 2 & 2 \\ \infty & 0 & \infty & 1 & \infty \\ \infty & \infty & 0 & \infty & 1 \\ \infty & \infty & \infty & 0 & \infty \\ \infty & \infty & \infty & \infty & 0 \end{bmatrix}$$

D is the distance matrix.

13.

$$D = \begin{bmatrix} 0 & 1 & 2 & 1 \\ \infty & 0 & \infty & \infty \\ \infty & \infty & 0 & \infty \\ \infty & \infty & 1 & 0 \end{bmatrix}$$

The distance sum of A is $1 + 2 + 1 = 4$, of B is 0, of C is 0, and of D is $0 + 1 = 1$.

15.

$$D = \begin{bmatrix} 0 & 1 & 1 & 2 & 2 \\ \infty & 0 & \infty & 1 & \infty \\ \infty & \infty & 0 & \infty & 1 \\ \infty & \infty & \infty & 0 & \infty \\ \infty & \infty & \infty & \infty & 0 \end{bmatrix}$$

The distance sum of A is $1 + 1 + 2 + 2 = 6$, of B is 1, of C is 1, of D is 0, and of E is 0.

17.

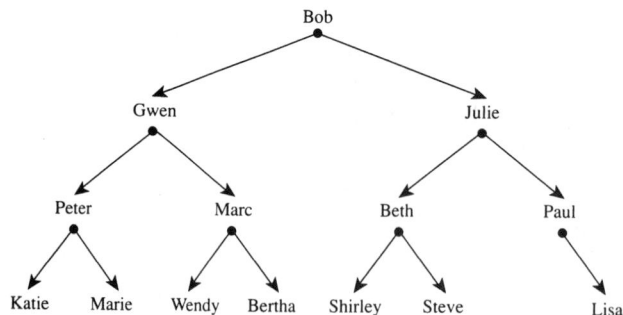

	Bob	Gwen	Julie	Peter	Marc	Beth	Paul	Katie	Marie	Wendy	Bertha	Shirl	Steve	Lisa
Bob	0	1	1	2	2	2	2	3	3	3	3	3	3	3
Gwen	∞	0	∞	1	1	∞	∞	2	2	2	2	∞	∞	∞
Julie	∞	∞	0	∞	∞	1	1	∞	∞	∞	∞	2	2	2
Peter	∞	∞	∞	0	∞	∞	∞	1	1	∞	∞	∞	∞	∞
Marc	∞	∞	∞	∞	0	∞	∞	∞	∞	1	1	∞	∞	∞
Beth	∞	∞	∞	∞	∞	0	∞	∞	∞	∞	∞	1	1	∞
Paul	∞	∞	∞	∞	∞	∞	0	∞	∞	∞	∞	∞	∞	1
Katie	∞	∞	∞	∞	∞	∞	∞	0	∞	∞	∞	∞	∞	∞
Marie	∞	∞	∞	∞	∞	∞	∞	∞	0	∞	∞	∞	∞	∞
Wendy	∞	∞	∞	∞	∞	∞	∞	∞	∞	0	∞	∞	∞	∞
Bertha	∞	∞	∞	∞	∞	∞	∞	∞	∞	∞	0	∞	∞	∞
Shirley	∞	∞	∞	∞	∞	∞	∞	∞	∞	∞	∞	0	∞	∞
Steve	∞	∞	∞	∞	∞	∞	∞	∞	∞	∞	∞	∞	0	∞
Lisa	∞	∞	∞	∞	∞	∞	∞	∞	∞	∞	∞	∞	∞	0

The status of Bob is $2(1) + 4(2) + 3(7) = 31$, of Gwen is $2(1) + 4(2) = 10$, of Julie is $2(1) + 3(2) = 8$, of Peter is 2, of Marc and Beth is 2, of Paul is 1 and of the other 7 individuals is 0.

19.

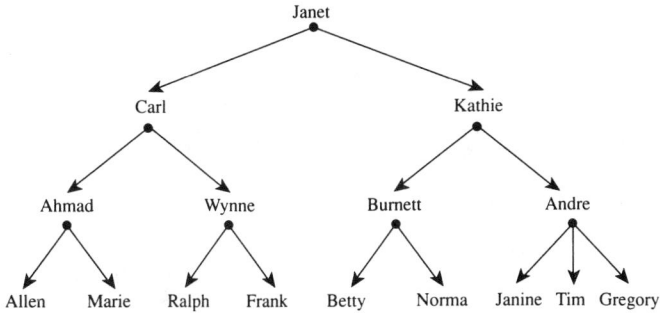

The distance matrix for the individuals with non-zero status follows.

		Janet	Carl / Kathie	Ahmad / Wynne	Burnett	Andre
	Janet	0	1 1	2 2	2	2
	Carl	∞	0 ∞	1 1	∞	∞
	Kathie	∞	∞ 0	∞ ∞	1	1
$D =$	Ahmad	∞	∞ ∞	0 ∞	∞	∞
	Wynne	∞	∞ ∞	∞ 0	∞	∞
	Burnett	∞	∞ ∞	∞ ∞	0	∞
	Andre	∞	∞ ∞	∞ ∞	∞	0

The status of Janet is $10 + 9(3) = 37$, of Carl is $2 + 2(4) = 10$,
of Kathie is $2 + 2(5) = 12$, of Ahmad, Wynne and Burnett is 2, of Andre is 3,
and of the other nine individuals is 0.

21. (a) The directed paths from S to H are S, A, D, H and S, B, F, G, H.
The sum for S, A, D, H is 12.   The sum for S, B, F, G, H is 15.
The critical path from S to H is S, B, F, G, H.

(b) The paths for the project are S, A, D, C, I, E with sum 18;
S, A, D, H, I, E with sum 17;   and S, B, F, G, H, I, E with sum 20.
The critical path for the project is S, B, F, G, H, I, E.

23.

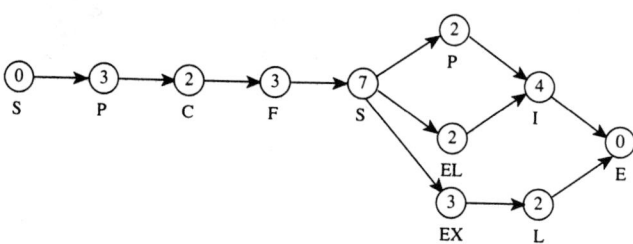

A critical path is S, P, C, F, S, P, I, E.

(a) The schedule is purchase land immediately, clear land in 3 weeks, build foundation in
5 weeks, build structure in 8 weeks, plumbing in 15 weeks, electrical work in 15
weeks, finishing interior in 17 weeks, finishing exterior in 15 weeks, and landscaping
in 18 weeks.

(b) the house will be completed in $3 + 2 + 3 + 7 + 2 + 4 = 21$ weeks.

25. (a) The critical path is S, A, D, H, I, E.

Task	Schedule
A	immediately
B	immediately
C	5 months hence
D	5 months hence
F	6 months hence
G	6 months hence
H	12 months hence
I	16 months hence

(b) The project will be completed in 24 months.

27.

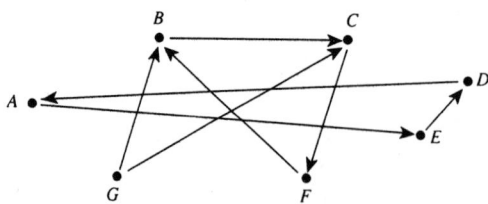

There is no directed path from A to B, or from B to A, or from C to A, or from D to B,
or from E to B, or from F to A, or from G to A.   Thus, the disease could not have spread
within the group and another person on campus with the disease has not been identified.

EXERCISE SET 7.8   CHAPTER REVIEW

1. (a) R has 3 choices and C has 4 choices.

  (b) $a_{21} = -4$.   So, R pays \$4 to C.

  (c) $a_{33} = 6$.   So, C pays \$6 to R.

3.
$$\begin{bmatrix} 1 & 2 & 3 & 0 \\ 3 & -3 & 2 & -2 \\ 1 & 2 & -2 & -1 \end{bmatrix}$$

We circle the smallest entry in each row and box the largest entry in each column.   0 is both smallest in its row and largest in its column.   So, it is a saddle point. R should play row 1 and C should play column 4.

5.
$$\begin{bmatrix} -3 & 2 & 4 \\ 2 & 0 & 3 \\ 5 & 4 & -3 \end{bmatrix}$$

We circle the smallest entry in each row and the largest entry in each column.   No entry is in a circle and a box.   So, the game is not strictly determined.

7. (a)
$$\begin{bmatrix} 2 & 3 & 0 \\ 4 & 1 & 5 \end{bmatrix}$$

There are no recessive rows or columns.   No entry is the smallest in its row and largest in its column.   So, the game is not strictly determined.

  (b)
$$\begin{bmatrix} 1 & 3 & -1 \\ 2 & 3 & 2 \\ 1 & 2 & 3 \\ -1 & 0 & 2 \end{bmatrix}$$

  (a)   Rows 1 and 4 are recessive.   After removing those rows, columns 2 and 3 are recessive.   Removing these columns we get $\begin{bmatrix} 2 \\ 1 \end{bmatrix}$.   Clearly, 2 is the saddle joint.

  (b)   With respect to the original matrix, R should play row 2 and C should play column 1.

  (c)   The value of the game is 2.

9. Let chain X be player R and chain Y be player C.

$$\begin{array}{c@{\quad}c@{\quad}c@{\quad}c} & A & B & C \\ A & \begin{bmatrix} 13 & -16.134 & -11.25 \\ -3.6 & 13 & -6.75 \\ 1.35 & -.15 & 13 \end{bmatrix} \end{array}$$

We entered 13 for a_{11}, a_{22}, a_{33} since when both stores are built in the same city, X gets 63% of the total business, which is 13 points over 50%. If X builds in A and Y in B, then X gets 80% of the business in A $((.8)(.23) = .184$ or 18.4% of the total business), 20% of the business in B $(.2(.33) = .066$ or 6.6% of the total) and 35% of the business in C $(.35(.44) = .154$ or 15.4% of the total). So, X gets $18.4 + .066 + 15.4 = 33.866\%$ of the total business. So, we enter -16.134 in row 1 column 2 since 33.866 in 16.134 points below 50. the other entries are obtained similarly.

This game is not strictly determined.

11.
$$\begin{bmatrix} -4 & 5 & 11 & -4 & -8 \\ 8 & 6 & 5 & 3 & 4 \\ 0 & -3 & 0 & 1 & 5 \\ 9 & -8 & 1 & -1 & 0 \end{bmatrix}$$

3 in row 2 column 4 is the saddle point. So, the Raccoons should use play #2 and the Cats should use defensive alignment #4.

13.
$$\begin{bmatrix} \frac{1}{3} & \frac{2}{3} \end{bmatrix} \begin{bmatrix} 45 & -90 & 150 \\ 60 & -75 & 180 \end{bmatrix} \begin{bmatrix} \frac{2}{5} \\ \frac{1}{5} \\ \frac{2}{5} \end{bmatrix} = \begin{bmatrix} 55 & -80 & 170 \end{bmatrix} \begin{bmatrix} \frac{2}{5} \\ \frac{1}{5} \\ \frac{2}{5} \end{bmatrix} = [74]$$

The value of the game is $74.

15.
$$A = \begin{bmatrix} 6 & 3 \\ -2 & 8 \end{bmatrix}$$ We use theorem 7.5. Here $a = 6$, $b = 3$, $c - 2$, $d = 8$.

So, $D = (6 + 8) - (3 + -2) = 14 - 1 = 13$.

$$p_1{}^* = \frac{8 - (-2)}{13} = \frac{10}{13}, \; p_2{}^* = 1 - \frac{10}{13} = \frac{3}{13} \text{ and } p^* = \begin{bmatrix} \frac{10}{13} & \frac{3}{13} \end{bmatrix}$$

$$q_1{}^* = \frac{8 - 3}{13} = \frac{5}{13}, \; q_2{}^* = 1 - \frac{5}{13} = \frac{8}{13} \text{ and } q^* = \begin{bmatrix} \frac{5}{13} \\ \frac{8}{13} \end{bmatrix}.$$

So, the optimal strategy for R is $\begin{bmatrix} \frac{10}{13} & \frac{3}{13} \end{bmatrix}$ and for C is $\begin{bmatrix} \frac{5}{13} \\ \frac{8}{13} \end{bmatrix}$.

17.
$$A = \begin{bmatrix} -4 & -2 & -1 & 3 \\ 6 & 1 & 3 & 5 \\ -2 & 0 & -6 & 1 \end{bmatrix}$$

The number 1 is the smallest element in row 2 and the largest element in column 2. So, it is a saddle point. The game is strictly determined. Its value is 1. R should always choose row 2 and C should always use column 2.

19. Let Ron be the row player and Carl be the column player. The matrix is

$$
\begin{array}{c c}
 & \begin{array}{c c} H & T \end{array} \\
\begin{array}{c} H \\ T \end{array} &
\begin{bmatrix} 2 & -2 \\ -2 & 2 \end{bmatrix}
\end{array}
$$

This game is not strictly determined. We use theorem 7.5 with $a = 2$, $b = -2$, $c = -2$, $d = 2$. So, $D = (2 + 2) - (-2 + -2) = 8$.

$$
p_1{}^* = \frac{2 - (-2)}{8} = \frac{1}{2}, \; p_2{}^* = \frac{1}{2}. \; \text{Thus, } p^* = \begin{bmatrix} \frac{1}{2} & \frac{1}{2} \end{bmatrix}.
$$

Also, $q_1{}^* = \dfrac{2 - (-2)}{8} = \dfrac{1}{2}$, $q_2{}^* = \dfrac{1}{2}$, and $q^* = \begin{bmatrix} \frac{1}{2} \\ \frac{1}{2} \end{bmatrix}$

$$
\begin{bmatrix} \frac{1}{2} & \frac{1}{2} \end{bmatrix}
\begin{bmatrix} 2 & -2 \\ -2 & 2 \end{bmatrix}
\begin{bmatrix} \frac{1}{2} \\ \frac{1}{2} \end{bmatrix}
$$

$$
= \begin{bmatrix} 0 & 0 \end{bmatrix}
\begin{bmatrix} \frac{1}{2} \\ \frac{1}{2} \end{bmatrix}
= \begin{bmatrix} 0 \end{bmatrix}. \quad \text{The value of the game is 0.}
$$

21.

$$
\begin{bmatrix}
5 & 8 & -8 & 2 \\
6 & 6 & -4 & 12 \\
-6 & 4 & 2 & -2 \\
5 & 7 & 9 & 11
\end{bmatrix}
$$

(a) First delete row 3 (it is recessive). Then delete column 2 and 4 (they are recessive). Finally, delete row 1, which is now recessive. We get

$$
A = \begin{bmatrix} 6 & -4 \\ 5 & 9 \end{bmatrix}
$$

We use theorem 7.5 with $a = 6$, $b = -4$, $c = 5$, $d = 9$.
So, $D = (6 + 9) - (5 + -4) = 14$.

$$
p_1{}^* = \frac{9 - 5}{14} = \frac{4}{14} = \frac{2}{7}, \text{ and } p_2{}^* = 1 - \frac{2}{7} = \frac{5}{7}. \; \text{Therefore, } p^* = \begin{bmatrix} \frac{2}{7} & \frac{5}{7} \end{bmatrix}.
$$

Similarly, $q_1{}^* = \dfrac{9 - (-4)}{14} = \dfrac{13}{14}$, $q_2{}^* = 1 - \dfrac{13}{14} = \dfrac{1}{14}$ and $q^* = \begin{bmatrix} \frac{13}{14} \\ \frac{1}{14} \end{bmatrix}$.

In terms of the original matrix, the optimal strategies are

$$\begin{bmatrix} 0 & \frac{2}{7} & 0 & \frac{5}{7} \end{bmatrix} \text{ for } R \text{ and } \begin{bmatrix} \frac{13}{14} \\ 0 \\ \frac{1}{14} \\ 0 \end{bmatrix} \text{ for } C.$$

$$\begin{bmatrix} 0 & \frac{2}{7} & 0 & \frac{5}{7} \end{bmatrix} \begin{bmatrix} 5 & 8 & -8 & 2 \\ 6 & 6 & -4 & 12 \\ -6 & 4 & 2 & -2 \\ 5 & 7 & 9 & 11 \end{bmatrix} \begin{bmatrix} \frac{13}{14} \\ 0 \\ \frac{1}{14} \\ 0 \end{bmatrix}$$

$$\begin{bmatrix} \frac{37}{7} & \frac{47}{7} & \frac{37}{7} & 2 \end{bmatrix} \begin{bmatrix} \frac{13}{14} \\ 0 \\ \frac{1}{14} \\ 0 \end{bmatrix} = \begin{bmatrix} \frac{37}{7} \end{bmatrix} \quad \text{The value of the game is } \frac{37}{7}.$$

(b)

$$\begin{bmatrix} 0 & 0 & 0 & 1 \end{bmatrix} \begin{bmatrix} 5 & 8 & -8 & 2 \\ 6 & 6 & -4 & 12 \\ -6 & 4 & 2 & -2 \\ 5 & 7 & 9 & 11 \end{bmatrix} \begin{bmatrix} \frac{13}{14} \\ 0 \\ \frac{1}{14} \\ 0 \end{bmatrix}$$

$$= \begin{bmatrix} 5 & 7 & 9 & 11 \end{bmatrix} \begin{bmatrix} \frac{13}{14} \\ 0 \\ \frac{1}{14} \\ 0 \end{bmatrix} = \begin{bmatrix} \frac{74}{14} \end{bmatrix} = \begin{bmatrix} \frac{37}{7} \end{bmatrix}$$

(c)

$$\begin{bmatrix} 0 & \frac{2}{7} & 0 & \frac{5}{7} \end{bmatrix} \begin{bmatrix} 5 & 8 & -8 & 2 \\ 6 & 6 & -4 & 12 \\ -6 & 4 & 2 & -2 \\ 5 & 7 & 9 & 11 \end{bmatrix} \begin{bmatrix} 0 \\ 1 \\ 0 \\ 0 \end{bmatrix}$$

$$= \begin{bmatrix} 0 & \frac{2}{7} & 0 & \frac{5}{7} \end{bmatrix} \begin{bmatrix} 8 \\ 6 \\ 4 \\ 7 \end{bmatrix} = \begin{bmatrix} \frac{32}{7} \end{bmatrix} \quad \text{The expected value of the game is } \frac{32}{7}.$$

23. Let Ron be the row player and Carol be the column player. The matrix is shown below.

$$
\begin{array}{c}
\\
\text{2 fingers} \\
\\
\text{3 fingers}
\end{array}
\begin{array}{cc}
\overset{2}{\underset{\text{fingers}}{}} & \overset{3}{\underset{\text{fingers}}{}} \\
\left[\begin{array}{cc}
4 & -5 \\
-5 & 6
\end{array}\right]
\end{array}
\quad \text{Let } k = 6 \text{ and obtain } B =
\left[\begin{array}{cc}
10 & 1 \\
1 & 12
\end{array}\right]
$$

<u>Problem 1:</u>   Minimize $z = x_1 + x_2$ subject to

$$
\begin{cases}
10x_1 + x_2 \geq 1 \\
x_1 + 12x_2 \geq 1 \\
\quad\quad x_1 \geq 0, \ x_2 \geq 0
\end{cases}
$$

<u>Problem 2:</u>   Maximize $Z = y_1 + y_2$ subject to

$$
\begin{cases}
10y_1 + y_2 \geq 1 \\
y_1 + 12y_2 \geq 1 \\
\quad\quad y_1 \geq 0, \ y_2 \geq 0
\end{cases}
$$

We solve the problem geometrically (See Figures 1 and 2).

The minumum value of Z is $\dfrac{11}{119} + \dfrac{9}{119} = \dfrac{20}{119}$ and occurs at $x_1 = \dfrac{11}{119}, \ x_2 = \dfrac{9}{119}$.

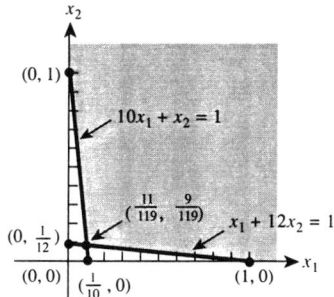

Figure 1

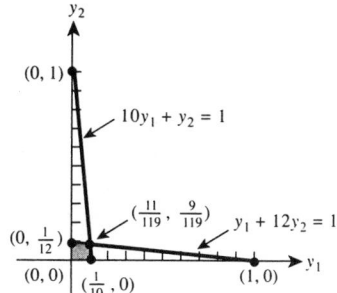

Figure 2

The maximum value of Z is $\dfrac{20}{119}$ and occurs at $y_1 = \dfrac{11}{119}, \ y_2 = \dfrac{9}{119}$. So, $v = \dfrac{1}{z} = \dfrac{119}{20}$,

$p_1{}^* = x_1 v = \dfrac{11}{119}\left(\dfrac{119}{20}\right) = \dfrac{11}{20}, \ p_2{}^* = x_2 v = \dfrac{9}{119}\left(\dfrac{119}{20}\right) = \dfrac{9}{20}.$ So, $p^* = \left[\begin{array}{cc} \dfrac{11}{20} & \dfrac{9}{20} \end{array}\right].$

Similarly, $q_1^* = y_1 v = \frac{11}{119}(\frac{119}{20}) = \frac{11}{20}$, $q_2^* = y_2 v = \frac{9}{119}(\frac{119}{20}) = \frac{9}{20}$.

Therefore, $q^* = \begin{bmatrix} \frac{11}{20} \\ \frac{9}{20} \end{bmatrix}$.

Check: $\begin{bmatrix} \frac{11}{20} & \frac{9}{20} \end{bmatrix} \begin{bmatrix} 10 & 1 \\ 1 & 12 \end{bmatrix} \begin{bmatrix} \frac{11}{20} \\ \frac{9}{20} \end{bmatrix} = \begin{bmatrix} \frac{119}{20} & \frac{119}{20} \end{bmatrix} \begin{bmatrix} \frac{11}{20} \\ \frac{9}{20} \end{bmatrix} = \begin{bmatrix} \frac{119}{20} \end{bmatrix}$.

The value of the original game is $\frac{119}{20} - 6 = \frac{-1}{20}$. So, the expected value of the game for Ron is $\$\frac{-1}{20}$.

25.
$$\begin{bmatrix} 6 & 9 & -7 & 3 \\ 7 & 11 & -3 & 13 \\ -5 & 5 & 4 & -1 \\ 6 & 8 & 10 & 12 \end{bmatrix}$$

Column 2 is recessive. We delete it. Row 1 is recessive. We delete it. Now column 3 is recessive. We delete it. finally, row 2 is recessive. We delete it. So, we consider

$A = \begin{bmatrix} 7 & -3 \\ 6 & 10 \end{bmatrix}$ Let $k = 4$ and obtain $B = \begin{bmatrix} 11 & 1 \\ 10 & 14 \end{bmatrix}$

Problem 1: Minimize $z = x_1 + x_2$ subject to
$$\begin{cases} 11x_1 + 10x_2 \geq 1 \\ x_1 + 14x_2 \geq 1 \\ x_1 \geq 0, \, x_2 \geq 0 \end{cases}$$

Problem 2: Maximize $Z = y_1 + y_2$ subject to
$$\begin{cases} 11y_1 + y_2 \leq 1 \\ 10y_1 + 14y_2 \leq 1 \\ y_1 \geq 0, \, y_2 \geq 0 \end{cases}$$

We solve the problem geometrically (See figures).

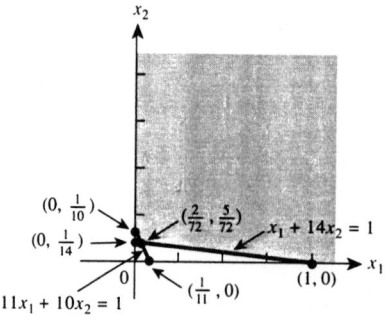
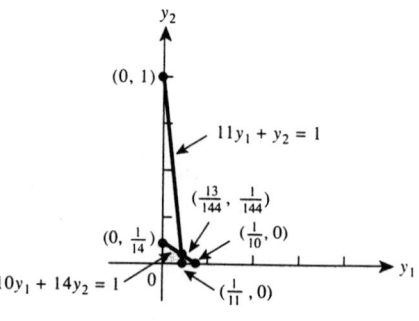

The minimum value of z is $\frac{7}{72}$. It occurs at $x_1 = \frac{2}{72}$, $x_2 = \frac{5}{72}$. Also, the maximum value of Z is $\frac{14}{144}$. It occurs when $y_1 = \frac{13}{144}$, $y_2 = \frac{1}{144}$. Now, $v = \frac{1}{z} = \frac{72}{7}$.

$$p_1{}^* = x_1 v = \frac{2}{72}\left(\frac{72}{7}\right) = \frac{2}{7}, \ p_2{}^* = x_2 v = \frac{5}{72}\left(\frac{72}{7}\right) = \frac{5}{7}. \text{ and } p^* = \left[\begin{array}{cc} \frac{2}{7} & \frac{5}{7} \end{array}\right].$$

Similarly, $q_1{}^* = y_1 v = \frac{13}{144}\left(\frac{72}{7}\right) = \frac{13}{14}, \ q_2{}^* = y_2 v = \frac{1}{144}\left(\frac{72}{7}\right) = \frac{1}{14}.$

Therefore, $q^* = \begin{bmatrix} \frac{13}{14} \\ \frac{1}{14} \end{bmatrix}.$

The value of the game is $\frac{72}{7} - 4 = \frac{44}{7}$. In terms of the original matrix

$$p^* = \left[\begin{array}{cccc} 0 & \frac{2}{7} & 0 & \frac{5}{7} \end{array}\right] \text{ and } q^* = \begin{bmatrix} \frac{13}{14} \\ 0 \\ \frac{1}{14} \\ 0 \end{bmatrix}$$

Store R should not use the radio or mail, use TV $\frac{2}{7}$ of the time and the paper $\frac{5}{7}$ of the time. Store C should use the radio $\frac{13}{14}$ of the time, mail $\frac{1}{14}$ of the time, and never use TV or paper.

27.

$$A = \begin{bmatrix} 1 & 2 & -1 & 6 \\ -2 & 4 & -3 & 4 \\ -1 & 1 & 2 & -3 \end{bmatrix}$$

Column 2 is recessive. Delete it. Then row 2 is recessive. Delete it. We get

$$\begin{bmatrix} 1 & -1 & 6 \\ -1 & 2 & -3 \end{bmatrix} \quad \text{Let } k = 4 \text{ and obtain } B = \begin{bmatrix} 5 & 3 & 10 \\ 3 & 6 & 1 \end{bmatrix}$$

Problem 1. Minimize $z = x_1 + x_2$ subject to

$$5x_1 + 3x_2 \geq 1$$
$$3x_1 + 6x_2 \geq 1$$
$$10x_1 + x_2 \geq 1$$
$$x_1 \geq 0, \ x_2 \geq 0$$

Problem 2: Maximize $Z = y_1 + y_2 + y_3$ subject to

$$5y_1 + 3y_2 + 10y_3 \leq 1$$
$$3y_1 + 6y_2 + y_3 \leq 1$$
$$y_1 \geq 0, \ y_2 \geq 0, \ y_3 \geq 0$$

We solve the second problem, using x_1, x_2 as slack variables.

$$\begin{cases} 5y_1 + 3y_2 + 10y_3 + x_1 & = 1 \\ 3y_1 + 6y_2 + y_3 & + x_2 = 1 \\ Z - y_1 - y_2 - y_3 & = 0 \end{cases}$$

The first tableau is

Tableau 1

Basis	y_1	y_2	y_3	x_1	x_2	Values of basic variables
x_1	5	3	(10)	1	0	1
x_2	3	6	1	0	1	1
Z	-1	-1	-1	0	0	0

$\uparrow$

Choose column 3 and row 1 as our first pivot column and row, respectively. Pivoting, we get

Tableau 2

Basis	y_1	y_2	y_3	x_1	x_2	Values of basic variables
y_3	.5	.3	1	.1	0	.1
x_2	2.5	(5.7)	0	$-.1$	1	.9
Z	$-.5$	$-.7$	0	.1	0	.1

$\uparrow$

Pivoting, we get

Tableau 3

Basis	y_1	y_2	y_3	x_1	x_2	Values of basic variables
y_3	$\left(\frac{7}{19}\right)$	0	1	$\frac{2}{19}$	$\frac{-1}{19}$	$\frac{1}{19}$
y_2	$\frac{25}{57}$	1	0	$\frac{-1}{57}$	$\frac{10}{57}$	$\frac{3}{19}$
Z	$\frac{-11}{57}$	0	0	$\frac{5}{57}$	$\frac{7}{57}$	$\frac{4}{19}$

$\uparrow$

Pivoting, we get

Tableau 4

Basis	y_1	y_2	y_3	x_1	x_2	Values of basic variables
y_1	1	0	$\frac{19}{7}$	$\frac{2}{7}$	$\frac{-1}{7}$	$\frac{1}{7}$
y_2	0	1	$\frac{-25}{21}$	$\frac{-1}{7}$	$\frac{5}{21}$	$\frac{2}{21}$
Z	0	0	$\frac{11}{21}$	$\frac{1}{7}$	$\frac{2}{21}$	$\frac{5}{21}$

There are no negative indicators in the bottom row. The maximum value of Z (and the minimum value of z) $\frac{5}{21}$ has been attained. It occurs when $x_1 = \frac{1}{7}$, $x_2 = \frac{2}{21}$

$y_1 = \frac{1}{7}$, $y_2 = \frac{2}{21}$, $y_3 = 0$. Now $v = \frac{1}{2} = \frac{21}{5}$. Also, $p_1^* = x_1 v = \frac{1}{7} \cdot \frac{21}{5} = \frac{3}{5}$, $p_2^* = \frac{2}{21} \cdot \frac{21}{5} = \frac{2}{5}$

so that $p^* = \begin{bmatrix} \frac{3}{5} & \frac{2}{5} \end{bmatrix}$. Similarly, $q_1^* = x_1 v = \frac{1}{7} \cdot \frac{21}{5} = \frac{3}{5}$, $q_2^* = \frac{2}{21} \cdot \frac{21}{5} = \frac{2}{5}$,

$q_3 v = 0(\frac{21}{5}) = 0$, so that $q^* = \begin{bmatrix} \frac{3}{5} \\ \frac{2}{5} \\ 0 \end{bmatrix}$. The value of the game is $\frac{21}{5} - 4 = \frac{1}{5}$.

In terms of the original matrix $p^* = \begin{bmatrix} \frac{3}{5} & 0 & \frac{2}{5} \end{bmatrix}$ and $q^* = \begin{bmatrix} \frac{3}{5} \\ 0 \\ \frac{2}{5} \\ 0 \end{bmatrix}$.

29. (a) The graph is traceable. The path E, B, A, C, B, D, C, E, D contains every vertex and every edge and no edge is repeated.

(b) The graph is traceable. All vertices are even. The path A, B. C, E, A, D, F, E, D, C, A contains every vertex and every edge and no edge is repeated.

(c) The graph is not traceable. Verticies B, C, D and E are odd.

(d) The graph is not connected and therefore not traceable.

31. $107 + 175 + 200 + 399 = 881$.

33. (a) 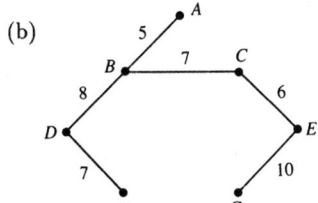 (b)

35.

37. (R, T), (T, V) and (R, U), (U, V) are directed paths of length 2 from R to V. (R, U), (U, S), (S, V) is a directed path of length 3 from R to V.
(b) yes (c) no (d) $d(R, V) = 2$, $d(T, U) = \infty$.

39.

41.

$$M = \begin{array}{c c} & \begin{array}{cccccc} A & B & C & D & E & F \end{array} \\ \begin{array}{c} A \\ B \\ C \\ D \\ E \\ F \end{array} & \left[\begin{array}{cccccc} 0 & 0 & 1 & 1 & 0 & 0 \\ 1 & 0 & 0 & 0 & 0 & 1 \\ 0 & 1 & 0 & 0 & 1 & 1 \\ 0 & 0 & 1 & 0 & 1 & 1 \\ 1 & 0 & 0 & 0 & 0 & 1 \\ 1 & 0 & 0 & 0 & 0 & 0 \end{array} \right] \end{array}$$

$$M^2 = \left[\begin{array}{cccccc} 0 & 1 & 1 & 0 & 2 & 2 \\ 1 & 0 & 1 & 1 & 0 & 0 \\ 3 & 0 & 0 & 0 & 0 & 2 \\ 2 & 1 & 0 & 0 & 1 & 2 \\ 1 & 0 & 1 & 1 & 0 & 0 \\ 0 & 0 & 1 & 1 & 0 & 0 \end{array} \right]$$

Committee member	Amount of one-or two-stage influence
Andre	$2 + 6 = 8$
Betty	$2 + 3 = 5$
Charles	$3 + 5 = 8$
Debbie	$3 + 6 = 9$
Ed	$2 + 3 = 5$
Frank	$1 + 2 = 3$

Debbie is the most influential member of the Committee.

43.

$$\begin{array}{c c} & \begin{array}{ccccc} R & S & T & U & V \end{array} \\ \begin{array}{c} R \\ S \\ T \\ U \\ V \end{array} & \left[\begin{array}{ccccc} 0 & 1 & 1 & 2 & 3 \\ \infty & 0 & \infty & 1 & 2 \\ \infty & \infty & 0 & 1 & 2 \\ \infty & \infty & \infty & 0 & 1 \\ \infty & \infty & \infty & \infty & 0 \end{array} \right] \end{array}$$

45.

$$D = \left[\begin{array}{cccc} 0 & \infty & 1 & 2 \\ 2 & 0 & 1 & 2 \\ 1 & \infty & 0 & 1 \\ \infty & \infty & \infty & 0 \end{array} \right]$$

The distance sum of A is $1 + 2 = 3$, of B is $2 + 1 + 2 = 5$, of C is $1 + 1 = 2$ and of D is 0.

47. (a) Directed paths from S to V are S, P, T, V with sum 20, S, Q, U, W, V with sum 26, and S, P, R, V with sum 24.   The critical path from S to V is S, Q, U, W, V.

(b) Directed paths for the project are S, P, R, V, E with sum 24; S, P, T, V, E with sum 20; $S, Q, U, W V, E$ with sum 26, and S, Q, U, W, X, Y, E with sum 31. The critical path for the project is S, Q, U, W, X, Y, E.

49. (a)

Task	Schedule
A	immediately
B	immediately
C	8 weeks hence
D	8 weeks hence
F	7 weeks hence
G	20 weeks hence
H	28 weeks hence
I	15 weeks hence
J	28 weeks hence

(b) The project will be completed in 40 weeks.

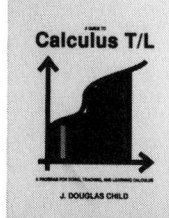

FOLD HERE

BUSINESS REPLY MAIL

FIRST CLASS   PERMIT NO. 358   PACIFIC GROVE, CA

POSTAGE WILL BE PAID BY ADDRESSEE

ATT: _____

Brooks/Cole Publishing Company
511 Forest Lodge Road
Pacific Grove, California 93950-9968

FOLD HERE